Electrically Contactless Online Impedance Measurement

IEEE Press
445 Hoes Lane
Piscataway, NJ 08854

Electrically Contactless Online Impedance Measurement

Theory and Applications

Zhenyu Zhao
Department of Electrical and Computer Engineering
National University of Singapore, Singapore

Kye Yak See
School of Electrical and Electronic Engineering
Nanyang Technological University, Singapore

IEEEPress
WILEY

Published by John Wiley & Sons, Inc., Hoboken, New Jersey.

Library of Congress Cataloging-in-Publication Data has been applied for:

Hardback ISBN: 9781394322732
ePDF ISBN: 9781394322756
ePub ISBN: 9781394322749
oBook ISBN: 9781394322763

Cover Design: Wiley
Cover Image: DrPixel/Getty Images

Contents

List of Figures

List of Tables

About the Authors

Dr. Zhenyu Zhao (Senior Member, IEEE) received the BEng degree in electrical engineering from Huazhong University of Science and Technology, Wuhan, China, in 2015, and the PhD degree in electrical engineering from Nanyang Technological University (NTU), Singapore, in 2021.

He is currently a Lecturer (Principal Investigator) in the Department of Electrical and Computer Engineering, National University of Singapore. His research interests include electromagnetic-based noncontact online impedance measurement; electromagnetic compatibility (EMC) of power electronics and transportation systems; noninvasive electromagnetic/ultrasonic monitoring of implantable devices and batteries; electromagnetic information leakage assessment and protection; and physics-informed, data-efficient AI methodologies for condition monitoring, testing, and diagnosis in complex systems, with applications to transportation systems, semiconductor devices, and human health. In these areas, he has published over 80 papers in leading international journals, including 13 ESI Highly Cited and Hot Papers.

Dr. Zhao received the Best Student Paper Award at the Joint IEEE International Symposium on Electromagnetic Compatibility and Asia-Pacific International Symposium on Electromagnetic Compatibility (Joint IEEE EMC & APEMC) in 2018, the Young Scientist Award and Best Paper Award at APEMC in 2022, the Young Scientist Award from the International Union of Radio Science (URSI) in 2023, the Emerging Leader Award from Measurement Science and Technology in 2024, the NUS College Teaching Excellence Award in AY2024/2025, the Best Paper Award at APEMC in 2026, etc. He was recognized as a Young Professional Ambassador of the IEEE EMC Society in 2025 and has been listed in the Stanford–Elsevier Top 2% Scientists List since 2025. He has served as the Technical Program Committee (TPC) Chair of IEEE APEMC 2026, TPC Co-Chair of ACES-China 2026 and EENES 2025, Publicity Chair of IEEE WiPDA-Asia 2025 and SWA 2024 and 2025. He is currently an Associate Editor of *IEEE Transactions on Transportation Electrification*, *IEEE Transactions on Vehicular Technology*, and *Measurement*, among others. Since 2024, he has served as Chair and Vice Chair of the IEEE EMC Society Singapore Chapter.

Professor Kye Yak See earned his BEng with First Class Honours from the National University of Singapore in 1986, followed by a PhD from Imperial College London, UK, in 1997.

Before his transition into academia, he spent nearly 10 years (1986–1994) in the private sector, holding senior technical and management positions across Singapore, the UK, and Hong Kong. After obtaining his PhD in 1997, he joined the School of Electrical and Electronic Engineering, NTU, Singapore, as a faculty member.

He is a distinguished academic and researcher, specializing in EMC, power electronics, and advanced online impedance measurement techniques. With a career spanning four decades across industry and academia, he has been a pivotal figure in shaping Singapore's electromagnetic research landscape and its application to critical urban infrastructure, particularly in rail transportation. Since 2007, Professor See has served as the Director of the Electromagnetic Effects Research Laboratory (EMERL). He also demonstrated significant leadership in public–private research partnership as the Director of the SMRT-NTU Smart Urban Rail Corporate Laboratory from 2015 to 2021, where he oversaw innovations in smart rail technology and reliability.

He has published nearly 300 peer-reviewed international journal and conference papers, co-authored three books, and secured five patents in his fields. His current research focuses on real-time condition monitoring of electrical systems and noninvasive measurement techniques, bridging the gap between theoretical electromagnetics and practical industrial safety.

Professor See has been very active in the global EMC community. He is the Founding Chair of both the IEEE EMC Society Singapore Chapter and the IEEE Joint AES/GRSS Singapore Chapter. His editorial contributions include serving as the Managing Editor of *IEEE Electromagnetic Compatibility Magazine* since 2011 and Associate Editor of *IEEE Transactions on Electromagnetic Compatibility* since 2024.

He has chaired several prestigious international conferences, most recently serving as the General Chair for the 2026 Asia-Pacific International Symposium on Electromagnetic Compatibility. In recognition of his service to the nation and his engineering achievements in the rail sector, he was awarded the 2021 National Day Public Administration Medal (Bronze) by the Prime Minister's Office of Singapore.

Beyond his academic and institutional roles, Professor See is an active proponent of technology commercialization. He co-founded a specialized technology startup alongside his former PhD students, focusing on the development of cutting-edge solutions for partial discharge monitoring and diagnosis.

Preface

Impedance is a fundamental quantity in electrical engineering, governing the behavior of circuits, components, and systems across a broad range of frequencies and operating conditions. While conventional impedance measurement techniques are well established under offline conditions, many modern electrical and electronic systems operate continuously at high voltage, high power, or in safety-critical environments where interrupting operation or making direct electrical contact is impractical, unsafe, or even impossible. In such contexts, the ability to measure impedance online, i.e., under actual operating conditions, becomes both technically challenging and critically important.

This book focuses on a particular class of online impedance measurement techniques: electrically contactless online impedance measurement based on inductive coupling. By exploiting electromagnetic induction, these techniques enable in-circuit impedance characterization of energized systems without direct electrical contact, thereby offering significant advantages in safety, deployment flexibility, and measurement fidelity. Over the past decade, electrically contactless impedance measurement has evolved from a niche research topic into a versatile and enabling technology with demonstrated value in power electronics, transportation systems, and power and energy infrastructures. This book aims to provide the first comprehensive and systematic treatment of this emerging field.

The primary objective of this book is to bridge theory, measurement methodology, and real-world application. Rather than presenting isolated techniques, the book develops a unified framework for electrically contactless online impedance measurement, beginning with fundamental principles and progressing through systematic measurement configurations to practical engineering case studies. Both frequency-domain and time-domain approaches are covered, with careful attention paid to calibration strategies, error mitigation, and implementation challenges encountered in real operating environments.

The book is organized into 10 chapters. Chapters 1 and 2 introduce the concept of online impedance, establish its significance across multiple engineering domains, and compare contact-based and contactless measurement approaches. Inductive coupling is identified as a particularly effective solution for non-contact online measurement, forming the foundation for the remainder of the book. Chapters 3–5 develop the core measurement methodologies, including two-probe, single-probe, and multi-probe configurations. These chapters present detailed analytical formulations, equivalent circuit models, cascaded two-port network representations, and both frequency-domain and time-domain implementations, enabling readers to understand not only how these methods work, but why they work. Chapter 6 provides a systematic comparison of the different contactless measurement setups, highlighting their respective strengths, limitations, and suitable application scenarios. This comparative perspective is intended to guide practitioners in selecting an appropriate configuration based on practical constraints such as signal-to-noise ratio (SNR), installation complexity, scalability, and measurement objectives. Chapters 7–9 demonstrate the practical impact of electrically contactless impedance measurement through representative applications in power electronics, transport electrification, and power and energy systems. These chapters illustrate how online impedance information enables capabilities such as electromagnetic interference (EMI) filter design, component and system health diagnostics, fault detection, and infrastructure condition monitoring, which are difficult or impossible to achieve using conventional offline methods. Finally, Chapter 10 summarizes the key contributions of the book and outlines future research directions, including advanced probe design, enhanced signal processing, portable measurement systems, artificial intelligence (AI) integration, and standardization for industrial deployment.

This book is intended for a broad readership, including graduate students, researchers, and practicing engineers working in power electronics, electromagnetic compatibility (EMC), transport electrification, and power and energy systems. A solid background in circuit theory and basic electromagnetics is assumed; however, the material is presented in a self-contained and progressive manner to support both learning and practical implementation. Wherever appropriate, theoretical developments are complemented by experimental validation and real-world case studies. It is our hope that this book will serve not only as a comprehensive reference on electrically contactless online impedance measurement, but also as a catalyst for further innovation and application of impedance-based sensing and diagnostic techniques in next-generation electrical and electronic systems.

Acknowledgments

The authors gratefully acknowledge Dr. Kangrong Li, Dr. Fei Fan, Dr. Arjuna Weesinghe, Dr. Sooriya Bandara, Dr. Manish Prajapati, and Dr. Huizhen Yang, among others, for their research contributions, which have influenced the development and technical content of this book. The authors also thank Dr. Kewei Deng for his assistance in the final preparation of the manuscript. The support provided by IEEE-Wiley throughout the publication process is also gratefully acknowledged.

1

Introduction

1.1 Background and Significance

Impedance, in electrical engineering, is a complex quantity that combines resistance and reactance to characterize the total opposition that a circuit or component presents to current flow [1]. This fundamental parameter exhibits distinct behavior under nonoperational (offline) and operational (online) conditions [2]. Offline impedance is measured when a device or system is powered off and is therefore unaffected by its actual operating state. In contrast, online impedance, also referred to as in-circuit impedance, is measured while the device or system is energized and interacting with the power source. As a result, online impedance provides real-time information that more accurately reflects the true electrical characteristics of the system under practical working conditions.

The significance of online impedance measurement spans a wide range of engineering domains, highlighting its essential role in improving reliability, efficiency, stability, and safety across diverse applications. In power electronics, online impedance measurement provides essential information for enhancing device reliability and operational efficiency, supporting system diagnostics and design improvements [3–5]. In the transportation sector, it facilitates condition monitoring of motor drive systems (MDS) [6–8], and plays a vital role in diagnosing and monitoring critical components in rail transit infrastructures, such as light rail transit (LRT) [9] and mass rapid transit (MRT) systems [10]. Moreover, online impedance information enables the systematic design of common-mode (CM) electromagnetic interference (EMI) filters for MDS [11], leading to more effective EMI suppression under real operating conditions. In power and energy systems, online impedance measurement is indispensable for the condition monitoring of critical assets such as batteries [12–14], power transformers [15], and high-power DC feeders [16]. Photovoltaic (PV) systems also benefit from online impedance measurement, which enhances their operational reliability [17, 18]. Furthermore, online impedance supports stability analysis of power electronic systems and grid-connected converters [19, 20], underscoring its expansive applicability and essential role in modern engineering.

Online impedance measurement can generally be implemented using three primary approaches: the voltage–current (V–I) approach, the capacitive coupling approach, and the inductive coupling approach [21]. The V–I approach determines the online impedance by measuring the voltage across and the current flowing through an energized system under test (SUT), followed by digital signal processing based on Ohm's Law [22–24]. The capacitive coupling approach employs coupling capacitors to block low-frequency or direct current (DC) power components while allowing high-frequency test signals to pass, enabling impedance extraction using an impedance analyzer (IA) or a vector network analyzer (VNA) [25–27]. However, both the V–I and capacitive coupling approaches require some form of direct electrical contact with the energized SUT, which increases implementation complexity and might introduce electrical safety risks in high-voltage and high-power environments. In contrast, the inductive coupling approach eliminates direct electrical contact with the energized SUT, thereby simplifying on-site implementation and enhancing safety [2]. Owing to its noncontact nature, it is also known as the electrically contactless online impedance measurement approach.

Based on the inductive coupling theory, a series of noncontact online impedance measurement setups has been developed to address different measurement requirements. These include the frequency-domain two-probe setup (FD-TPS) [28], time-domain two-probe setup (TD-TPS) [29], frequency-domain single-probe setup (FD-SPS) [30], and time-domain single-probe setup (TD-SPS) [31]. In addition, the frequency-domain three-probe setup (FD-THPS) [32] and time-domain multi-probe setup (TD-MPS) [33] have been developed to simultaneously measure the online impedances of SUTs across two or multiple branches powered by the same source. Despite being rooted in the inductive coupling approach, these setups exhibit respective characteristics in design theory, principle, and applicability. Each configuration is tailored to meet unique measurement challenges, offering solutions for time-variant measurements, multifrequency simultaneous measurements, varying measurement efficiency and complexity in implementation, etc. This array of noncontact measurement setups underscores the flexibility and adaptability of the inductive coupling approach, establishing it as a versatile and effective solution for contemporary challenges in online impedance measurement.

1.2 Organization of This Book

This book provides a comprehensive exploration of the theory, technological developments, and practical applications of electrically contactless online impedance measurement. It is designed to bridge the gap between traditional impedance measurement techniques and modern noncontact approaches, presenting both foundational principles and recent innovations that have shaped the field. The chapters are organized as follows:

This chapter discusses the evolution and importance of online impedance measurement, emphasizing the scope and significance of noncontact methods across diverse applications, preparing the reader for the detailed technical exploration to follow.

Chapter 2 presents a detailed comparison of three prevalent online impedance measurement approaches: the *V–I* approach, the capacitive coupling approach, and the inductive coupling approach. This chapter establishes a fundamental basis for understanding the advantages and limitations of each method, with particular emphasis on the benefits of inductive coupling for noncontact measurement.

Chapter 3 explores two main types of two-probe setups (i.e., FD-TPS and TD-TPS) based on inductive coupling, detailing the effects of probe-to-probe coupling and the calibration techniques used to mitigate these effects.

Chapter 4 analyzes two types of single-probe setups (i.e., FD-SPS and TD-SPS) utilizing the inductive coupling approach, covering their theoretical foundations, configurations, practical implementations, and necessary calibration procedures.

Chapter 5 elaborates on two types of complex multi-probe setups (i.e., FD-THPS and TD-MPS) and their different unique objectives compared to two-probe and single-probe setups.

Chapter 6 provides a detailed comparative analysis of single-probe, two-probe, and multi-probe setups, evaluating their advantages, limitations, and suitability for different application criteria, aiding readers in selecting an appropriate technology for specific needs.

Chapter 7 presents applications of noncontact online impedance measurement in power electronics, including EMI filter design for switched-mode power supplies, as well as health diagnosis and voltage-dependent parasitic capacitance extraction of power semiconductor devices.

Chapter 8 examines the use of noncontact online impedance measurement in transportation systems, including MDS, LRT, and MRT. This chapter focuses on their systematic CM EMI filter design, fault detection, and health monitoring, respectively.

Chapter 9 discusses applications in power and energy systems, such as transformer condition monitoring, defect detection in high-power DC feeders, and EMI estimation of PV systems. These examples demonstrate how noncontact techniques strengthen system reliability and maintenance strategies.

Chapter 10 concludes the book by summarizing key insights and projecting future research directions and technological advancements in noncontact online impedance measurement, suggesting paths forward for this evolving field.

References

1 L. Callegaro. *Electrical impedance: principles, measurement, and applications.* Boca Raton, FL, USA: CRC Press, 2012, pp. 5–8.

2 Z. Zhao, "Measurement setup consideration and implementation for inductively coupled online impedance extraction," Ph.D. thesis, School Electr. Electron. Eng., Nanyang Tech. Univ., Singapore, 2021.

3 V. Tarateeraseth, K. Y. See, F. G. Canavero et al., "Systematic electromagnetic interference filter design based on information from in-circuit impedance measurements," *IEEE Trans. Electromagn. Compat.*, vol. 52, no. 3, pp. 588–598, Aug. 2010.

4 H. C. Yang, K. Y. See, R. Simanjorang et al., "Offline health diagnosis of power device based on nonintrusive inductively coupled approach," *IEEE J. Emerg. Sel. Top. Power Electron.*, vol. 6, no. 4, pp. 2053–2059, Mar. 2018.

5 Z. Zhao, K. Y. See, W. Wang et al., "Voltage-dependent capacitance extraction of SiC power MOSFETs using inductively coupled in-circuit impedance measurement technique," *IEEE Trans. Electromagn. Compat.*, vol. 61, no. 4, pp. 1322–1328, Aug. 2019.

6 Z. Zhao, F. Fan, W. Wang et al., "Detection of stator inter-turn short-circuit faults in inverter-fed induction motors by online common-mode impedance monitoring," *IEEE Trans. Instrum. Meas.*, vol. 70, pp. 1–10, 2021.

7 D. Zheng, G. Lu, and P. Zhang, "A noninvasive interturn insulation condition monitoring method based on the common-mode impedance spectrum of inverter-fed machines," *IEEE Trans. Ind. Appl.*, vol. 57, no. 5, pp. 4786–4795, Sep.–Oct. 2021.

8 D. Zheng, G. Lu, and P. Zhang, "An improved online stator insulation monitoring method based on common-mode impedance spectrum considering the effect of aging position," *IEEE Trans. Ind. Appl.*, vol. 58, no. 3, pp. 3558–3566, May–Jun. 2022.

9 Z. Zhao, E. K. Chua, T. Svimonishvili et al., "Current collector health monitoring of LRT trains powered by three-phase AC power rail based on inductive coupling method," *IEEE Trans. Ind. Electron.*, vol. 71, no. 3, pp. 3103–3112, Mar. 2024.

10 H. Jie, Y. Chang, Z. Zhao et al., "Defect detection and classification of railway track system for in-service MRT in tropical regions using a contactless TBMS and adaptive-DBSCAN," *IEEE Trans. Intell. Transp. Syst.*, vol. 26, no. 11, pp. 18695–18707, 2025.

11 F. Fan, K. Y. See, X. Liu et al., "Systematic common-mode filter design for inverter-driven motor system based on in-circuit impedance extraction," *IEEE Trans. Electromagn. Compat.*, vol. 62, no. 5, pp. 1711–1722, Oct. 2020.

12 S. Rodrigues, N. Munichandraiah, and A. Shukla, "A review of state-of-charge indication of batteries by means of AC impedance measurements," *J. Power Sources*, vol. 87, pp. 12–20, 2000.

13 Y. Fu, J. Xu, M. Shi et al., "A fast impedance calculation-based battery state-of-health estimation method," *IEEE Trans. Ind. Electron.*, vol. 69, no. 7, pp. 7019–7028, Jul. 2022.

14 L. Raijmakers, D. Danilov, J. Van Lammeren et al., "Sensorless battery temperature measurements based on electrochemical impedance spectroscopy," *J. Power Sources*, vol. 247, pp. 539–544, 2014.

15 S. B. Rathnayaka, K. Y. See, and K. Li, "On-line impedance monitoring of transformer based on inductive coupling approach," *IEEE Trans. Dielectr. Electr. Insul.*, vol. 24, no. 2, pp. 1273–1279, Apr. 2017.

16 K.-R. Li, K. Y. See, F. Fan et al., "Defects detection of high-power DC feeder with multiple parallel cables using inductive coupling technique," *IEEE Trans. Power Del.*, vol. 35, no. 2, pp. 787–793, Apr. 2019.

17 M. Prajapati and K. Y. See, "Extraction of equivalent noise source model from photovoltaic systems," *IEEE Trans. Electromagn. Compat.*, vol. 61, no. 3, pp. 903–910, Jun. 2019.

18 M. Prajapati, F. Fan, Z. Zhao et al., "Estimation of radiated emissions from PV system through black box approach," *IEEE Trans. Instrum. Meas.*, vol. 70, pp. 1–4, 2021.

19 M. Liserre, R. Teodorescu, and F. Blaabjerg, "Stability of photovoltaic and wind turbine grid-connected inverters for a large set of grid impedance values," *IEEE Trans. Power Electron.*, vol. 21, no. 1, pp. 263–272, Jan. 2006.

20 Y. Liao and X. Wang, "Impedance-based stability analysis for interconnected converter systems with open-loop RHP poles," *IEEE Trans. Power Electron.*, vol. 35, no. 4, pp. 4388–4397, Apr. 2020.

21 Z. Zhao, F. Fan, H. Jie et al., "In-circuit impedance measurement setups of inductive coupling approach: a review," in *Proc. IEEE Asia-Pacific Symp. Electromagn. Compat. (APEMC)*, Beijing, China, 2022.

22 L. Huang, C. Zhang, W. Feng et al., "Online impedance measurement of operational batteries utilizing Sinc function signal injection via DC-DC power converter," *Measurement*, vol. 225, p. 114058, Feb. 2024.

23 H. Hu, P. Pan, Y. Song et al., "A novel controlled frequency band impedance measurement approach for single-phase railway traction power system," *IEEE Trans. Ind. Electron.*, vol. 67, no. 1, pp. 244–253, 2020.

24 M. Li, H. Nian, B. Hu et al., "Design method of multi-sine signal for broadband impedance measurement considering frequency coupling characteristic," *IEEE J. Emerg. Sel. Top. Power Electron.*, vol. 10, no. 1, pp. 532–543, Feb. 2022.

25 C. Gonzalez, J. Pleite, V. Valdivia et al., "An overview of the on line application of frequency response analysis (FRA)," in *Proc. IEEE Int. Symp. Ind. Electron.*, Vigo, Spain, 2007, pp. 1294–1299.

26 W. Su, H. Li, Z. Zhao et al., "In-circuit impedance extraction of individual components in power converter using capacitive coupling fixture adapters method for electromagnetic interference estimation," *IEEE Trans. Ind. Electron.*, vol. 72, no. 7, pp. 7616–7627, 2025.

27 X. Shang, D. Su, H. Xu et al., "A noise source impedance extraction method for operating SMPS using modified LISN and simplified calibration procedure," *IEEE Trans. Power Electron.*, vol. 32, no. 6, pp. 4132–4139, Jun. 2017.

28 K. Li, K. Y. See, and X. Li, "Inductive coupled in-circuit impedance monitoring of electrical system using two-port ABCD network approach," *IEEE Trans. Instrum. Meas.*, vol. 64, no. 9, pp. 2489–2495, Sep. 2015.

29 Z. Zhao, K. Y. See, E. K. Chua et al., "Time-variant in-circuit impedance monitoring based on the inductive coupling method," *IEEE Trans. Instrum. Meas.*, vol. 68, no. 1, pp. 169–176, 2019.

30 A. Weerasinghe, Z. Zhao, N. B. Narampanawe et al., "Single-probe inductively coupled in-circuit impedance measurement," *IEEE Trans. Electromagn. Compat.*, vol. 64, no. 1, pp. 2–10, Feb. 2022.

31 A. Weerasinghe, Z. Zhao, Q. Sun et al., "A novel single-probe setup for multifrequency simultaneous measurement of in-circuit impedance," *IEEE Trans. Ind. Electron.*, vol. 70, no. 9, pp. 9538–9549, Sep. 2023.

32 K. Li, A. Videt, and N. Idir, "Multiprobe measurement method for voltage-dependent capacitances of power semiconductor devices in high voltage," *IEEE Trans. Power Electron.*, vol. 28, no. 11, pp. 5414–5422, Nov. 2013.

33 Z. Zhao and K. Y. See, "A multiprobe inductive coupling method for online impedance measurement of electrical devices distributed in multibranch cables," *IEEE Trans. Instrum. Meas.*, vol. 69, no. 9, pp. 5975–5977, Sep. 2020.

2

Contact and Contactless Online Measurement

This chapter reviews three prevalent approaches for online impedance measurement, namely, the voltage–current (V–I) approach, the capacitive coupling approach, and the inductive coupling approach. These methods reflect the evolution from traditional contact-based measurement techniques toward noncontact solutions that emphasize safety, ease of deployment, and adaptability across diverse operating environments. By examining the operating principles, circuit models, implementation considerations, and inherent limitations of each approach, this chapter establishes a foundation for understanding why inductively coupled configurations have become the preferred choice for electrically contactless online impedance measurement. The concepts introduced in this chapter provide the theoretical basis for the detailed measurement setups discussed in Chapters 3–5.

2.1 V–I Approach

The V–I approach is a straightforward method in which the online impedance of an electrical SUT, denoted as Z_{SUT}, is obtained by measuring the test signal voltage across the SUT and the corresponding test signal current passing through it. Z_{SUT} is then determined using Ohm's Law.

Based on the characteristics of the test signals, the V–I approach can be further divided into passive methods and active methods. The passive methods employ the existing harmonics present in the energized SUT as the test signal [1]. However, when the background noise is significant, passive methods may suffer from a low SNR, which can degrade impedance measurement accuracy [2]. In addition, these methods can only measure impedances at harmonic frequencies. In contrast, active methods perform online impedance measurement by injecting a transient or steady-state test signal using a dedicated signal generation device, such as a current transformer [3], a voltage transformer [3], or a converter connected to the

SUT [1]. By properly selecting the amplitude and frequency content of the injected test signal, the online impedance at frequencies of interest can be measured accurately.

Figure 2.1 shows the basic measurement setup of the *V–I* approach using a current transformer as the SID, where an SG is used to generate the test signal. The test signal is injected into the wire connected between the power source and the SUT through the current transformer (i.e., SID). Subsequently, the induced signal voltage across the SUT and the current passing through it are measured using a current sensor and a voltage sensor, respectively. Finally, an ADC samples the analog voltage and current of the test signals, and a DSP algorithm is applied to obtain Z_{SUT}.

Since Ohm's Law is used for impedance extraction, various DSP algorithms have been developed for the *V–I* approach, among which the discrete Fourier transform (DFT) is widely used [4]. However, inter-harmonic distortions and transient events can affect the performance of the DFT due to spectral leakage and picket fence effects. To overcome these limitations, the CWT with an adjustable window has been proposed [5]. Owing to its controllable selection of scale ranges, CWT is effective for identifying resonance peak points in the SUT impedance. Nevertheless, CWT produces high redundancy for real-time applications, and its computational complexity is significant. In addition, the DWPT and its improved variants have also been used in this approach [6].

In summary, the *V–I* approach enables online impedance measurement by directly relating the measured voltage and current responses of the SUT to injected or inherent test signals. By employing either passive harmonic components or actively generated steady-state or transient excitations, this method allows

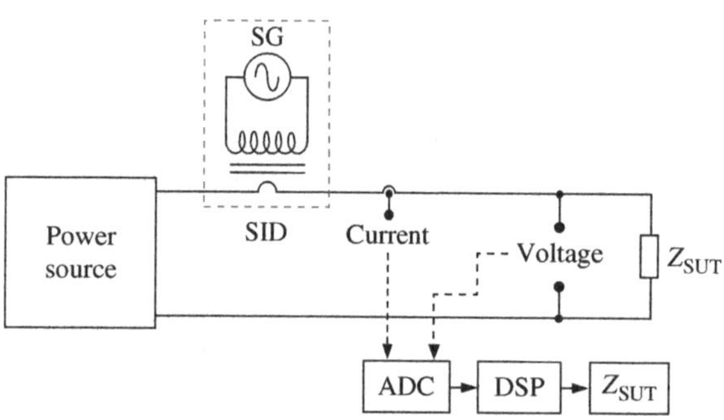

Figure 2.1 Basic measurement setup of the *V–I* approach: an illustration of using a current transformer as the signal injection device. *Source*: Adapted from [3].

online impedance characterization at selected frequencies of interest. The use of established signal processing techniques, such as DFT-, CWT-, and DWPT-based methods, further supports accurate online impedance extraction under practical operating conditions.

2.2 Capacitive Coupling Approach

Figure 2.2 shows the basic measurement setup of the capacitive coupling approach for extracting the online impedance of an SUT, which is powered by either a DC or a low-frequency (e.g., 50/60 Hz) alternating current (AC) power source. To extract Z_{SUT}, two coupling capacitors (C_1 and C_2), two inductors (L_1 and L_2), and an impedance analyzer (IA) or a vector network analyzer (VNA) form the basic measurement setup [7–9]. In this configuration, C_1 and C_2 are used to block the DC or low-frequency power signal from entering the measurement instrument (IA or VNA), while allowing the high-frequency test signal to enter the SUT. Conversely, L_1 and L_2 block the high-frequency test signal from entering the power source, but present very low impedance to the DC or low-frequency power signal.

Figure 2.3 shows the simplification of the measurement configuration in Figure 2.2 for the high-frequency test signal, where L_1 and L_2 can be regarded as open circuits, and Z_{C1} and Z_{C2} are the equivalent impedances of C_1 and C_2, respectively. Similarly, Figure 2.4 shows the simplification of the measurement configuration in Figure 2.2 for the DC or low-frequency power signal, where C_1 and C_2 can be regarded as open circuits.

As shown in Figure 2.3, if an IA is used as the measurement instrument [7, 8], the resultant impedance of the SUT together with C_1 and C_2 (i.e., $Z_{SUT} + Z_{C1} + Z_{C2}$) can be directly measured by the IA. Once the resultant impedance is obtained, Z_{SUT} can be extracted by de-embedding ($Z_{C1} + Z_{C2}$) from the resultant impedance.

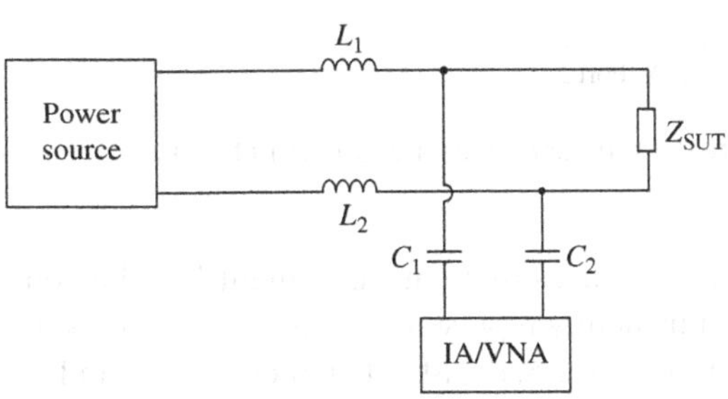

Figure 2.2 Basic measurement setup of the capacitive coupling approach. *Source*: Adapted from [7].

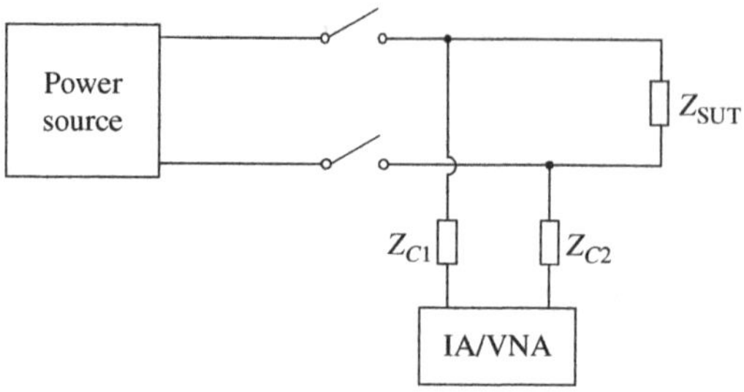

Figure 2.3 Simplification of Figure 2.2 for the high-frequency test signal.
Source: Adapted from [7].

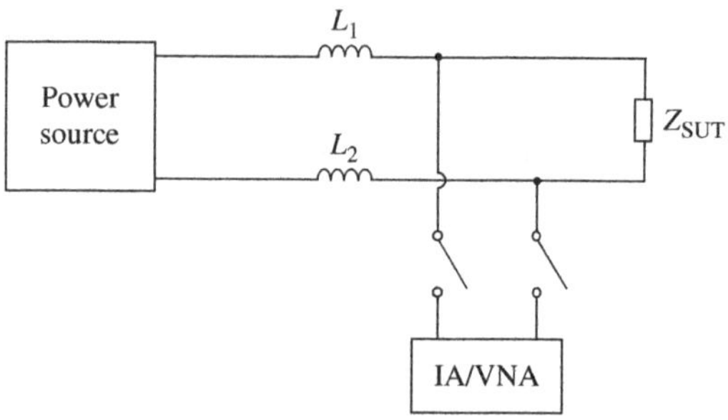

Figure 2.4 Simplification of Figure 2.2 for the DC or low-frequency power signal.
Source: Adapted from [7].

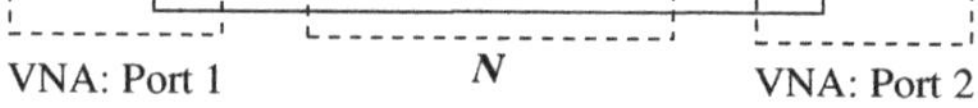

Figure 2.5 Equivalent model of Figure 2.3 represented as a two-port network.
Source: Adapted from [9].

Alternatively, if a VNA is used as the measurement instrument [9], the configuration shown in Figure 2.2 is equivalently represented as a two-port network model, as illustrated in Figure 2.5. In Figure 2.5, the SUT together with C_1 and C_2 can be equated as a two-port network $\boldsymbol{N}$, where $\boldsymbol{N}$ is expressed in terms of transmission ($ABCD$) parameters,

$$\boldsymbol{N} = \begin{bmatrix} A & B \\ C & D \end{bmatrix} \tag{2.1}$$

Since $B = Z_{SUT} + Z_{C1} + Z_{C2}$ [10], the scattering parameters (S-parameters) of $\boldsymbol{N}$ can be directly measured using a VNA and then converted into the parameter B through Equation (2.2) [10],

$$B = 50 \cdot \frac{(1 + S_{11})(1 + S_{22}) - S_{12}S_{21}}{2S_{21}} \tag{2.2}$$

After B is obtained, Z_{SUT} can be finally obtained by de-embedding ($Z_{C1} + Z_{C2}$) from B.

Overall, the capacitive coupling approach provides an effective means for online impedance measurement by separating high-frequency test signals from DC or low-frequency power components using appropriately arranged coupling capacitors and inductors. Both IA-based and VNA-based implementations have been described, together with their equivalent circuit and two-port network representations. These formulations establish a clear analytical framework for de-embedding the coupling network and extracting the desired SUT's impedance, forming a theoretical basis for the capacitive coupling-based online impedance measurement techniques.

2.3 Inductive Coupling Approach

Sections 2.1 and 2.2 reviewed two contact-based online impedance measurement approaches. In both methods, the voltage sensor used in the V–I approach and the coupling capacitor employed in the capacitive coupling approach require direct electrical connections to the SUT for online impedance extraction. These physical connections pose electrical safety hazards to service personnel involved in instrument installation, operation, and maintenance. Moreover, when the SUT operates at high voltage, the coupling capacitor and the voltage sensor are subjected to significant dielectric and thermal stresses, necessitating regular maintenance and replacement and potentially causing unnecessary downtime of the SUT.

In contrast, measurement setups based on the inductive coupling approach do not require any direct electrical connections to the energized SUT. The clamp-on inductive probes used in this approach can be easily mounted on or removed from the insulated wire or cable supplying power to the SUT, thereby simplifying on-site implementation and significantly enhancing operational safety. Owing to their nonintrusive nature, inductively coupled configurations are particularly attractive for online impedance measurements in high-voltage, high-power, or safety-critical systems.

The earliest work on inductive coupling for online impedance measurement was reported in [11], which introduced a FD-TPS. This configuration was later improved by incorporating the cascaded two-port network concept to simplify the impedance extraction procedure [12]. Over time, this approach evolved into a broader family of noncontact online impedance measurement configurations, encompassing both frequency-domain and time-domain techniques. These configurations include two-probe, single-probe, and multi-probe setups, each designed to address specific measurement objectives, operating conditions, and practical constraints.

The following chapters provide a detailed treatment of these inductive-coupling-based measurement configurations. Specifically, Chapters 3–5 examine the theoretical foundations, calibration procedures, and key characteristics of frequency-domain and time-domain two-probe, single-probe, and multi-probe setups, respectively, thereby establishing a comprehensive framework for electrically contactless online impedance measurement. Chapter 6 compares the advantages and limitations of these contactless configurations and introduces selection criteria based on application requirements. Chapters 7–9 present representative applications of inductively coupled impedance measurement techniques in power electronics, transport electrification, and power and energy systems. Finally, Chapter 10 discusses potential future research directions in this field.

2.4 Chapter Summary

This chapter reviewed the prevalent approaches for online impedance measurement, tracing their evolution from traditional contact-based techniques to electrically contactless solutions. The V–I approach and the capacitive coupling approach were first introduced, with emphasis on their operating principles, measurement configurations, and impedance extraction methods. While these techniques enable accurate online impedance characterization in their respective applications, their reliance on direct electrical connections to energized systems introduces safety concerns and practical limitations in high-voltage and high-power environments.

To address these challenges, the inductive coupling approach was presented as a nonintrusive alternative that enables electrically contactless online impedance measurement. By employing clamp-on inductive probes, this approach eliminates direct electrical contact with the SUT, thereby enhancing operational safety and deployment flexibility. The chapter also outlined the historical development and classification of inductively coupled measurement configurations, establishing a conceptual foundation for the detailed methodologies and applications discussed in subsequent chapters.

References

1 Z. Liu, C. Zhang, Z. Huang et al., "On-line battery impedance measurement using uniform measurement control sequence via a DC–DC power converter," *IEEE Trans. Ind. Electron.*, vol. 72, no. 12, pp. 14787–14796, Dec. 2025.

2 D. K. Alves, R. L. A. Ribeiro, F. B. Costa et al., "Real-time wavelet-based grid impedance estimation method," *IEEE Trans. Ind. Electron.*, vol. 66, no. 10, pp. 8263–8265, Oct. 2019.

3 C. J. Kikkert, "An on-line PLC frequency impedance analyzer," in *Proc. IEEE Int. Conf. Smart Grid Commun.*, Vancouver, BC, Canada, 2013.

4 H. Chen, Y. Yan, and H. Zhao, "Extraction of common-mode impedance of an inverter-fed induction motor," *IEEE Trans. Electromagn. Compat.*, vol. 58, no. 2, pp. 599–606, Apr. 2016.

5 M. Sumner, A. Abusorrah, D. Thomas et al., "Real time parameter estimation for power quality control and intelligent protection of grid-connected power electronic converters," *IEEE Trans. Smart Grid*, vol. 5, no. 4, pp. 1602–1607, Jul. 2014.

6 D. K. Alves, R. L. de Araujo Ribeiro, F. B. Costa et al., "Wavelet-based monitor for grid impedance estimation of three-phase networks," *IEEE Trans. Ind. Electron.*, vol. 68, no. 3, pp. 2564–2574, Mar. 2021.

7 C. González, J. Pleite, V. Valdivia et al., "An overview of the online application of frequency response analysis (FRA)," in *Proc. IEEE Int. Conf. Ind. Electron.*, Vigo, Spain, 2007.

8 T. Funaki, N. Phankong, T. Kimoto et al., "Measuring terminal capacitance and its voltage dependency for high-voltage power devices," *IEEE Trans. Power Electron.*, vol. 24, no. 6, pp. 1486–1493, Jun. 2009.

9 X. Shang, D. Su, H. Xu et al., "A noise source impedance extraction method for operating SMPS using modified LISN and simplified calibration procedure," *IEEE Trans. Power Electron.*, vol. 32, no. 6, pp. 4132–4139, Jun. 2017.

10 D. M. Pozar, *Microwave engineering*. Hoboken, NJ, USA: Wiley, 2012, pp. 165–227.

11 R. A. Southwick and W. C. Dolle, "Line impedance measuring instrumentation utilizing current probe coupling," *IEEE Trans. Electromagn. Compat.*, vol. EMC-13, no. 4, pp. 31–36, Nov. 1971.

12 K. Li, K. Y. See, and X. Li, "Inductive coupled in-circuit impedance monitoring of electrical system using two-port ABCD network approach," *IEEE Trans. Instrum. Meas.*, vol. 64, no. 9, pp. 2489–2495, Sep. 2015.

3

Two-probe Setups

This chapter presents two-probe configurations for electrically contactless online impedance measurement based on the inductive coupling approach. The two-probe setup was first reported in [1], where separate injecting and receiving probes were employed to enable noncontact online impedance extraction. This configuration was later enhanced by incorporating the cascaded two-port network concept to simplify the impedance extraction procedure [2]. Subsequently, the two-probe framework was extended to support time-variant online impedance monitoring, enabling dynamic characterization of operating systems [3].

Building on these developments, this chapter systematically introduces the key components and methodologies of two-probe measurement systems. The inductive probes and their electrical characteristics are first described, followed by the introduction of the cascaded two-port network concept. FD-TPS and TD-TPS are then presented in detail. Since two inductive probes are simultaneously employed, probe-to-probe coupling effects are considered, and corresponding calibration techniques are introduced to mitigate their impact on measurement accuracy. Finally, practical considerations for deploying two-probe setups in harsh industrial environments, such as significant electrical noise and power surges, are discussed.

3.1 Inductive Probes

Inductive probes, commonly referred to as AC current probes, have been widely used for current measurement in metering and diagnostic applications [4, 5]. Current probes can be broadly classified into two categories: probes designed for DC measurement and probes designed for AC measurement. DC current probes are typically based on the Hall effect, enabling the measurement of both DC and low-frequency AC currents, whereas AC current probes operate on the principle of electromagnetic induction [6]. For AC current probes, they can be used either to inject an AC excitation signal into a conductor (e.g., a cable or wire) or to receive an AC signal induced by the current flowing through the conductor. These probes are usually implemented in a clamp-on configuration, allowing them to be mounted onto a cable or wire without direct electrical contact. As a result, they introduce minimal disturbance to the electrical SUT. A wide range of commercial inductive probes is available with varying bandwidths, current ratings, and aperture sizes, as illustrated in Figure 3.1.

The operating principle of an inductive probe is similar to that of a transformer, where the primary winding is formed by the current-carrying conductor being clamped, and the secondary winding consists of an n-turn copper coil wound on a magnetic core. The corresponding equivalent circuit model is shown in Figure 3.2 [6].

As illustrated in Figure 3.2a, an inductive probe clamped onto a current-carrying conductor can be modeled as a one-turn primary winding coupled to an n-turn secondary winding. The secondary winding is terminated with a burden resistor R. For a more comprehensive equivalent circuit, shown in Figure 3.2b, parasitic components must be considered, including the primary and secondary stray capacitances C_P and C_S. The magnetic core is modeled by a magnetizing inductance L_m in parallel with a core-loss resistance R_C. The parameters R_S and L_l represent the secondary winding resistance and leakage inductance, respectively, while L_R denotes the equivalent series inductance of the burden resistor.

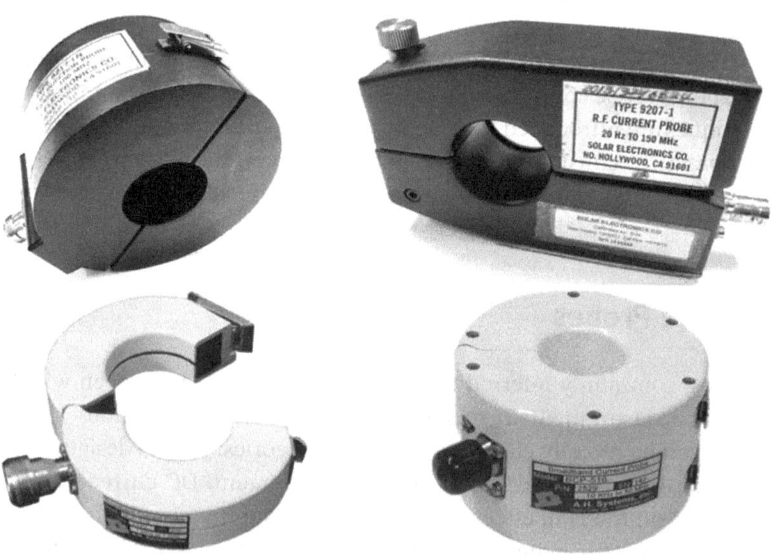

Figure 3.1 Examples of commercial inductive (AC current) probes.

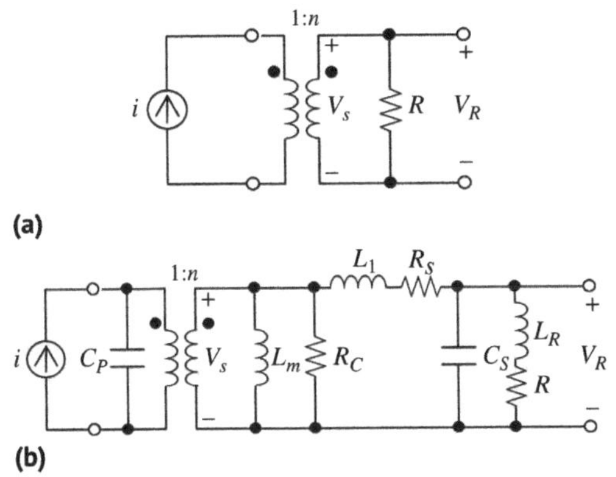

Figure 3.2 Equivalent circuit models of an inductive probe. (a) Basic transformer-based model. (b) Comprehensive model including parasitic elements.

Figure 3.3 Photo of a circular transmission-line test fixture (Solar 9125-1) used for measuring the transfer impedance of an inductive probe.

In practice, the transfer impedance of the inductive probe is required to convert the measured secondary voltage V_R into the corresponding primary conductor current. The transfer impedance of inductive probes can be determined using a circular cross-section transmission-line test fixture (e.g., Solar 9125-1), as illustrated in Figure 3.3.

During the measurement, the inductive probe is clamped onto the inner conductor of the fixture, while the outer conductor serves as a low-impedance reference ground for the vector network analyzer (VNA), as shown in Figure 3.4. One end of the fixture is connected to VNA Port 1, and the other end is terminated with a 50 Ω load.

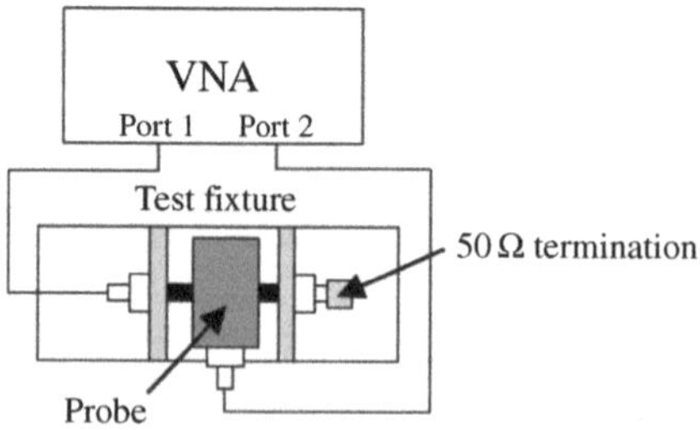

Figure 3.4 Measurement configuration for extracting the transfer impedance of an inductive probe using the test fixture shown in Figure 3.3.

The transfer impedance of an inductive probe is defined as

$$Z_T = \frac{V_2^-}{I_1} \tag{3.1}$$

where V_2^- is the reflected voltage wave measured at the output port of the inductive probe, and I_1 is the current flowing through the 50 Ω termination. The current I_1 can be expressed as

$$I_1 = \frac{V_1}{50} = \frac{V_1^+(1 + S_{11})}{50} \tag{3.2}$$

where V_1^+ denotes the incident voltage wave at Port 1, and S_{11} is the input-port voltage reflection coefficient. By substituting Equation (3.2) into Equation (3.1), the transfer impedance of the inductive probe can be expressed in terms of S-parameters as

$$Z_T = \frac{50V_2^-}{V_1^+(1 + S_{11})} = \frac{50S_{21}}{1 + S_{11}} \tag{3.3}$$

where S_{21} represents the forward voltage transmission coefficient.

As a result, the frequency-dependent transfer impedance of an inductive probe can be directly determined from measured S-parameters. Figure 3.5 presents an example of the measured transfer impedance of a Solar 9217-1N inductive probe.

Overall, inductive probes provide a nonintrusive and versatile means of interfacing with energized conductors for current sensing and signal coupling without direct electrical contact. By exploiting electromagnetic induction, these probes can function both as receivers of current-induced signals and as injectors of controlled excitation, making them well suited for online impedance measurement applications. The transformer-based operating principle and equivalent circuit representations clarify the key factors governing probe behavior, while the transfer impedance offers a convenient and quantitative characterization for converting measured probe voltages into conductor current. The ability to accurately

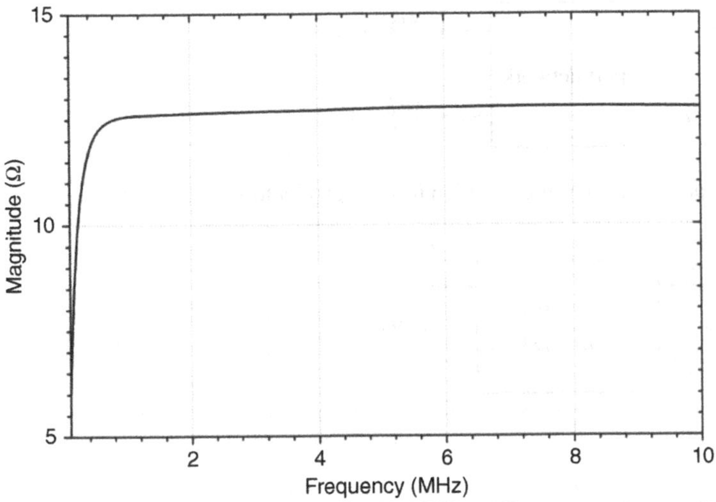

Figure 3.5 Measured transfer impedance versus frequency of a Solar 9217-1N inductive probe.

determine probe transfer impedance using standardized test fixtures and S-parameter measurements establishes a reliable foundation for the subsequent development of inductively coupled online impedance measurement setups.

3.2 Cascaded Two-port Network Concept

This section introduces the cascaded two-port network concept employed in two-probe setups for online impedance measurement, which was first reported in [2]. The two-port transmission network described by $ABCD$ parameters is adopted because it can be conveniently cascaded, making it particularly suitable for systematic network analysis. Since measurements obtained from a VNA are expressed in terms of S-parameters, the conversion between S-parameters and $ABCD$ parameters is also presented for completeness and practical relevance. In general, S-parameters describe a two-port network in terms of incident, reflected, and transmitted voltage waves, whereas impedance- or admittance-based descriptions directly relate voltages and currents. For circuit-level impedance extraction and network cascading, the $ABCD$ representation provides a more intuitive and mathematically convenient framework.

Figure 3.6 illustrates a typical two-port network expressed in terms of S-parameters, where V_1^+ and V_1^- denote the amplitudes of the incident and reflected voltage waves at Port 1, respectively, and V_2^+ and V_2^- denote those at

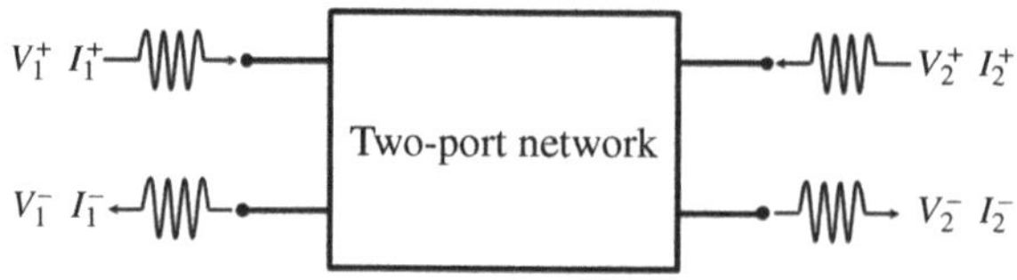

Figure 3.6 Two-port network representation using S-parameters.

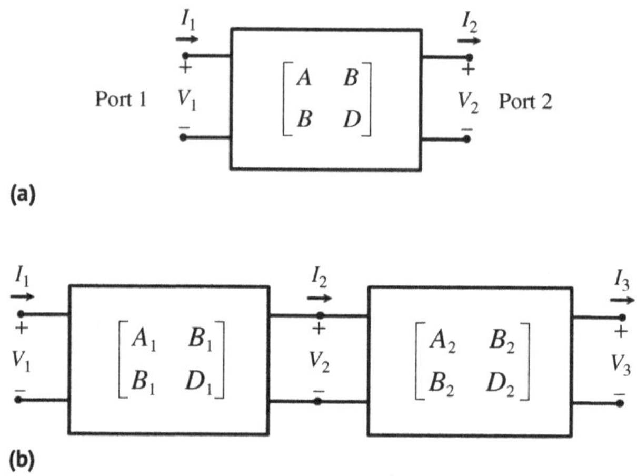

Figure 3.7 (a) Single two-port network represented by $ABCD$ parameters. (b) Two cascaded two-port networks represented by $ABCD$ parameters.

Port 2. The relationship between these waves can be expressed using the S-parameter matrix as

$$\begin{bmatrix} V_1^- \\ V_2^- \end{bmatrix} = \begin{bmatrix} S_{11} & S_{12} \\ S_{21} & S_{22} \end{bmatrix} \begin{bmatrix} V_1^+ \\ V_2^+ \end{bmatrix} \tag{3.4}$$

where S_{11}, S_{12}, S_{21}, and S_{22} represent the input-port reflection coefficient, reverse transmission coefficient, forward transmission coefficient, and output-port reflection coefficient, respectively.

Alternatively, a two-port network can be described using $ABCD$ parameters, as illustrated in Figure 3.7a. In this representation, the input and output voltages and currents are related as

$$\begin{bmatrix} V_1 \\ I_1 \end{bmatrix} = \begin{bmatrix} A & B \\ C & D \end{bmatrix} \begin{bmatrix} V_2 \\ I_2 \end{bmatrix} \tag{3.5}$$

A key advantage of the $ABCD$ formulation is its suitability for cascaded networks. As shown in Figure 3.7b, if two two-port networks are cascaded and described by the $ABCD$ matrices in Equations (3.6) and (3.7), respectively,

$$\begin{bmatrix} V_1 \\ I_1 \end{bmatrix} = \begin{bmatrix} A_1 & B_1 \\ C_1 & D_1 \end{bmatrix} \begin{bmatrix} V_2 \\ I_2 \end{bmatrix} \tag{3.6}$$

$$\begin{bmatrix} V_2 \\ I_2 \end{bmatrix} = \begin{bmatrix} A_2 & B_2 \\ C_2 & D_2 \end{bmatrix} \begin{bmatrix} V_3 \\ I_3 \end{bmatrix} \tag{3.7}$$

then substituting Equation (3.7) into Equation (3.6) yields

$$\begin{bmatrix} V_1 \\ I_1 \end{bmatrix} = \begin{bmatrix} A_1 & B_1 \\ C_1 & D_1 \end{bmatrix} \begin{bmatrix} A_2 & B_2 \\ C_2 & D_2 \end{bmatrix} \begin{bmatrix} V_3 \\ I_3 \end{bmatrix} \tag{3.8}$$

This result shows that the overall *ABCD* matrix of cascaded networks is obtained by the matrix product of the individual *ABCD* matrices, which forms the theoretical basis for modeling the two-probe impedance measurement setup as a cascade of multiple network elements. Table 3.1 summarizes the *ABCD* parameters of several commonly used circuit models.

If the two-port network measurement is performed using a VNA, the *ABCD* parameters can be obtained by converting the measured *S*-parameters according to the following relationships [7]:

$$A = \frac{(1 + S_{11})(1 - S_{22}) + S_{12}S_{21}}{2S_{21}} \tag{3.9}$$

$$B = Z_0 \frac{(1 + S_{11})(1 + S_{22}) - S_{12}S_{21}}{2S_{21}} \tag{3.10}$$

$$C = \frac{(1 - S_{11})(1 - S_{22}) - S_{12}S_{21}}{2Z_0S_{21}} \tag{3.11}$$

Table 3.1 *ABCD* parameters of commonly used network models.

Series Impedance Model

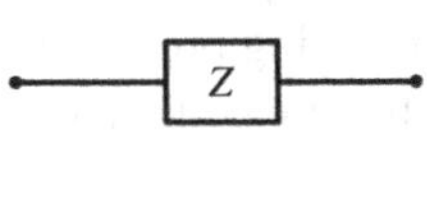

$$\begin{bmatrix} A & B \\ C & D \end{bmatrix} = \begin{bmatrix} 1 & Z \\ 0 & 1 \end{bmatrix}$$

Transmission Line Model

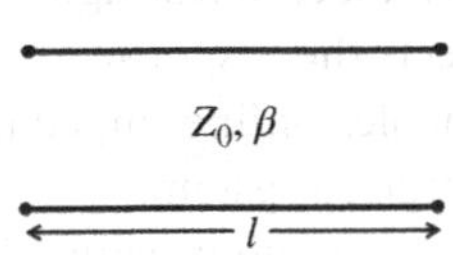

$$\begin{bmatrix} A & B \\ C & D \end{bmatrix} = \begin{bmatrix} \cos \beta l & jZ_0 \sin \beta l \\ jY_0 \sin \beta l & \cos \beta l \end{bmatrix}$$

$$D = \frac{(1 - S_{11})(1 + S_{22}) + S_{12}S_{21}}{2S_{21}} \tag{3.12}$$

where Z_0 denotes the characteristic impedance of the VNA, typically 50 Ω.

These conversion formulas enable the direct transformation of experimentally measured S-parameters into $ABCD$ parameters, thereby allowing the measured two-port network to be seamlessly incorporated into cascaded network analysis. This capability is useful for modeling the overall measurement chain in two-probe inductive impedance measurement setups, where multiple network elements (including probes and SUT) will be represented and analyzed in a unified framework.

Overall, this section establishes the cascaded two-port network framework as an available analytical tool for two-probe online impedance measurement. By introducing the $ABCD$ parameter representation and its relationship with S-parameters measured by a VNA, it provides a systematic and scalable method for modeling measurement chains. The ability to cascade individual network elements through matrix multiplication forms the basis for accurate impedance extraction in subsequent two-probe configurations.

3.3 Setup for Frequency-domain Measurement

This section introduces the FD-TPS for online impedance measurement based on the inductive coupling approach. By combining two clamp-on inductive probes with VNA measurements and cascaded two-port network concept, the FD-TPS enables nonintrusive extraction of system online impedance under normal operating conditions. The measurement principle, equivalent network representation, and probe pre-characterization requirements are presented in detail.

Figure 3.8 illustrates a basic FD-TPS for online impedance measurement. The DC or low-frequency (e.g., 50/60 Hz) power supply is represented by a source voltage V_S and a source impedance Z_S. The electrical system powered by this supply is characterized by an equivalent impedance Z_E. To perform online impedance measurement, the setup employs a VNA together with two clamp-on inductive probes. Port 1 of the VNA injects a test signal into the system through the injecting probe, while Port 2 detects the response of the same signal via the receiving probe. This configuration enables online impedance extraction without direct electrical contact with the energized system.

The overall measurement setup shown in Figure 3.8 can be modeled by an equivalent network consisting of three cascaded two-port networks, denoted

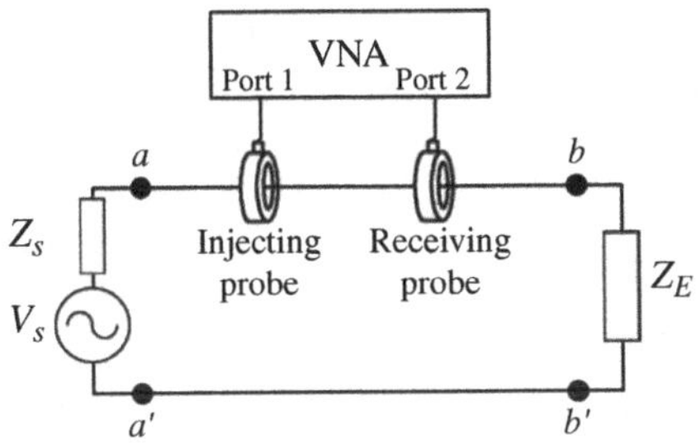

Figure 3.8 FD-TPS for online impedance measurement.

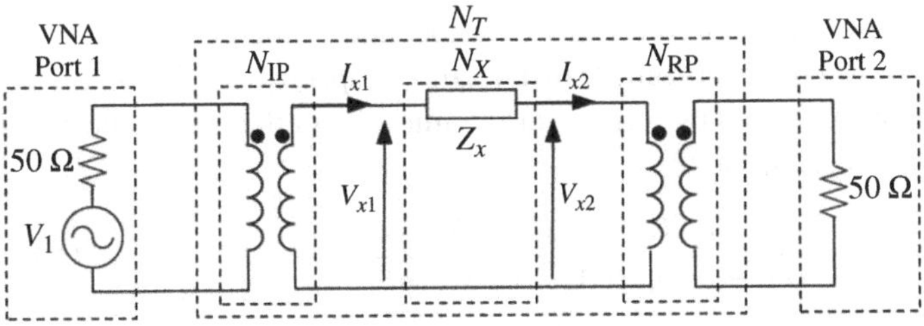

Figure 3.9 Equivalent cascaded two-port network representation of the setup shown in Figure 3.8.

as $\boldsymbol{N}_{\mathbf{IP}}$, $\boldsymbol{N}_{\boldsymbol{X}}$, and $\boldsymbol{N}_{\mathbf{RP}}$, as illustrated in Figure 3.9, where $\boldsymbol{N}_{\mathbf{IP}}$, $\boldsymbol{N}_{\boldsymbol{X}}$, and $\boldsymbol{N}_{\mathbf{RP}}$ correspond to the two-port representations of the injecting probe, the unknown online impedance, and the receiving probe, respectively. The unknown impedance to be extracted, denoted as Z_x, is the resultant series impedance composed of Z_S, the wiring impedance Z_W between nodes a–b and, a'–b' and Z_E (i.e., $Z_x = Z_S + Z_W + Z_E$).

In practical systems, Z_S and Z_W are often small compared to Z_E, and thus, Z_x is usually dominated by Z_E. The resultant *ABCD* matrix of the overall cascaded two-port network $\boldsymbol{N}_{\boldsymbol{T}}$, as observed by VNA Ports 1 and 2, is given by

$$\begin{bmatrix} A & B \\ C & D \end{bmatrix}_T = \begin{bmatrix} A & B \\ C & D \end{bmatrix}_{\mathrm{IP}} \begin{bmatrix} A & B \\ C & D \end{bmatrix}_X \begin{bmatrix} A & B \\ C & D \end{bmatrix}_{\mathrm{RP}} \tag{3.13}$$

where the matrices $\begin{bmatrix} A & B \\ C & D \end{bmatrix}_{\mathrm{IP}}$, $\begin{bmatrix} A & B \\ C & D \end{bmatrix}_X$, and $\begin{bmatrix} A & B \\ C & D \end{bmatrix}_{\mathrm{RP}}$ denote the *ABCD* parameters of $\boldsymbol{N}_{\mathbf{IP}}$, $\boldsymbol{N}_{\boldsymbol{X}}$, and $\boldsymbol{N}_{\mathbf{RP}}$, respectively.

Based on the voltage–current relationships of the two-port network corresponding to the unknown impedance, the *ABCD* parameters of $\boldsymbol{N_X}$ are defined as

$$\begin{bmatrix} A & B \\ C & D \end{bmatrix}_X = \begin{bmatrix} \left.\frac{V_{x1}}{V_{x2}}\right|_{I_{x2}=0} & \left.\frac{V_{x1}}{I_{x2}}\right|_{V_{x2}=0} \\ \left.\frac{I_{x1}}{V_{x2}}\right|_{I_{x2}=0} & \left.\frac{I_{x1}}{I_{x2}}\right|_{V_{x2}=0} \end{bmatrix}_X \tag{3.14}$$

where V_{x1} and I_{x1} are the input voltage and current at Port 1 of $\boldsymbol{N_X}$, and V_{x2} and I_{x2} are the output voltage and current at Port 2 of $\boldsymbol{N_X}$, as indicated in Figure 3.9.

By definition, the parameters A, B, C, and D correspond to the open-circuit voltage transfer ratio, short-circuit transfer impedance, open-circuit transfer admittance, and short-circuit current transfer ratio, respectively. The unknown impedance Z_x can be extracted by solving for parameter B of the *ABCD* matrix of $\boldsymbol{N_X}$. Specifically, Z_x is given by

$$Z_x = B_X = \frac{A_{\text{RP}}(D_{\text{IP}}B_T - B_{\text{IP}}D_T) - B_{\text{RP}}(A_T D_{\text{IP}} - C_T B_{\text{IP}})}{(A_{\text{IP}}D_{\text{IP}} - C_{\text{IP}}B_{\text{IP}})\,(A_{\text{RP}}D_{\text{RP}} - C_{\text{RP}}B_{\text{RP}})} \tag{3.15}$$

Before determining Z_x, the two-port characteristics of both inductive probes must be accurately pre-characterized. For an ideal inductive probe, the *ABCD* parameters are given by

$$\begin{bmatrix} A & B \\ C & D \end{bmatrix} = \begin{bmatrix} 1/n & 0 \\ 0 & n \end{bmatrix} \tag{3.16}$$

where n is the number of turns of the secondary winding on the magnetic core.

In practice, inductive probes are nonideal due to effects such as leakage inductance, core loss, and inter-winding capacitance. Consequently, their actual *ABCD* parameters must be carefully characterized over the frequency range of interest to ensure accurate impedance extraction. For FD-TPS measurements, the injecting and receiving probes are typically characterized using a circular cross-section transmission-line test fixture. To ensure measurement accuracy and repeatability, the voltage standing wave ratio (VSWR) of the fixture should be less than 1.2, ensuring that most of the injected signal is delivered to the load with minimal reflection. Figure 3.10 shows the VSWR measurement setup, where VNA Port 1 is connected to one port of the fixture and the other port is terminated with 50 Ω. As an example, Figure 3.11 presents the measured VSWR of a Solar 9125-1 fixture, where the maximum VSWR is 1.04 at 10 MHz, satisfying the required criterion.

To characterize an inductive probe, the probe is clamped onto the inner conductor of the test fixture, while the outer conductor serves as the common ground

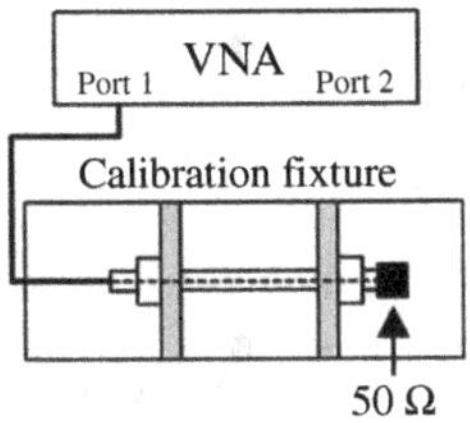

Figure 3.10 VSWR measurement setup for the test fixture.

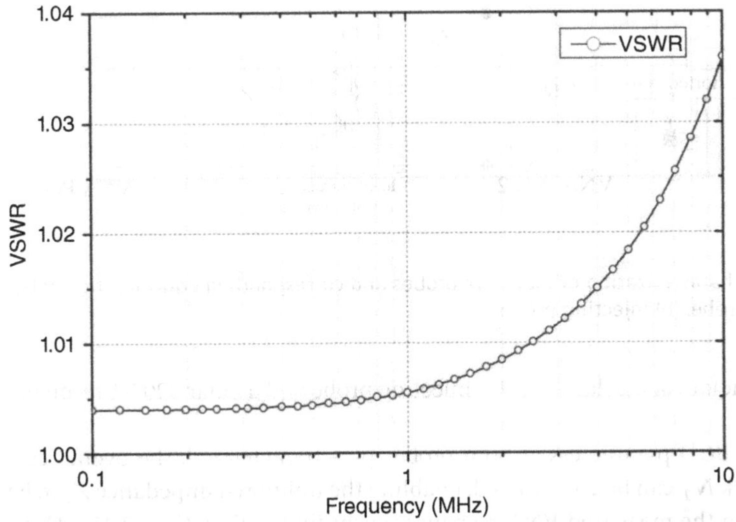

Figure 3.11 Measured VSWR versus frequency for the Solar 9125-1 fixture.

reference for the VNA. One end of the fixture is connected to a VNA port, and the other end is terminated with a short circuit. For the receiving probe, VNA Port 1 injects a test signal into the test fixture. The resulting current flowing through the inner conductor induces a voltage at the probe output, which is measured by VNA Port 2, as illustrated in Figure 3.12a. For the injecting probe, the connections to VNA Ports 1 and 2 are interchanged, as shown in Figure 3.12b. The corresponding equivalent circuits are derived based on the inductive probe model illustrated in Figure 3.12.

Once the two-port S-parameters of the receiving and injecting probes are measured, they are converted into the corresponding $ABCD$ parameters of $\boldsymbol{N}_{\mathbf{IP}}$ and $\boldsymbol{N}_{\mathbf{RP}}$, respectively. As illustrative examples, Figure 3.13a,b show the measured

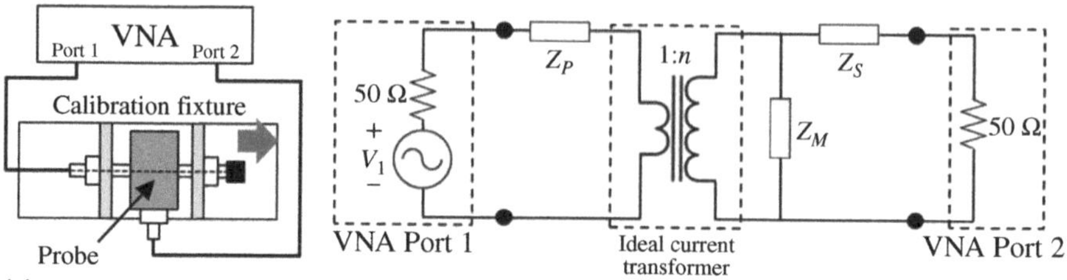

(a)

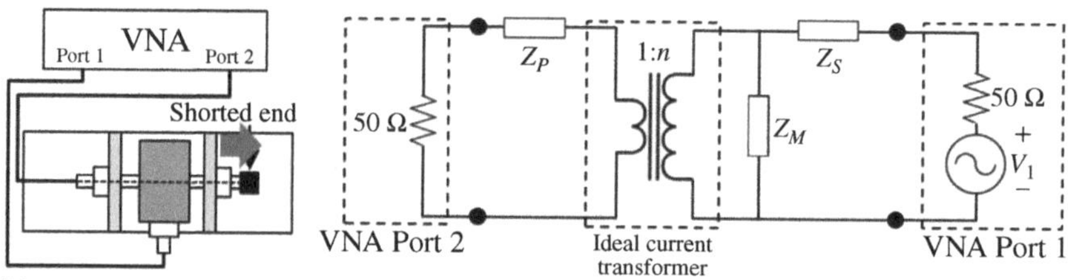

(b)

Figure 3.12 Characterization of inductive probes and corresponding equivalent circuits. (a) Receiving probe. (b) Injecting probe.

ABCD parameters of a Solar 9217-1N injecting probe and a Solar 9207-1 receiving probe.

With the *ABCD* parameters of both probes pre-characterized, the overall cascaded network $\boldsymbol{N_T}$ can be constructed, enabling the unknown impedance Z_x to be extracted from the measured VNA responses according to Equation (3.15). Thus, Z_E can be obtained by de-embedding $(Z_S + Z_W)$ from Z_x:

$$Z_E = Z_x - (Z_S + Z_W) \tag{3.17}$$

In some practical applications where Z_S and Z_W are relatively small compared with Z_E, Z_x provides a good approximation of Z_E.

Overall, this section presents a systematic frequency-domain two-probe measurement framework (i.e., FD-TPS) for electrically contactless online impedance extraction based on inductive coupling. By modeling the FD-TPS using cascaded two-port networks, the impedance under test can be analytically extracted from measured *ABCD* parameters. The characterization procedures for injecting and receiving probes are established. These results validate the effectiveness of the FD-TPS and provide an analytical foundation for online impedance extraction under normal operating conditions.

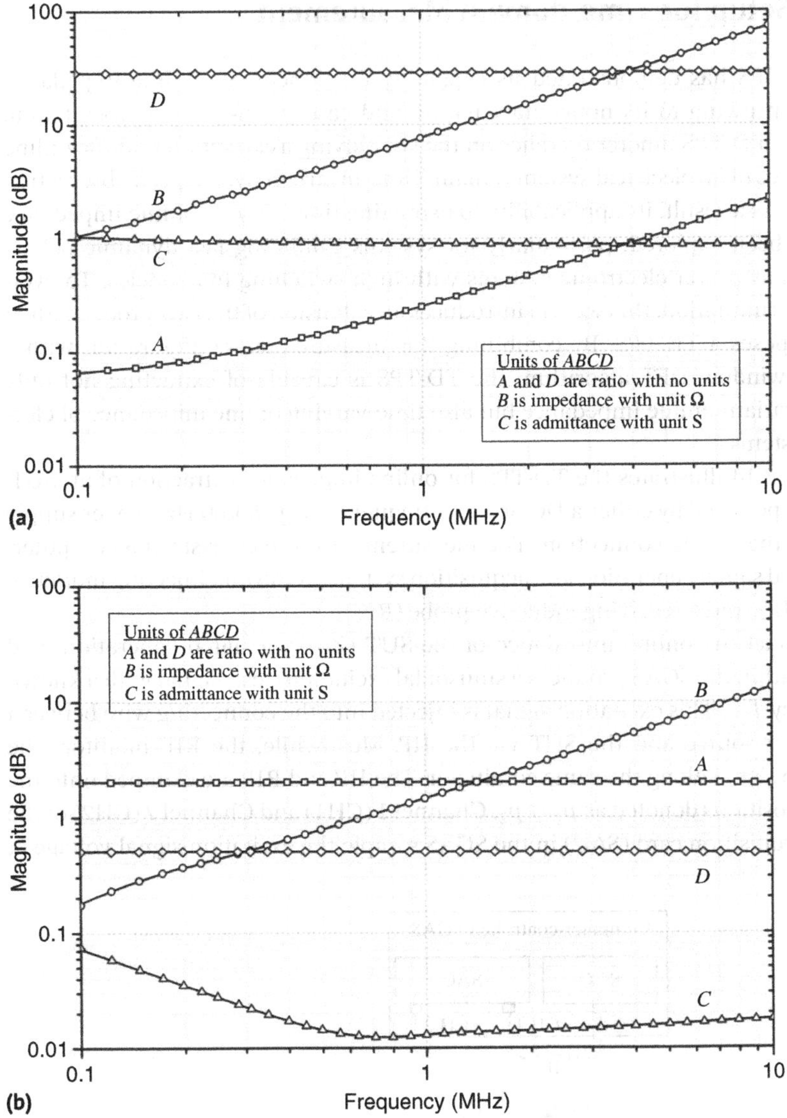

Figure 3.13 Examples of measured *ABCD* parameters of inductive probes. (a) Receiving probe. (b) Injecting probe.

3.4 Setup for Time-domain Measurement

The FD-TPS has demonstrated its capability for noncontact online impedance extraction owing to its noncontact nature and ease of on-site implementation. However, FD-TPS inherently relies on the simplifying assumption that the online impedance of an electrical system remains time-invariant over a given observation interval. As a result, its applicability to extracting time-varying online impedance has not been explored, particularly for systems exhibiting fast dynamic behavior, such as power electronic systems with high switching frequencies. To overcome this limitation, this section introduces an extension of the two-probe method and proposes a TD-TPS. By combining this measurement configuration with a moving-window DFT algorithm, the TD-TPS is capable of extracting not only time-invariant online impedance but also time-varying online impedance of electrical systems.

Figure 3.14 illustrates the TD-TPS for online impedance extraction of an SUT, which is powered by either a DC or a low-frequency (e.g., 50/60 Hz) power supply through the wiring connection. The measurement setup consists of a computer-controlled signal generation and acquisition system (SGAS), an injecting inductive probe (IIP), and a receiving inductive probe (RIP).

To extract the online impedance of the SUT (Z_{SUT}), a signal generation card (SGC) within the SGAS produces a sinusoidal excitation (test) signal with a known frequency f_{sig}. This excitation signal is injected into the connecting wire between the power source and the SUT via the IIP. Meanwhile, the RIP monitors the response signal along the same conductor. The IIP and RIP are clamped onto the wire at positions denoted as $p_1 - p_2$. Channel 1 (CH1) and Channel 2 (CH2) of the signal acquisition card (SAC) in the SGAS sample the excitation signal voltage at

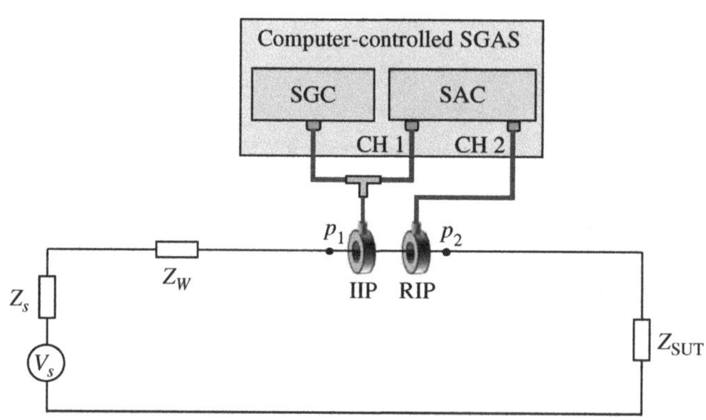

Figure 3.14 TD-TPS for online impedance measurement.

the input of the IIP and the response signal voltage at the output of the RIP, respectively. V_S and Z_S represent the equivalent source voltage and internal impedance of the power source. Z_W denotes the equivalent impedance of the wiring connection, excluding the part clamped by the IIP and RIP. Consequently, the resultant impedance (denoted as Z_X) of the power source, the wiring connection excluding the part being clamped, and the SUT can be expressed as

$$Z_X = Z_S + Z_W + Z_{SUT} \tag{3.18}$$

Based on the cascaded two-port network concept introduced in Section 3.2, Figure 3.15 presents the equivalent circuit model corresponding to the TD-TPS shown in Figure 3.14. V_{SGC} and Z_{SGC} are the equivalent source voltage and internal impedance of the SGC, respectively. Z_{CH1} and Z_{CH2} are the input impedances of CH1 and CH2 of the SAC, respectively. V_1 and I_1 denote the excitation signal voltage and current at the input port of the IIP, respectively, while V_2 and I_2 represent the response signal voltage and current at the output port of the RIP, respectively. $\boldsymbol{N_I}$is the two-port network of the IIP with the wire being clamped, where L_{k1} and C_{p1} represent the leakage inductance and parasitic capacitance between the winding of the IIP and its frame, respectively. Similarly, $\boldsymbol{N_R}$ denotes the two-port network of the RIP, with L_{k2} and C_{p2} representing the corresponding leakage inductance and parasitic capacitance. $\boldsymbol{N_X}$ is the two-port network corresponding to the resultant impedance Z_X to be measured.

Since the frequency of the sinusoidal test signal f_{sig} generated by the SGC is much higher than the frequency of the DC or low-frequency power source, the source voltage V_S can be treated as an AC short circuit at f_{sig}. By expressing the three cascaded two-port networks ($\boldsymbol{N_I}$, $\boldsymbol{N_X}$, and $\boldsymbol{N_R}$) in terms of their respective

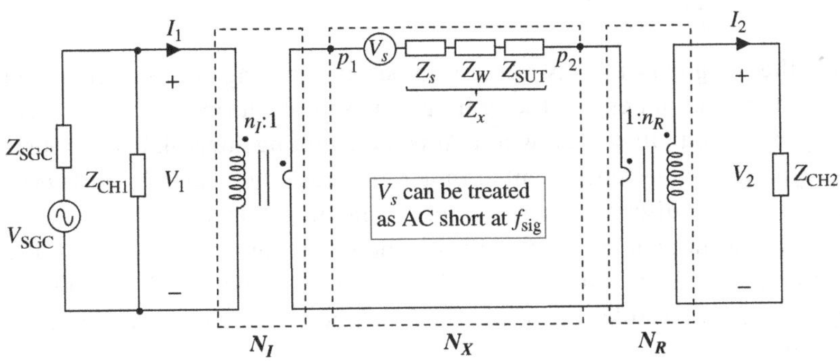

Figure 3.15 Equivalent circuit model of Figure 3.14 represented by cascaded two-port networks.

ABCD parameters, the relationship between the input port of the IIP and the output port of the RIP can be written as

$$\begin{bmatrix} V_1 \\ I_1 \end{bmatrix} = \begin{bmatrix} A_I & B_I \\ C_I & D_I \end{bmatrix} \begin{bmatrix} A_X & B_X \\ C_X & D_X \end{bmatrix} \begin{bmatrix} A_R & B_R \\ C_R & D_R \end{bmatrix} \begin{bmatrix} V_2 \\ I_2 \end{bmatrix} \tag{3.19}$$

Since $I_2 = V_2/Z_{CH2}$ and $\boldsymbol{N_X}$ can be expressed as [7]

$$\boldsymbol{N_X} = \begin{bmatrix} 1 & Z_x \\ 0 & 1 \end{bmatrix} \tag{3.20}$$

Substituting Equation (3.20) into Equation (3.19) yields the expression for extracting Z_x:

$$Z_x = \frac{1}{A_I(C_R + D_R/Z_{CH2})} \cdot \frac{V_1}{V_2} - \frac{A_R + B_R/Z_{CH2}}{C_R + D_R/Z_{CH2}} - \frac{B_I}{A_I} \tag{3.21}$$

As mentioned earlier, the *ABCD* parameters of $\boldsymbol{N_I}$ and $\boldsymbol{N_R}$ can be pre-characterized individually using the setup shown in Figure 3.12. The input impedance Z_{CH2} can be directly obtained from the datasheet of the SAC [8, 9]. Therefore, the resultant impedance Z_x can be determined from Equation (3.21) through the measurement of V_1 and V_2. Once Z_x is obtained, the online impedance of the system under test can be extracted by de-embedding $(Z_S + Z_W)$ from Z_x. To obtain the dynamic voltage equivalents at different time instants, a moving-window DFT is applied [10]. The moving window has a fixed size, and the DFT within the moving window is defined as

$$V_k(m) = \sum_{n=m-N+1}^{m} v(n) e^{-j2\pi k[n-(m-N+1)]/N} \tag{3.22}$$

where j is the imaginary unit, N is the number of sampling points in the fixed moving window, and m denotes the mth discrete sample during the entire measurement process, with $m = t/\Delta t$, where Δt is the sampling interval. $V_k(m)$ represents the dynamic voltage equivalent at time t and frequency index k, while $v(n)$ denotes the $(n+1)$th discrete voltage sample in the time domain.

Unlike the conventional DFT with a fixed window ranging from 0 to $N-1$, the summation range of the moving-window DFT varies with the sampling index m. By applying Euler's formula, Equation (3.22) can be rewritten in trigonometric form as follows:

$$V_k(m) = \sum_{n=m-N+1}^{m} v(n)\left[\cos\left(2\pi k \frac{n-m-1}{N}\right) - j \cdot \sin\left(2\pi k \frac{n-m-1}{N}\right)\right] \tag{3.23}$$

Since the discrete voltage sequences $v_1(n)$ and $v_2(n)$ are directly sampled by channels CH1 and CH2 of the SAC, respectively, the frequency-domain voltage equivalents V_1 and V_2 at time t and frequency index k can be obtained using Equation (3.23). Consequently, the resultant impedance Z_x at time t and frequency index k can be calculated using Equation (3.21), and the corresponding online impedance Z_{SUT} can be determined through de-embedding as described earlier.

Figure 3.16 illustrates the procedure for extracting Z_{SUT} using the moving-window DFT algorithm. The procedure begins by sampling the voltage signals $v_1(m)$ and $v_2(m)$ at $t = 0$, where m corresponds to the $(m + 1)$th discrete sample and is related to time by $m = t/\Delta t$. When $m \geq N - 1$, the DFT is performed to obtain the voltage equivalents V_1 and V_2 at time t and frequency index k. These voltage equivalents are computed using the most recent N samples in the time domain, ranging from $(m - N + 1)\Delta t$ to $m\Delta t$. Based on the extracted V_1 and V_2, the resultant impedance Z_x at time t and frequency index k is calculated using Equation (3.21), followed by the extraction of Z_{SUT} through de-embedding. The algorithm then checks whether the current sampling time t has reached the final sampling time t_{end}. If not, the sampling time is updated to $t = t + \Delta t$, and the same process is repeated until $t \geq t_{end}$.

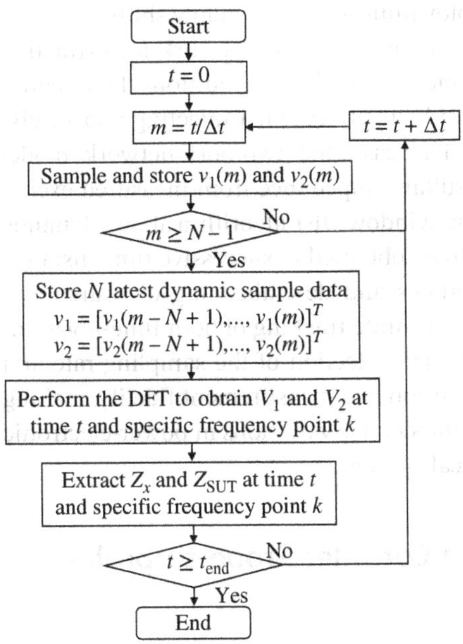

Figure 3.16 Procedure for extracting Z_{SUT} using the moving-window DFT algorithm.

For the moving-window DFT algorithm, the selection of the sampling rate f_s (where $f_s = 1/\Delta t$) and the window size N has a significant impact on the measurement performance. To ensure sufficient sampling resolution within each cycle of the sinusoidal test signal, f_s is recommended to be at least 10 times the excitation frequency f_{sig}, i.e.,$f_s = 10f_{\text{sig}}$, which provides 10 sampling points per signal cycle. In addition, to achieve reliable DFT accuracy, each DFT operation should include data spanning at least 10 complete cycles of the sinusoidal excitation. Accordingly, the window size is set to $N = 100$. For time-variant SUT, the selection of f_s must also consider the characteristic frequency of impedance variation f_{ch}. To ensure that the impedance can be regarded as quasi-stationary within each moving window, the duration of the window, N/f_s, should be much smaller than $1/f_{\text{ch}}$, leading to the condition

$$f_s \gg N f_{\text{ch}} \tag{3.24}$$

Considering that the sampling rate of commercially available SACs can reach several gigahertz [9], the TD-TPS is capable of extracting the time-variant online impedance of most electrical systems. Furthermore, since the moving window is advanced by one sample at a time, the algorithm provides superior temporal resolution and improved sensitivity to dynamic impedance variations compared with conventional DFT approaches that employ multi-sample window shifts.

This section extends the FD-TPS to a time-domain framework for extracting time-variant online impedance under practical operating conditions. By integrating a computer-controlled SGAS with inductively coupled injecting and receiving probes, a TD-TPS is established. The cascaded two-port network model enables systematic derivation of the resultant impedance from measured excitation and response signals, while a moving-window DFT algorithm allows dynamic frequency-domain voltage equivalents to be obtained at successive time instants. Through pre-characterization of the probes and de-embedding of source and wiring impedances, the TD-TPS enables accurate tracking of both time-invariant and time-variant system impedances. Proper selection of the sampling rate and window size ensures high temporal resolution and measurement fidelity, making the TD-TPS suitable for monitoring fast impedance variations in power-electronic and other dynamically operating electrical systems.

3.5 Three-term Calibration to Consider Probe-to-probe Coupling Effects

3.5.1 Three-port Network Concept

Although the TPS has been extended to handle time-variant online impedance extraction, the potential impact of probe-to-probe coupling between the IIP and the RIP on the extraction accuracy has not yet been investigated. This effect becomes particularly significant in practical applications with severe space constraints, where the two probes must be placed in close proximity, leading to nonnegligible mutual coupling effects [11]. To mitigate the resulting measurement errors, this section introduces two three-term calibration methods aimed at compensating for probe-to-probe coupling and improving the overall accuracy of online impedance extraction. In previous TPS-based analyses, the interaction between the IIP and RIP has been implicitly neglected. However, when the probes are positioned close to each other, the induced electromagnetic coupling may significantly distort the measured signals and cannot be ignored. To explicitly account for both probe-to-wire coupling and probe-to-probe coupling, a comprehensive equivalent circuit model based on the three-port network concept is introduced [7]. Accordingly, the TD-TPS shown in Figure 3.14 is reformulated as a three-port network, as illustrated in Figure 3.17.

In Figure 3.17, V_{SGC} and Z_{SGC} denote the equivalent source voltage and internal impedance of the SGC, respectively. Z_{CH1} and Z_{CH2} represent the input impedances of channels CH1 and CH2 of the SAC. V_1 and I_1 are the excitation signal voltage and current at the input port of the IIP, respectively, where V_1 is directly measured by CH1 of the SAC. Similarly, V_2 and I_2 are the response signal voltage and current at the output port of the RIP, respectively, where V_2 is directly measured by CH2 of the SAC and $I_2 = V_2/Z_{CH2}$. The voltage V_3 is the induced signal voltage between nodes p_1 and p_2, and I_3 is the induced signal current flowing through the resultant impedance Z_x to be measured. Since the frequency f_{sig} of the sinusoidal test signal generated by the SGC is much higher than that of the power source, the source voltage V_S can be treated as an AC short circuit at f_{sig}. Consequently, V_3 and I_3 are related by $I_3 = V_3/Z_x$. Under this formulation, both the probe-to-wire coupling

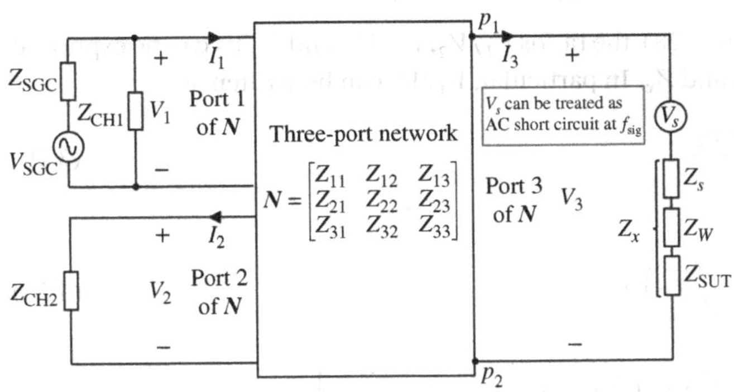

Figure 3.17 Equivalent circuit model of Figure 3.14 represented as a three-port network.

and the probe-to-probe coupling are fully captured by a three-port network, which can be described by the impedance matrix

$$\boldsymbol{N} = \begin{bmatrix} Z_{11} & Z_{12} & Z_{13} \\ Z_{21} & Z_{22} & Z_{23} \\ Z_{31} & Z_{32} & Z_{33} \end{bmatrix} \tag{3.25}$$

where Z_{ij} $(i, j = 1, 2, 3)$ denotes the impedance parameter of the three-port network $\boldsymbol{N}$. Port 1 corresponds to the input port of the IIP, Port 2 corresponds to the output port of the RIP, and Port 3 corresponds to the induced port between nodes p_1 and p_2. The corresponding port voltages and currents are related by

$$\begin{bmatrix} V_1 \\ V_2 \\ V_3 \end{bmatrix} = \begin{bmatrix} Z_{11} & Z_{12} & Z_{13} \\ Z_{21} & Z_{22} & Z_{23} \\ Z_{31} & Z_{32} & Z_{33} \end{bmatrix} \begin{bmatrix} I_1 \\ I_2 \\ I_3 \end{bmatrix} \tag{3.26}$$

Since $I_2 = V_2/Z_{\mathrm{CH2}}$ and $I_3 = V_3/Z_x$, Equation (3.26) can be rewritten as

$$\begin{bmatrix} V_1 \\ V_2 \\ V_3 \end{bmatrix} = \begin{bmatrix} Z_{11} & Z_{12} & Z_{13} \\ Z_{21} & Z_{22} & Z_{23} \\ Z_{31} & Z_{32} & Z_{33} \end{bmatrix} \begin{bmatrix} I_1 \\ V_2/Z_{\mathrm{CH2}} \\ V_3/Z_x \end{bmatrix} \tag{3.27}$$

Dividing both sides of Equation (3.27) by V_2 yields

$$\begin{bmatrix} V_1/V_2 \\ 1 \\ V_3/V_2 \end{bmatrix} = \begin{bmatrix} Z_{11} & Z_{12} & Z_{13} \\ Z_{21} & Z_{22} & Z_{23} \\ Z_{31} & Z_{32} & Z_{33} \end{bmatrix} \begin{bmatrix} I_1/V_2 \\ 1/Z_{\mathrm{CH2}} \\ (V_3/V_2)/Z_x \end{bmatrix} \tag{3.28}$$

By solving Equation (3.28), the ratios V_1/V_2, V_3/V_2, and I_1/V_2 can be expressed in terms of Z_{ij}, Z_{CH2}, and Z_x. In particular, V_1/V_2 can be written as

$$\frac{V_1}{V_2} = \frac{a_1 \cdot Z_x + a_2}{Z_x + a_3} \tag{3.29}$$

where

$$a_1 = \frac{Z_{11}}{Z_{21}} \cdot \left(1 - \frac{Z_{22}}{Z_{\mathrm{CH2}}}\right) + \frac{Z_{12}}{Z_{\mathrm{CH2}}} \tag{3.30}$$

$$\begin{aligned} a_2 = {} & \left(Z_{13} - \frac{Z_{11}Z_{23}}{Z_{21}}\right) \cdot \left[\frac{Z_{31}}{Z_{21}} \cdot \left(1 - \frac{Z_{22}}{Z_{\mathrm{CH2}}}\right) + \frac{Z_{32}}{Z_{\mathrm{CH2}}}\right] \\ & - \left(Z_{33} - \frac{Z_{31}Z_{23}}{Z_{21}}\right) \cdot \left[\frac{Z_{11}}{Z_{21}} \cdot \left(1 - \frac{Z_{22}}{Z_{\mathrm{CH2}}}\right) + \frac{Z_{12}}{Z_{\mathrm{CH2}}}\right] \end{aligned} \tag{3.31}$$

$$a_3 = \frac{Z_{31}Z_{23}}{Z_{21}} - Z_{33} \tag{3.32}$$

The detailed derivation of Equation (3.29) is provided as follows. From Equation (3.28), we obtain

$$\frac{V_1}{V_2} = Z_{11} \cdot \frac{I_1}{V_2} + \frac{Z_{12}}{Z_{CH2}} + \frac{Z_{13}}{Z_x} \cdot \frac{V_3}{V_2} \tag{3.33}$$

$$1 = Z_{21} \cdot \frac{I_1}{V_2} + \frac{Z_{22}}{Z_{CH2}} + \frac{Z_{23}}{Z_x} \cdot \frac{V_3}{V_2} \tag{3.34}$$

$$\frac{V_3}{V_2} = Z_{31} \cdot \frac{I_1}{V_2} + \frac{Z_{32}}{Z_{CH2}} + \frac{Z_{33}}{Z_x} \cdot \frac{V_3}{V_2} \tag{3.35}$$

From Equation (3.34), the ratio I_1/V_2 can be expressed as

$$\frac{I_1}{V_2} = \frac{1}{Z_{21}}\left(1 - \frac{Z_{22}}{Z_{CH2}} - \frac{Z_{23}}{Z_x} \cdot \frac{V_3}{V_2}\right) \tag{3.36}$$

Substituting Equation (3.36) into Equation (3.35), the ratio V_3/V_2 is obtained as

$$\frac{V_3}{V_2} = \frac{\frac{Z_{31}}{Z_{21}}\left(1 - \frac{Z_{22}}{Z_{CH2}}\right) + \frac{Z_{32}}{Z_{CH2}}}{1 + \frac{Z_{31}Z_{23}}{Z_{21}Z_x} - \frac{Z_{33}}{Z_x}} \tag{3.37}$$

Substituting Equations (3.36) and (3.37) into Equation (3.33), the final expression for V_1/V_2 is obtained and corresponds to Equation (3.29). Consequently, Z_x can be explicitly expressed as a function of the measured voltage ratio V_1/V_2 as

$$Z_x = \frac{a_3 \cdot \frac{V_1}{V_2} - a_2}{-\frac{V_1}{V_2} + a_1} \tag{3.38}$$

3.5.2 Open-short-load Calibration

In this section, a three-term calibration technique is introduced and elaborated. By performing this calibration, the probe-to-probe coupling influence on the extracted online impedance can be effectively eliminated. From Equation (3.38), Z_x can be determined from the voltage ratio V_1/V_2 once the coefficients a_1, a_2, and a_3 are known, where V_1 and V_2 can be directly extracted from the measurement setup shown in Figure 3.14. By examining Equations (3.30)–(3.32), it can be observed that, for a given SAC with a fixed input impedance Z_{CH2}, the coefficients a_1, a_2, and a_3 depend solely on the impedance parameters of the three-port network $\boldsymbol{N}$ (i.e., Z_{ij}). These parameters are determined by the physical characteristics of the inductive probes and the test signal frequency f_{sig}. In addition, the spacing between the inductive probes, denoted as d, also affects the elements Z_{ij}, since the strength of probe-to-probe coupling varies with distance. To determine the values of a_1, a_2, and a_3, the unknown impedance Z_x is replaced by three known but distinct termination conditions prior to actual online measurement. As standard industrial calibration procedures commonly employ open, short, and 50 Ω resistive load terminations, the same calibration standards are adopted here.

Based on Equation (3.38), when Port 3 of the three-port network $\boldsymbol{N}$ in Figure 3.17 is left open (i.e., $Z_x = \infty$, denoted as R_∞), the following relation is obtained:

$$a_1 = \left.\frac{V_1}{V_2}\right|_{R_\infty} \tag{3.39}$$

When Port 3 of $\boldsymbol{N}$ is short-circuited (i.e., $Z_x = 0$, denoted as R_0), the following expression holds:

$$a_2 = a_3 \cdot \left.\frac{V_1}{V_2}\right|_{R_0} \tag{3.40}$$

When Port 3 of $\boldsymbol{N}$ is terminated with a 50 Ω resistive load (i.e., $Z_x = 50\ \Omega$, denoted as R_{50}), the following is obtained:

$$R_{50} = \frac{a_3 \cdot \left.\frac{V_1}{V_2}\right|_{R_{50}} - a_2}{-\left.\frac{V_1}{V_2}\right|_{R_{50}} + a_1} \tag{3.41}$$

In Equations (3.39)–(3.41), $V_1/V_2\,|_t$ represents the voltage ratio measured at the input of the IIP and the output of the RIP, respectively, where the subscript t denotes the termination condition (R_∞, R_0, or R_{50}). Since $V_1/V_2\,|_t$ can be directly obtained from the measurement setup, the coefficients a_2 and a_3 can be determined from Equations (3.42) and (3.43), respectively,

$$a_2 = R_{50} \cdot \left.\frac{V_1}{V_2}\right|_{R_0} \cdot \frac{\left(\left.\frac{V_1}{V_2}\right|_{R_\infty} - \left.\frac{V_1}{V_2}\right|_{R_{50}}\right)}{\left(\left.\frac{V_1}{V_2}\right|_{R_{50}} - \left.\frac{V_1}{V_2}\right|_{R_0}\right)} \tag{3.42}$$

$$a_3 = R_{50} \cdot \frac{\left(\left.\frac{V_1}{V_2}\right|_{R_\infty} - \left.\frac{V_1}{V_2}\right|_{R_{50}}\right)}{\left(\left.\frac{V_1}{V_2}\right|_{R_{50}} - \left.\frac{V_1}{V_2}\right|_{R_0}\right)} \tag{3.43}$$

Substituting Equations (3.39), (3.42), and (3.43) into Equation (3.38), the unknown impedance Z_x can be determined as

$$Z_x = R_{50} \cdot \frac{\left(\left.\frac{V_1}{V_2}\right|_{R_\infty} - \left.\frac{V_1}{V_2}\right|_{R_{50}}\right)\left(\frac{V_1}{V_2} - \left.\frac{V_1}{V_2}\right|_{R_0}\right)}{\left(\left.\frac{V_1}{V_2}\right|_{R_{50}} - \left.\frac{V_1}{V_2}\right|_{R_0}\right)\left(\left.\frac{V_1}{V_2}\right|_{R_\infty} - \frac{V_1}{V_2}\right)} \tag{3.44}$$

It should be noted that a prerequisite for extracting Z_x using Equation (3.44) is that the voltages $V_1|_t$ and $V_2|_t$ under each termination condition must be measurable. Since $V_1|_t$ corresponds to the injected signal, it is typically well above the noise floor of the measurement system and therefore exhibits a high SNR. Under the conditions $Z_x = 0$ or $Z_x = R_{50}$, the voltages $V_2|_{R_0}$ and $V_2|_{R_{50}}$ are also generally well above the noise floor. When $Z_x = \infty$, $V_2|_{R_\infty}$ remains measurable if the probe-to-probe coupling is sufficiently strong. Conversely, if the coupling is weak, $V_2|_{R_\infty}$ may fall below the noise floor, in which case the probe-to-probe coupling has a negligible influence on the extracted impedance. Under such conditions, the simpler two-probe formulation in Equation (3.21) remains sufficiently accurate for extracting Z_x.

To simplify the expression of Equation (3.21), it can be rewritten as

$$Z_x = k \cdot \frac{V_1}{V_2} + b \tag{3.45}$$

where

$$k = \frac{1}{A_I\left(C_R + \frac{D_R}{Z_{CH2}}\right)} \tag{3.46}$$

$$b = -\frac{A_R + \frac{B_R}{Z_{CH2}}}{C_R + \frac{D_R}{Z_{CH2}}} - \frac{B_I}{A_I} \tag{3.47}$$

In addition to the characterization technique shown in Figure 3.12, the calibration technique can also be used to determine the values of k and b. When Z_x is replaced by a short circuit ($Z_x = 0\ \Omega$), Equation (3.45) becomes

$$0 = k \cdot \left.\frac{V_1}{V_2}\right|_{R_0} + b \tag{3.48}$$

Similarly, when Z_x is replaced by a 50 Ω resistive load ($Z_x = R_{50}$), Equation (3.45) becomes

$$R_{50} = k \cdot \left.\frac{V_1}{V_2}\right|_{R_{50}} + b \tag{3.49}$$

By solving Equations (3.48) and (3.49), k and b can be obtained as

$$k = \frac{R_{50}}{\left(\left.\frac{V_1}{V_2}\right|_{R_{50}} - \left.\frac{V_1}{V_2}\right|_{R_0}\right)} \tag{3.50}$$

$$b = \frac{-R_{50} \cdot \left.\frac{V_1}{V_2}\right|_{R_0}}{\left(\left.\frac{V_1}{V_2}\right|_{R_{50}} - \left.\frac{V_1}{V_2}\right|_{R_0}\right)} \tag{3.51}$$

Substituting Equations (3.50) and (3.51) into Equation (3.45), Z_x can finally be determined as

$$Z_x = \frac{R_{50} \cdot \left(\frac{V_1}{V_2} - \left.\frac{V_1}{V_2}\right|_{R_0}\right)}{\left(\left.\frac{V_1}{V_2}\right|_{R_{50}} - \left.\frac{V_1}{V_2}\right|_{R_0}\right)} \tag{3.52}$$

Figure 3.18 summarizes the online impedance measurement process based on the OSL calibration technique and can be described as follows:

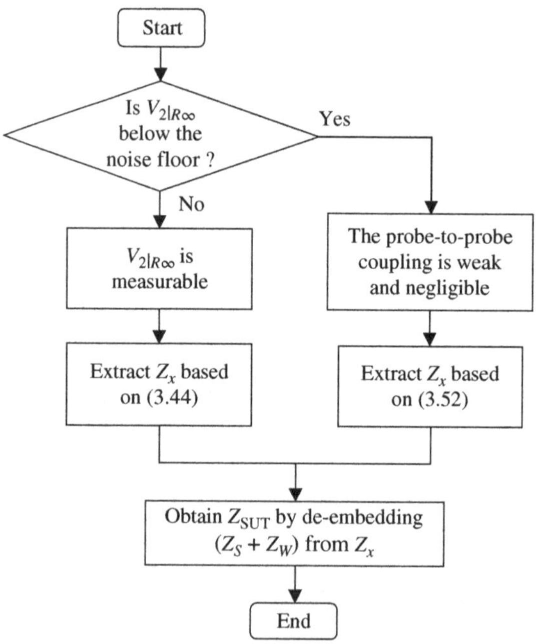

Figure 3.18 Online impedance measurement process based on the OSL calibration technique.

- If $V_2|_{R_\infty}$ is above the noise floor of the measurement system, the probe-to-probe coupling is significant and $V_2|_{R_\infty}$ is measurable. In this case, Equation (3.44) should be used to determine Z_x.
- If $V_2|_{R_\infty}$ is below the noise floor of the measurement system, the probe-to-probe coupling is weak and can be neglected. Therefore, Equation (3.52) can be applied to determine Z_x.
- After extracting Z_x, the online impedance of the system under test, Z_{SUT}, can be obtained by de-embedding $(Z_s + Z_W)$ from Z_x.

3.5.3 Improved Three-term Calibration

As explained earlier, to determine the coefficients a_1, a_2, and a_3 of a given TPS, the OSL calibration technique requires three calibration components corresponding to an open circuit, a short circuit, and a 50 Ω resistive load. In this calibration, the impedances of the calibration components are assumed to be sufficiently close to their ideal values (i.e., ∞, 0, and 50 Ω, respectively) at the test frequency. This assumption imposes stringent requirements on the calibration components. However, due to inherent parasitic series inductances and shunt capacitances, the impedances of practical calibration components deviate from their ideal values, particularly at relatively high test frequencies. Consequently, the OSL calibration technique is only valid when the test frequency is relatively low such that the parasitic effects of the calibration components are negligible [11].

To overcome this limitation, an improved three-term calibration technique is introduced to enhance measurement accuracy even when the test frequency is relatively high, and the parasitic effects of the calibration components are significant. Instead of assuming ideal calibration standards, the improved calibration employs three distinct known calibration components (denoted as CC1, CC2, and CC3) with frequency-dependent impedances Z_{CC1}, Z_{CC2}, and Z_{CC3}, respectively, where $Z_{CC1} \neq Z_{CC2} \neq Z_{CC3}$. Unlike the OSL calibration, this calibration technique does not require the calibration components to exhibit ideal impedance characteristics. Prior to calibration, the frequency-dependent impedances Z_{CC1}, Z_{CC2}, and Z_{CC3} are accurately measured over the frequency range of interest using a precision IA. Based on Equation (3.38), when the terminals $p_1 - p_2$ are successively terminated with Z_{CC1}, Z_{CC2}, and Z_{CC3}, the following relationships are obtained:

$$Z_{CC1} = \frac{a_3(V_1/V_2|_{CC1}) - a_2}{-V_1/V_2|_{CC1} + a_1} \tag{3.53}$$

$$Z_{CC2} = \frac{a_3(V_1/V_2|_{CC2}) - a_2}{-V_1/V_2|_{CC2} + a_1} \tag{3.54}$$

$$Z_{CC3} = \frac{a_3(V_1/V_2|_{CC3}) - a_2}{-V_1/V_2|_{CC3} + a_1} \tag{3.55}$$

where $V_1/V_2\,|_{CCi}$ $(i = 1, 2, 3)$ can be directly measured using the TD-TPS.

Rearranging Equations (3.53)–(3.55) yields

$$Z_{CC1}a_1 + a_2 - (V_1/V_2|_{CC1})a_3 = (V_1/V_2|_{CC1})Z_{CC1} \tag{3.56}$$

$$Z_{CC2}a_1 + a_2 - (V_1/V_2|_{CC2})a_3 = (V_1/V_2|_{CC2})Z_{CC2} \tag{3.57}$$

$$Z_{CC3}a_1 + a_2 - (V_1/V_2|_{CC3})a_3 = (V_1/V_2|_{CC3})Z_{CC3} \tag{3.58}$$

Expressing Equations (3.56)–(3.58) in matrix form gives

$$\begin{bmatrix} Z_{CC1} & 1 & -V_1/V_2|_{CC1} \\ Z_{CC2} & 1 & -V_1/V_2|_{CC2} \\ Z_{CC3} & 1 & -V_1/V_2|_{CC3} \end{bmatrix} \begin{bmatrix} a_1 \\ a_2 \\ a_3 \end{bmatrix} = \begin{bmatrix} (V_1/V_2|_{CC1}) \cdot Z_{CC1} \\ (V_1/V_2|_{CC2}) \cdot Z_{CC2} \\ (V_1/V_2|_{CC3}) \cdot Z_{CC3} \end{bmatrix} \tag{3.59}$$

Using Cramer's rule, the calibration coefficients a_1, a_2, and a_3 can be solved as

$$a_1 = \begin{bmatrix} (V_1/V_2|_{CC1})(V_1/V_2|_{CC2})(Z_{CC1} - Z_{CC2}) \\ +(V_1/V_2|_{CC1})(V_1/V_2|_{CC3})(Z_{CC3} - Z_{CC1}) \\ +(V_1/V_2|_{CC2})(V_1/V_2|_{CC3})(Z_{CC2} - Z_{CC3}) \end{bmatrix} / \Lambda \tag{3.60}$$

$$a_2 = \begin{bmatrix} (V_1/V_2|_{CC1})(V_1/V_2|_{CC2})(Z_{CC2} - Z_{CC1})Z_{CC3} \\ +(V_1/V_2|_{CC1})(V_1/V_2|_{CC3})(Z_{CC1} - Z_{CC3})Z_{CC2} \\ +(V_1/V_2|_{CC2})(V_1/V_2|_{CC3})(Z_{CC3} - Z_{CC2})Z_{CC1} \end{bmatrix} / \Lambda \tag{3.61}$$

$$a_3 = \begin{bmatrix} (V_1/V_2|_{CC1})(Z_{CC2} - Z_{CC3})Z_{CC1} \\ +(V_1/V_2|_{CC2})(Z_{CC3} - Z_{CC1})Z_{CC2} \\ +(V_1/V_2|_{CC3})(Z_{CC1} - Z_{CC2})Z_{CC3} \end{bmatrix} / \Lambda \tag{3.62}$$

$$\begin{aligned} \Lambda = (V_1/V_2|_{CC1})(Z_{CC3} - Z_{CC2}) + (V_1/V_2|_{CC2}) \\ \cdot(Z_{CC1} - Z_{CC3}) + (V_1/V_2|_{CC3})(Z_{CC2} - Z_{CC1}) \end{aligned} \tag{3.63}$$

It should be noted that Λ must be nonzero to ensure the uniqueness of the solutions for a_1, a_2, and a_3.

This section addresses the influence of probe-to-probe coupling on the accuracy of two-probe online impedance measurements and presents systematic calibration strategies to mitigate its effects. By modeling the interaction between the injecting and receiving probes using a three-port network, the coupling mechanisms are incorporated into the measurement framework. A three-term calibration approach is first formulated to correct coupling-induced errors, followed by an OSL-based implementation for practical calibration. To overcome the limitations of the OSL calibration at higher frequencies, an improved three-term calibration method is further introduced, which relies on nonideal calibration standards with known impedances. Together, these methods enable robust and accurate online impedance extraction even when probes are placed in close proximity, significantly extending the applicability of the TPS under space-constrained and high-frequency measurement conditions.

3.6 Measurement Setup Consideration Under Significant Electrical Noise and Power Surge Conditions

A computer-controlled SGAS-based inductively coupled measurement setup, together with the three-term calibration techniques, has been introduced in the previous sections to enable the extraction of both time-invariant and time-variant online impedances of electrical systems. However, due to practical constraints, the measurement setup shown in Figure 3.14 cannot be directly applied to electrical systems operating under significant electrical noise and power surge conditions. In such environments, strong electrical noise can severely mask the injected test signal, leading to a degraded SNR, while high-amplitude power surges may cause unexpected damage to sensitive measurement instruments. To address these practical implementation challenges, the measurement setup in Figure 3.14 must be enhanced by improving its SNR through signal amplification and by increasing its robustness via appropriate surge protection measures. Accordingly, this section introduces an improved measurement setup that incorporates signal amplification and surge protection modules, enabling reliable online impedance measurements in harsh electrical environments.

3.6.1 Measurement Setup Incorporating Signal Amplification and Surge Protection Modules

Figure 3.19 illustrates an enhanced measurement setup that incorporates signal amplification and surge protection modules for the online impedance extraction of an SUT. The SUT is powered by either a DC source or a low-frequency (e.g., 50/60 Hz) power supply through the wiring connection. The overall measurement setup consists of a computer-controlled SGAS, a power amplifier (PA), two attenuators (AT1 and AT2), two surge protectors (SP1 and SP2), an IIP, and a RIP.

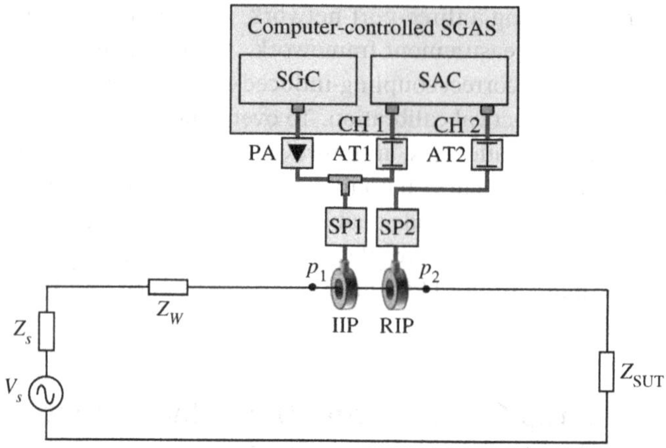

Figure 3.19 TD-TPS incorporating signal amplification and surge protection modules.

To extract the online impedance of the SUT (Z_{SUT}), the SGC in the SGAS produces a sinusoidal excitation (test) signal with a known frequency (f_{sig}). This excitation signal is amplified by the PA and then injected into the connecting wire through the IIP, while the RIP monitors the response signal. The IIP and RIP are clamped onto the connecting wire between locations p_1 and p_2, with a spacing distance denoted as d. Channel 1 (CH1) of the SAC in the SGAS measures the excitation signal voltage at the output of AT1, whereas Channel 2 (CH2) measures the response signal voltage at the output of AT2. The attenuators AT1 and AT2 are employed to ensure that the measured signal voltage levels remain within the permissible input range of the SAC. The surge protectors SP1 and SP2 are introduced to protect the measurement instruments from power surges originating from the electrical system. By incorporating signal amplification and surge protection, this measurement setup can be reliably applied to electrical systems operating under significant electrical noise and power surge conditions. As a result, the robustness and practical applicability of the inductive coupling-based online impedance measurement approach are substantially enhanced.

3.6.2 Equivalent Circuit Model Based on Two-port and Three-port Network Concepts

Based on the two-port and three-port network concepts [12], Figure 3.20 illustrates the equivalent circuit model corresponding to the measurement setup shown in Figure 3.19. In this model, V_{SGC} and Z_{SGC} denote the equivalent source voltage and

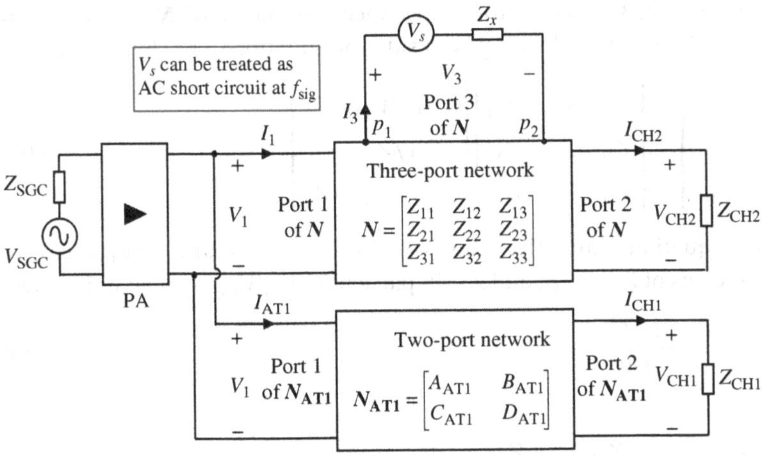

Figure 3.20 Equivalent circuit model of Figure 3.19 represented by two-port and three-port networks.

internal impedance of the SGC, respectively. Z_{CH1} and Z_{CH2} represent the input impedances of CH1 and CH2 of the SAC, respectively. V_1 and I_1 are the excitation signal voltage and current at the input port of surge protector SP1, respectively. V_{CH2} and I_{CH2} are the response signal voltage and current at the output port of attenuator AT2, respectively. V_{CH2} can be directly measured by CH2 of the SAC, and $I_{CH2} = V_{CH2}/Z_{CH2}$. V_3 denotes the induced signal voltage between points p_1 and p_2, while I_3 is the induced signal current flowing through the resultant impedance Z_x to be extracted.

With the inclusion of the PA, the test signal can be sufficiently amplified such that the induced signal level is significantly higher than the background noise, even under severe noise conditions. Since the frequency f_{sig} of the injected sinusoidal test signal is much higher than the frequency of the power source, V_S can be treated as an AC short circuit at f_{sig}. Consequently, V_3 and I_3 satisfy the relation $I_3 = V_3/Z_x$. The network $\boldsymbol{N}$ is modeled as a three-port network, where Port 1 corresponds to the input port of SP1, Port 2 corresponds to the output port of AT2, and Port 3 corresponds to the port between points p_1 and p_2. Therefore, the voltages and currents at the three ports of $\boldsymbol{N}$ are related by

$$\begin{bmatrix} V_1 \\ V_{CH2} \\ V_3 \end{bmatrix} = \begin{bmatrix} Z_{11} & Z_{12} & Z_{13} \\ Z_{21} & Z_{22} & Z_{23} \\ Z_{31} & Z_{32} & Z_{33} \end{bmatrix} \begin{bmatrix} I_1 \\ I_{CH2} \\ I_3 \end{bmatrix} \tag{3.64}$$

where Z_{ij} $(i, j = 1, 2, 3)$ denotes the impedance parameter of $\boldsymbol{N}$. Since $I_{CH2} = V_{CH2}/Z_{CH2}$ and $I_3 = V_3/Z_x$, dividing both sides of Equation (3.64) by V_{CH2} yields

$$\begin{bmatrix} V_1/V_{CH2} \\ 1 \\ V_3/V_{CH2} \end{bmatrix} = \begin{bmatrix} Z_{11} & Z_{12} & Z_{13} \\ Z_{21} & Z_{22} & Z_{23} \\ Z_{31} & Z_{32} & Z_{33} \end{bmatrix} \begin{bmatrix} I_1/V_{CH2} \\ 1/Z_{CH2} \\ (V_3/V_{CH2})/Z_x \end{bmatrix} \tag{3.65}$$

By solving Equation (3.65), the ratios V_1/V_{CH2}, V_3/V_{CH2}, and I_1/V_{CH2} can be expressed in terms of Z_{ij}, Z_{CH2}, and Z_x. In particular, V_1/V_{CH2} can be written as

$$\frac{V_1}{V_{CH2}} = \frac{a_1 \cdot Z_x + a_2}{Z_x + a_3} \tag{3.66}$$

where

$$a_1 = \frac{Z_{11}}{Z_{21}} \cdot \left(1 - \frac{Z_{22}}{Z_{CH2}}\right) + \frac{Z_{12}}{Z_{CH2}} \tag{3.67}$$

$$\begin{aligned} a_2 = &\left(Z_{13} - \frac{Z_{11}Z_{23}}{Z_{21}}\right) \cdot \left[\frac{Z_{31}}{Z_{21}} \cdot \left(1 - \frac{Z_{22}}{Z_{CH2}}\right) + \frac{Z_{32}}{Z_{CH2}}\right] \\ &-\left(Z_{33} - \frac{Z_{31}Z_{23}}{Z_{21}}\right) \cdot \left[\frac{Z_{11}}{Z_{21}} \cdot \left(1 - \frac{Z_{22}}{Z_{CH2}}\right) + \frac{Z_{12}}{Z_{CH2}}\right] \end{aligned} \tag{3.68}$$

$$a_3 = \frac{Z_{31}Z_{23}}{Z_{21}} - Z_{33} \tag{3.69}$$

In addition, attenuator AT1 can be modeled as a two-port network $\boldsymbol{N}_{\mathbf{AT1}}$, as shown in Figure 3.20. The voltages and currents at the ports of $\boldsymbol{N}_{\mathbf{AT1}}$ are related by

$$\begin{bmatrix} V_1 \\ I_{AT1} \end{bmatrix} = \begin{bmatrix} A_{AT1} & B_{AT1} \\ C_{AT1} & D_{AT1} \end{bmatrix} \begin{bmatrix} V_{CH1} \\ I_{CH1} \end{bmatrix} \tag{3.70}$$

where $A_{AT1}, B_{AT1}, C_{AT1}$, and D_{AT1} are the transmission parameters of $\boldsymbol{N}_{\mathbf{AT1}}$. I_{AT1} is the test signal current at the input port of the AT1; V_{CH1} and I_{CH1} are the excitation signal voltage and current at the output of AT1, respectively. Since V_{CH1} can be directly measured by CH1 of the SAC and $I_{CH1} = V_{CH1}/Z_{CH1}$, solving Equation (3.70) yields

$$V_1 = \left(A_{AT1} + \frac{B_{AT1}}{Z_{CH1}}\right) \cdot V_{CH1} \tag{3.71}$$

Substituting Equation (3.71) into Equation (3.66), the expression becomes

$$\frac{V_{CH1}}{V_{CH2}} = \frac{a_4 \cdot Z_x + a_5}{Z_x + a_3} \tag{3.72}$$

where

$$a_4 = \frac{a_1}{A_{\text{AT1}} + \frac{B_{\text{AT1}}}{Z_{\text{CH1}}}} \tag{3.73}$$

$$a_5 = -\frac{a_2}{A_{\text{AT1}} + \frac{B_{\text{AT1}}}{Z_{\text{CH1}}}} \tag{3.74}$$

Finally, the resultant impedance Z_x can be expressed as a function of the directly measurable voltage ratio $V_{\text{CH1}}/V_{\text{CH2}}$ as

$$Z_x = \frac{a_3 \cdot \frac{V_{\text{CH1}}}{V_{\text{CH2}}} - a_5}{-\frac{V_{\text{CH1}}}{V_{\text{CH2}}} + a_4} \tag{3.75}$$

Once a_3, a_4, and a_5 are determined, Z_x can be obtained from Equation (3.75), where V_{CH1} and V_{CH2} are directly measured by CH1 and CH2 of the SAC in the measurement setup, respectively. From Equations (3.67)–(3.69), (3.73), and (3.74), it can be observed that the coefficients a_3, a_4, and a_5 depend solely on the impedance parameters of $\boldsymbol{N}$, the transmission parameters of $\boldsymbol{N}_{\textbf{AT1}}$, and the input impedances of CH1 and CH2 of the SAC (i.e., Z_{CH1} and Z_{CH2}). Since $\boldsymbol{N}, \boldsymbol{N}_{\textbf{AT1}}, \boldsymbol{Z}_{\textbf{CH1}}$, and $\boldsymbol{Z}_{\textbf{CH2}}$ at a given frequency are uniquely determined by the inherent characteristics of the instruments in the measurement setup, the values of a_3, a_4, and a_5 at that frequency remain unchanged for a specific measurement configuration. To determine a_3, a_4, and a_5 for a given measurement setup at the test frequency f_{sig}, a three-term calibration technique, as described in Section 3.5, can be performed prior to the actual online measurement by sequentially replacing Z_x with three known but distinct termination conditions. After obtaining a_3, a_4, and a_5, the unknown Z_x can be calculated from Equation (3.75) using the measured values of V_{CH1} and V_{CH2}. Subsequently, Z_{SUT} can be extracted by de-embedding $(Z_S + Z_W)$ from Z_x.

This section addresses the practical challenges of applying the TD-TPS in electrical systems subject to significant electrical noise and power surge conditions. By incorporating signal amplification, attenuation, and surge protection modules into the measurement setup, the introduced configurations significantly enhance the SNR while ensuring the safety and robustness of the measurement instruments. Equivalent circuit models based on two-port and three-port network concepts are further developed to accurately characterize the modified setup and enable reliable impedance extraction under harsh operating environments. For the FD-TPS, similar enhancement strategies can be implemented following the same design principles. Overall, these enhancements extend the applicability of the two-probe measurement framework to real-world electrical systems with severe noise and transient disturbances.

3.7 Chapter Summary

This chapter systematically presents the theory, implementation, and practical considerations of two-probe setups for electrically contactless online impedance measurement. Beginning with the fundamental operating principles of inductive probes and the cascaded two-port network concept, the chapter establishes a unified analytical framework for impedance extraction using both frequency-domain and time-domain two-probe configurations. Practical measurement setups are then developed for time-invariant and time-variant systems, incorporating pre-characterization, de-embedding, and moving-window signal processing techniques. To address probe-to-probe coupling effects, three-term calibration methods based on three-port network theory are introduced, including conventional OSL calibration and an improved approach that accounts for the frequency-dependent parasitics of calibration components. Finally, enhanced measurement configurations incorporating signal amplification and surge protection are introduced to ensure robust operation under severe electrical noise and power surge conditions. Together, these developments provide a comprehensive and scalable two-probe measurement framework suitable for accurate online impedance extraction in realistic and challenging electrical environments.

References

1 R. A. Southwick, and W. C. Dolle, "Line impedance measuring instrumentation utilizing current probe coupling," *IEEE Trans. Electromagn. Compat.*, vol. EMC-13, no. 4, pp. 31–36, 1971.

2 K. R. Li, K. Y. See, and X. M. Li, "Inductive coupled in-circuit impedance monitoring of electrical system using two-port ABCD network approach," *IEEE Trans. Instrum. Meas.*, vol. 64, no. 9, pp. 2489–2495, 2015.

3 Z. Zhao, K. Y. See, E. K. Chua et al., "Time-variant in-circuit impedance monitoring based on the inductive coupling method," *IEEE Trans. Instrum. Meas.*, vol. 68, no. 1, pp. 169–176, 2019.

4 D. C. Smith, "Current probes, more useful than you think," in *Proc. IEEE Int. Symp. Electromagn. Compat.*, USA, Aug. 1998.

5 M. Yahyavi, F. Brojeni, and M. Vaziri, "Basic theory and practical considerations in a current transformer," in *Proc. IEEE Power Eng. Soc. Gen. Meeting*, USA, Jun. 2007.

6 N. Kondrath and M. K. Kazimierczuk, "Bandwidth of current transformers," *IEEE Trans. Instrum. Meas.*, vol. 58, no. 6, pp. 2008–2016, Oct. 2009.

7 D. M. Pozar, *Microwave engineering: theory and techniques*. Hoboken, NJ, USA: Wiley, 2021, pp. 166–193.

8 *PXI-5922 Oscilloscope*, Nat. Instrum., Austin, TX, USA, 2018.

9 *PXIe-5186 Digitizer*, Nat. Instrum., Austin, TX, USA, 2015.

10 J. A. Rosendo Macias and A. Gomez Exposito, "Efficient moving-window DFT algorithms," *IEEE Trans. Circuits Syst. II. Analog Digit. Signal Process. (1993–2003)*, vol. 45, no. 2, pp. 256–260, Feb. 1998.

11 Z. Zhao, A. Weerasinghe, W. Wang et al., "Eliminating the effect of probe-to-probe coupling in inductive coupling method for in-circuit impedance measurement," *IEEE Trans. Instrum. Meas.*, vol. 70, pp. 1–8, 2021.

12 D. M. Pozar, *Microwave engineering*. Hoboken, NJ, USA: Wiley, 2012, pp. 165–227.

4

Single-probe Setups

This chapter presents a systematic and comprehensive treatment of the single-probe inductive coupling approach for electrically contactless online impedance measurement. Building upon the fundamental principles introduced in the preceding chapters, the single-probe setup (SPS) simplifies the measurement hardware while maintaining accuracy and flexibility for in-circuit applications. Both frequency-domain and time-domain implementations are investigated, encompassing the theoretical formulation of online impedance extraction, pre-measurement characterization, systematic error cancellation, excitation signal design, and practical deployment considerations. Through detailed analytical derivations and experimental insights, this chapter establishes the SPS as a versatile and robust solution for online impedance measurement in energized electrical systems.

4.1 Setup for Frequency-domain Measurement

4.1.1 Basic Principle

Figure 4.1 illustrates the FD-SPS for online impedance measurement of an energized SUT [1]. The SUT consists of a load powered by an AC or DC power supply through connecting cables. The measurement plane at the load is denoted by l–l'. An inductive probe is clamped onto one of the power cables at a clamping location denoted by c–c'. The inductive probe is connected to an impedance measurement instrument, whose measurement reference plane is denoted by m–m'. The objective of the FD-SPS is to extract the online impedance of the SUT as observed at c–c', or equivalently at l–l' when the load termination is accessible. By establishing an analytical relationship between the measurement reference plane m–m' and the clamping plane c–c', the SUT impedance can be determined. Similarly, by establishing the relationship between m–m' and l–l', the load impedance can also be extracted.

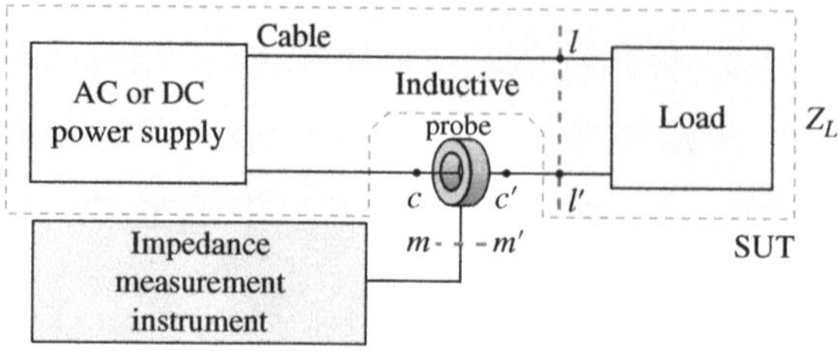

Figure 4.1 Basic FD-SPS configuration for online impedance measurement.

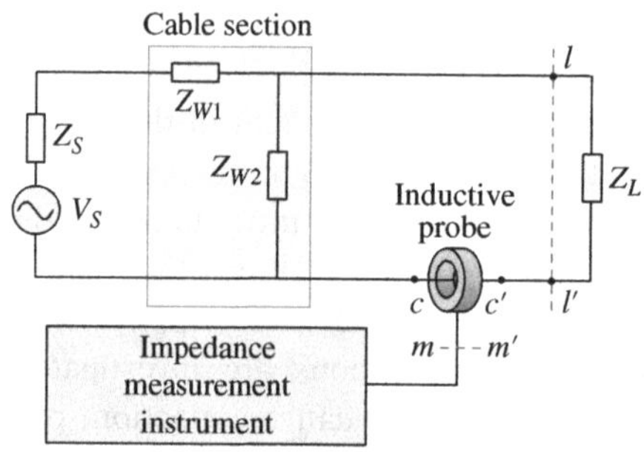

Figure 4.2 Equivalent circuit of the FD-SPS.

The equivalent circuit of the FD-SPS is shown in Figure 4.2. The power supply of the energized SUT is represented by its Thevenin equivalent voltage V_S and internal impedance Z_S, while the load is represented by its impedance Z_L. In the frequency range of interest, the physical length of the power cables is typically much shorter than the wavelength corresponding to the highest measurement frequency. Therefore, the cables can be modeled using lumped-element representations consisting of series and shunt impedances, denoted by Z_{W1} and Z_{W2}, respectively. These impedances capture the nonideal parasitic effects of the cables, which become increasingly significant at frequencies above approximately 10 MHz.

The FD-SPS can also be represented as two cascaded two-port networks, as shown in Figure 4.3. Each two-port network is characterized by its corresponding *ABCD* parameters. In this representation, the power supply together with the connecting cables is combined and modeled as a single two-port network, while the inductive probe constitutes the other two-port network. The impedance observed by the measurement instrument at the reference plane m–m' is denoted by Z_m. Similarly, the impedance seen by the inductive probe at the clamping plane c–c' is denoted by Z_c. The impedance measurement instrument can be either an IA or a VNA, as discussed in the subsequent sections.

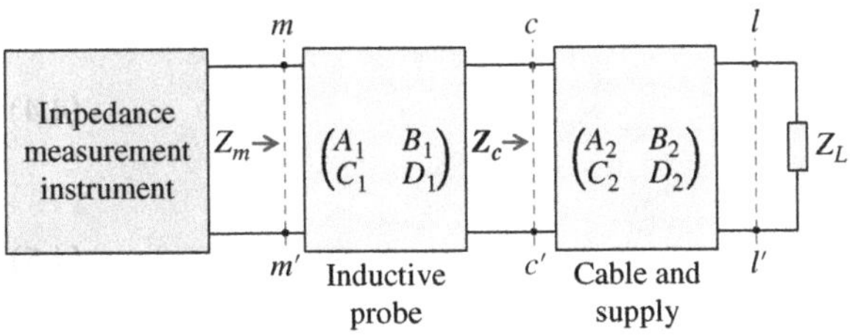

Figure 4.3 Cascaded two-port network representation of the FD-SPS.

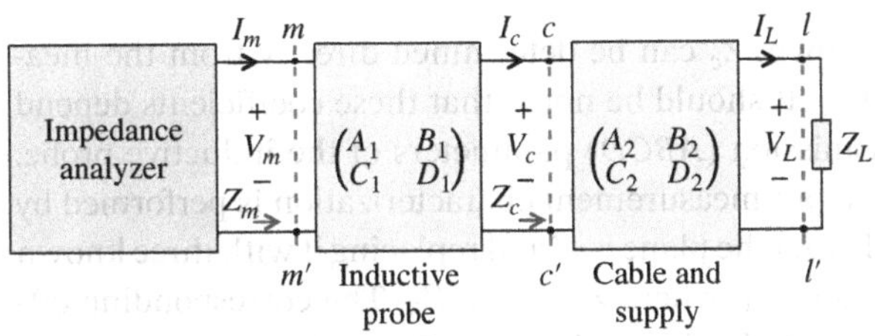

Figure 4.4 Online impedance measurement using the IA-based FD-SPS.

4.1.2 IA-based FD-SPS

Figure 4.4 illustrates the FD-SPS configuration when an IA is employed as the impedance measurement instrument. The voltage and current generated by the excitation of the IA at the measurement reference plane m–m' are denoted by V_m and I_m, respectively. The resulting voltage and current at the probe clamping location c–c' are denoted by V_c and I_c, respectively, while the corresponding voltage and current at the load reference plane l–l' are denoted by V_L and I_L, respectively. The impedance directly observed by the IA is denoted by Z_m.

Through network analysis, the relationship between the measurement planes m–m' and c–c' can be expressed as

$$Z_m = \frac{V_m}{I_m} \frac{A_1 V_c + B_1 I_c}{C_1 V_c + D_1 I_c} \tag{4.1}$$

Since $Z_c = V_c / I_c$, Equation (4.1) can be rewritten as

$$Z_m = \frac{A_1 Z_c + B_1}{C_1 Z_c + D_1} \tag{4.2}$$

Rearranging Equation (4.2) yields

$$Z_c = \frac{k_1 Z_m + k_2}{Z_m + k_3} \tag{4.3}$$

where

$$k_1 = -\frac{D_1}{C_1} \tag{4.4}$$

$$k_2 = \frac{B_1}{C_1} \tag{4.5}$$

$$k_3 = -\frac{A_1}{C_1} \tag{4.6}$$

Once k_1, k_2, and k_3 are known, Z_c can be determined directly from the measured Z_m using Equation (4.3). It should be noted that these coefficients depend solely on the two-port transmission ($ABCD$) parameters of the inductive probe. To determine k_1, k_2, and k_3, a pre-measurement characterization is performed by removing the energized SUT from the plane c–c' and replacing it with three known distinct impedances Z_1, Z_2, and Z_3, where $Z_1 \neq Z_2 \neq Z_3$. The corresponding off-circuit pre-measurement characterization is shown in Figure 4.5.

By substituting Z_1, Z_2, and Z_3 into Equation (4.3), the following expressions are obtained:

$$Z_1 = \frac{k_1 Z_{m_1} + k_2}{Z_{m_1} + k_3} \tag{4.7}$$

$$Z_2 = \frac{k_1 Z_{m_2} + k_2}{Z_{m_2} + k_3} \tag{4.8}$$

$$Z_3 = \frac{k_1 Z_{m_3} + k_2}{Z_{m_3} + k_3} \tag{4.9}$$

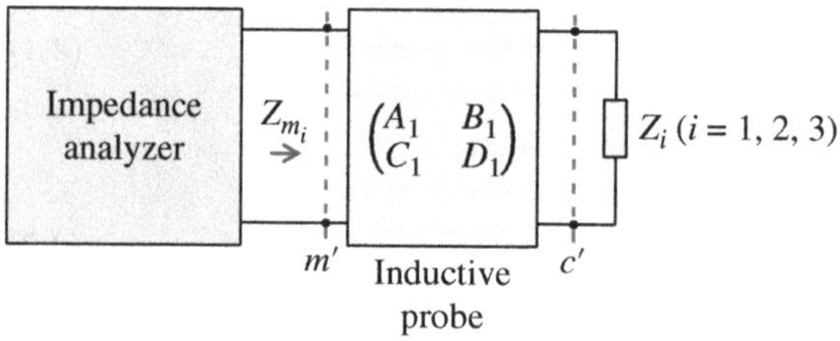

Figure 4.5 Pre-measurement characterization of the IA-based FD-SPS at c–c'.

where Z_{m_1}, Z_{m_2}, and Z_{m_3} are the impedances measured by the IA at m–m' corresponding to Z_1, Z_2, and Z_3, respectively. By combining Equations (4.3), (4.7), and (4.8), the following relationship is obtained:

$$\left(\frac{Z_c - Z_1}{Z_1 - Z_2}\right) = \left(\frac{\dfrac{k_1 Z_m + k_2}{Z_m + k_3} - \dfrac{k_1 Z_{m_2} + k_2}{Z_{m_2} + k_3}}{\dfrac{k_1 Z_{m_1} + k_2}{Z_{m_1} + k_3} - \dfrac{k_1 Z_{m_2} + k_2}{Z_{m_2} + k_3}}\right) \tag{4.10}$$

Similarly, combining Equations (4.3), (4.8), and (4.9) yields

$$\left(\frac{Z_2 - Z_3}{Z_3 - Z_c}\right) = \left(\frac{\dfrac{k_1 Z_{m_2} + k_2}{Z_{m_2} + k_3} - \dfrac{k_1 Z_{m_3} + k_2}{Z_{m_3} + k_3}}{\dfrac{k_1 Z_{m_2} + k_2}{Z_{m_3} + k_3} - \dfrac{k_1 Z_m + k_2}{Z_m + k_3}}\right) \tag{4.11}$$

By multiplying Equations (4.10) and (4.11), Z_c can be expressed as

$$Z_c = \frac{a + bZ_m}{c + dZ_m} \tag{4.12}$$

where

$$a = Z_1 Z_2 (Z_{m_2} - Z_{m_1}) Z_{m_3} + Z_1 Z_3 (Z_{m_1} - Z_{m_3}) Z_{m_2} + Z_2 Z_3 (Z_{m_3} - Z_{m_2}) Z_{m_1} \tag{4.13}$$

$$b = Z_1 Z_2 (Z_{m_1} - Z_{m_2}) + Z_1 Z_3 (Z_{m_3} - Z_{m_1}) + Z_2 Z_3 (Z_{m_2} - Z_{m_3}) \tag{4.14}$$

$$c = Z_1 (Z_{m_2} - Z_{m_3}) Z_{m_1} + Z_2 (Z_{m_3} - Z_{m_1}) Z_{m_2} + Z_3 (Z_{m_1} - Z_{m_2}) Z_{m_3} \tag{4.15}$$

$$d = Z_1 (Z_{m_3} - Z_{m_2}) + Z_2 (Z_{m_1} - Z_{m_3}) + Z_3 (Z_{m_2} - Z_{m_1}) \tag{4.16}$$

By comparing Equations (4.3) and (4.12), k_1, k_2, and k_3 can be obtained as follows:

$$k_1 = b/d \tag{4.17}$$

$$k_2 = a/d \tag{4.18}$$

$$k_3 = c/d \tag{4.19}$$

With the measured values Z_{m_1}, Z_{m_2}, and Z_{m_3} for the respective calibration impedances Z_1, Z_2, and Z_3, the coefficients k_1, k_2, and k_3 can be determined. Once these coefficients are known, the relationship between Z_c and Z_m is fully established, enabling any unknown Z_c to be extracted from the corresponding measured Z_m at m–m'. It should be noted that both k_1 and k_2 have units of ohms (Ω), whereas k_3 is dimensionless. Similarly, through network analysis, the relationship between m–m' and l–l' in Figure 4.4 can be expressed as follows:

$$Z_L = \frac{h_1 Z_m + h_2}{Z_m + h_3} \tag{4.20}$$

where

$$h_1 = -(B_2C_1 + D_1D_2)/(A_2C_1 + C_2D_1) \tag{4.21}$$

$$h_2 = (A_1B_2 + B_1D_2)/(A_2C_1 + C_2D_1) \tag{4.22}$$

$$h_3 = -(A_1A_2 + B_1C_2)/(A_2C_1 + C_2D_1) \tag{4.23}$$

Once $h_1, h_2,$ and h_3 are known, Z_L can be extracted from the measured Z_m. These coefficients depend on the *ABCD* parameters of the inductive probe, the power supply, and the connecting cables. To determine $h_1, h_2,$ and h_3, a pre-measurement characterization can be performed by removing Z_L (when accessible) from the plane l–l' and replacing it with three known distinct impedances Z_1, Z_2, and Z_3, where $Z_1 \neq Z_2 \neq Z_3$. The corresponding pre-measurement characterization setup is shown in Figure 4.6.

By substituting $Z_1, Z_2,$ and Z_3 into Equation (4.20), the following expressions are obtained:

$$Z_1 = \frac{h_1 Z_{m_1} + h_2}{Z_{m_1} + h_3} \tag{4.24}$$

$$Z_2 = \frac{h_1 Z_{m_2} + h_2}{Z_{m_2} + h_3} \tag{4.25}$$

$$Z_3 = \frac{h_1 Z_{m_3} + h_2}{Z_{m_3} + h_3} \tag{4.26}$$

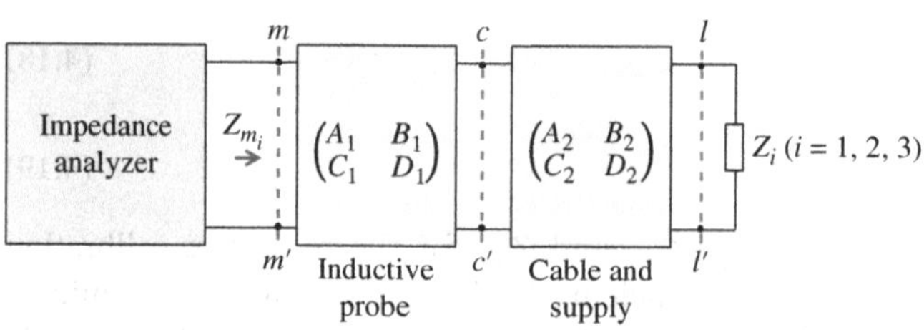

Figure 4.6 Pre-measurement characterization of the IA-based FD-SPS at l–l'.

where Z_{m_1}, Z_{m_2}, and Z_{m_3} are the impedances measured by the IA at m–m' for Z_1, Z_2, and Z_3, respectively. Combining Equations (4.20), (4.24), and (4.25) yields

$$\left(\frac{Z_L - Z_1}{Z_1 - Z_2}\right) = \left(\frac{\dfrac{h_1 Z_m + h_2}{Z_m + h_3} - \dfrac{h_1 Z_{m_1} + h_2}{Z_{m_1} + h_3}}{\dfrac{h_1 Z_{m_1} + h_2}{Z_{m_1} + h_3} - \dfrac{h_1 Z_{m_2} + h_2}{Z_{m_2} + h_3}}\right) \tag{4.27}$$

Similarly, combining Equations (4.20), (4.25), and (4.26) leads to

$$\left(\frac{Z_2 - Z_3}{Z_3 - Z_L}\right) = \left(\frac{\dfrac{h_1 Z_{m_2} + h_2}{Z_{m_2} + h_3} - \dfrac{h_1 Z_{m_3} + h_2}{Z_{m_3} + h_3}}{\dfrac{h_1 Z_{m_3} + h_2}{Z_{m_3} + h_3} - \dfrac{h_1 Z_m + h_2}{Z_m + h_3}}\right) \tag{4.28}$$

By multiplying Equations (4.27) and (4.28), the load impedance Z_L can be expressed as

$$Z_L = \frac{a' + b' Z_m}{c' + d' Z_m} \tag{4.29}$$

where

$$a' = Z_1 Z_2 (Z_{m_2} - Z_{m_1}) Z_{m_3} + Z_1 Z_3 (Z_{m_1} - Z_{m_3}) Z_{m_2} + Z_2 Z_3 (Z_{m_3} - Z_{m_2}) Z_{m_1} \tag{4.30}$$

$$b' = Z_1 Z_2 (Z_{m_1} - Z_{m_2}) + Z_1 Z_3 (Z_{m_3} - Z_{m_1}) + Z_2 Z_3 (Z_{m_2} - Z_{m_3}) \tag{4.31}$$

$$c' = Z_1 (Z_{m_2} - Z_{m_3}) Z_{m_1} + Z_2 (Z_{m_3} - Z_{m_1}) Z_{m_2} + Z_3 (Z_{m_1} - Z_{m_2}) Z_{m_3} \tag{4.32}$$

$$d' = Z_1 (Z_{m_3} - Z_{m_2}) + Z_2 (Z_{m_1} - Z_{m_3}) + Z_3 (Z_{m_2} - Z_{m_1}) \tag{4.33}$$

By comparing Equations (4.20) and (4.29), the coefficients h_1, h_2, and h_3 can be expressed as

$$h_1 = \frac{b'}{d'} \tag{4.34}$$

$$h_2 = \frac{a'}{d'} \tag{4.35}$$

$$h_3 = \frac{c'}{d'} \tag{4.36}$$

With the measured values of Z_{m_1}, Z_{m_2}, and Z_{m_3} corresponding to Z_1, Z_2, and Z_3, respectively, the coefficients h_1, h_2, and h_3 can be determined. Once these coefficients are known, the relationship between Z_L and Z_m is fully established, enabling the extraction of any unknown load impedance Z_L from the measured Z_m at m–m'. It should be noted that both h_1 and h_2 have units of ohms (Ω), whereas h_3 is dimensionless. Furthermore, any changes in the cabling or the power supply will require re-characterization of h_1, h_2, and h_3.

4.1.3 VNA-based FD-SPS

Figure 4.7 illustrates the FD-SPS configuration when a VNA is used as the impedance measurement instrument. The reflection coefficient measured by the VNA at the reference plane m–m' is denoted by Γ_m. The reflection coefficient and impedance observed at c–c' are denoted by Γ_c and Z_c, respectively. Similarly, the reflection coefficient and impedance of the load at l–l' are denoted by Γ_L and Z_L, respectively.

Through network analysis, the relationship between the reference planes m–m' and c–c'can be expressed as

$$\Gamma_m = \frac{\frac{A_1 Z_c + B_1}{C_1 Z_c + D_1} - Z_0}{\frac{A_1 Z_c + B_1}{C_1 Z_c + D_1} + Z_0} \tag{4.37}$$

where Z_0 is the reference impedance of the VNA, typically 50 Ω. Equation (4.37) can be rearranged as

$$Z_c = \frac{k_1 \Gamma_m + k_2}{\Gamma_m + k_3} \tag{4.38}$$

where

$$k_1 = -\frac{B_1 + D_1 Z_0}{A_1 + C_1 Z_0} \tag{4.39}$$

$$k_2 = -\frac{D_1 Z_0 - B_1}{A_1 + C_1 Z_0} \tag{4.40}$$

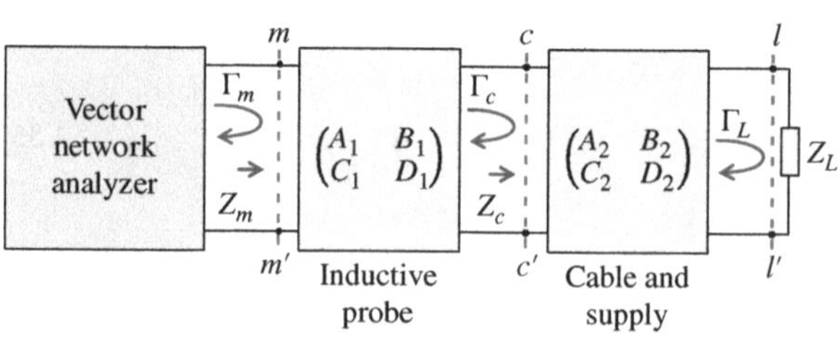

Figure 4.7 Online impedance measurement using the VNA-based FD-SPS.

$$k_3 = \frac{C_1 Z_0 - A_1}{A_1 + C_1 Z_0} \tag{4.41}$$

Once the coefficients k_1, k_2, and k_3 are determined, Z_c can be extracted directly from the measured Γ_m. It should be noted that these coefficients depend on the two-port *ABCD* parameters of the inductive probe as well as the reference impedance of the VNA (Z_0). To determine k_1, k_2, and k_3, a pre-measurement characterization can be performed by removing the SUT at c–c' and replacing it with three known and distinct impedances Z_1, Z_2, and Z_3, where $Z_1 \neq Z_2 \neq Z_3$. The corresponding pre-measurement characterization setup is shown in Figure 4.8.

By substituting Z_1, Z_2, and Z_3 into Equation (4.38), the following expressions are obtained:

$$Z_1 = \frac{k_1 \Gamma_{m_1} + k_2}{\Gamma_{m_1} + k_3} \tag{4.42}$$

$$Z_2 = \frac{k_1 \Gamma_{m_2} + k_2}{\Gamma_{m_2} + k_3} \tag{4.43}$$

$$Z_3 = \frac{k_1 \Gamma_{m_3} + k_2}{\Gamma_{m_3} + k_3} \tag{4.44}$$

where Γ_{m_1}, Γ_{m_2}, and Γ_{m_3} are the reflection coefficients measured by the VNA at m–m' corresponding to Z_1, Z_2, and Z_3 at c–c', respectively. Combining Equations (4.38), (4.42), and (4.43) yields

$$\left(\frac{Z_c - Z_1}{Z_1 - Z_2}\right) = \left(\frac{\dfrac{k_1\Gamma_m + k_2}{\Gamma_m + k_3} - \dfrac{k_1\Gamma_{m_1} + k_2}{\Gamma_{m_1} + k_3}}{\dfrac{k_1\Gamma_{m_1} + k_2}{\Gamma_{m_1} + k_3} - \dfrac{k_1\Gamma_{m_2} + k_2}{\Gamma_{m_2} + k_3}}\right) \tag{4.45}$$

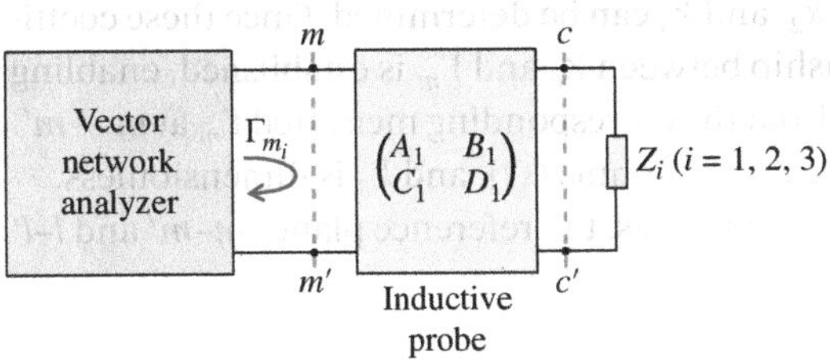

Figure 4.8 Pre-measurement characterization of the VNA-based FD-SPS at c–c'.

Similarly, combining Equations (4.38), (4.43), and (4.44) leads to

$$\left(\frac{Z_2 - Z_3}{Z_3 - Z_c}\right) = \left(\frac{\dfrac{k_1\Gamma_{m_2} + k_2}{\Gamma_{m_2} + k_3} - \dfrac{k_1\Gamma_{m_3} + k_2}{\Gamma_{m_3} + k_3}}{\dfrac{k_1\Gamma_{m_3} + k_2}{\Gamma_{m_3} + k_3} - \dfrac{k_1\Gamma_{m} + k_2}{\Gamma_{m} + k_3}}\right) \tag{4.46}$$

By multiplying Equations (4.45) and (4.46), Z_c can be expressed as

$$Z_c = \frac{a + b\Gamma_m}{c + d\Gamma_m} \tag{4.47}$$

where

$$a = \Gamma_{m_1}\Gamma_{m_3}Z_2(Z_3 - Z_1) + \Gamma_{m_2}\Gamma_{m_3}Z_1(Z_2 - Z_3) + \Gamma_{m_1}\Gamma_{m_2}Z_3(Z_1 - Z_2) \tag{4.48}$$

$$b = \Gamma_{m_1}Z_1(Z_2 - Z_3) + \Gamma_{m_2}Z_2(Z_3 - Z_1) + \Gamma_{m_3}Z_3(Z_1 - Z_2) \tag{4.49}$$

$$c = \Gamma_{m_1}Z_1(\Gamma_{m_2} - \Gamma_{m_3}) + \Gamma_{m_2}Z_2(\Gamma_{m_3} - \Gamma_{m_1}) + \Gamma_{m_3}Z_3(\Gamma_{m_1} - \Gamma_{m_2}) \tag{4.50}$$

$$d = Z_1(\Gamma_{m_3} - \Gamma_{m_2}) + Z_2(\Gamma_{m_1} - \Gamma_{m_3}) + Z_3(\Gamma_{m_2} - \Gamma_{m_1}) \tag{4.51}$$

By comparing Equations (4.3) and (4.47), the coefficients k_1, k_2, and k_3 can be expressed as

$$k_1 = \frac{b}{d} \tag{4.52}$$

$$k_2 = \frac{a}{d} \tag{4.53}$$

$$k_3 = \frac{c}{d} \tag{4.54}$$

With the measured Γ_{m_1}, Γ_{m_2}, and Γ_{m_3} corresponding to the known impedances Z_1, Z_2, and Z_3, the coefficients k_1, k_2, and k_3 can be determined. Once these coefficients are known, the full relationship between Z_c and Γ_m is established, enabling the extraction of any unknown Z_c from the corresponding measured Γ_m at $m - m'$. It is noted that both k_1 and k_2 have units of ohms (Ω), and k_3 is dimensionless.

If Z_L is accessible, through network analysis, the reference planes m–m' and l–l' in Figure 4.7 are related by

$$Z_L = \frac{h_1\Gamma_m + h_2}{\Gamma_m + h_3} \tag{4.55}$$

where

$$h_1 = \frac{-Z_0(B_2C_1 + D_1D_2) + A_1B_2 + B_1D_2}{Z_0(A_1C_1 + C_2D_1) + A_1A_2 + B_1C_2} \tag{4.56}$$

$$h_2 = \frac{-Z_0(B_2C_1 + D_1D_2) - A_1B_2 - B_1D_2}{Z_0(A_1C_1 + C_2D_1) + A_1A_2 + B_1C_2} \tag{4.57}$$

$$h_3 = \frac{Z_0(A_2C_1 + C_2D_1) - A_1A_2 - B_1C_2}{Z_0(A_1C_1 + C_2D_1) + A_1A_2 + B_1C_2} \tag{4.58}$$

Again, with known h_1, h_2, and h_3, Z_L can be determined from the measured Z_m. The coefficients h_1, h_2, and h_3 depend on the $ABCD$ parameters of the inductive probe, the power supply, the connecting cables, and the VNA reference impedance Z_0.

To determine h_1, h_2, and h_3, a pre-measurement characterization is performed, as shown in Figure 4.9, by replacing Z_L at l–l' (if accessible) with three known distinct impedances Z_1, Z_2 and Z_3, where $Z_1 \neq Z_2 \neq Z_3$.

By substituting Z_1, Z_2, and Z_3 into Equation (4.55), the following expressions are obtained:

$$Z_1 = \frac{h_1\Gamma_{m_1} + h_2}{\Gamma_{m_1} + h_3} \tag{4.59}$$

$$Z_2 = \frac{h_1\Gamma_{m_2} + h_2}{\Gamma_{m_2} + h_3} \tag{4.60}$$

$$Z_3 = \frac{h_1\Gamma_{m_3} + h_2}{\Gamma_{m_3} + h_3} \tag{4.61}$$

where Γ_{m_1}, Γ_{m_2}, and Γ_{m_3} are the reflection coefficients measured by the VNA at m–m' corresponding to Z_1, Z_2, and Z_3, respectively. Combining Equations (4.55), (4.59), and (4.60) yields

$$\left(\frac{Z_L - Z_1}{Z_1 - Z_2}\right) = \left(\frac{\dfrac{h_1\Gamma_m + h_2}{\Gamma_m + h_3} - \dfrac{h_1\Gamma_{m_1} + h_2}{\Gamma_{m_1} + h_3}}{\dfrac{h_1\Gamma_{m_1} + h_2}{\Gamma_{m_1} + h_3} - \dfrac{h_1\Gamma_{m_2} + h_2}{\Gamma_{m_2} + h_3}}\right) \tag{4.62}$$

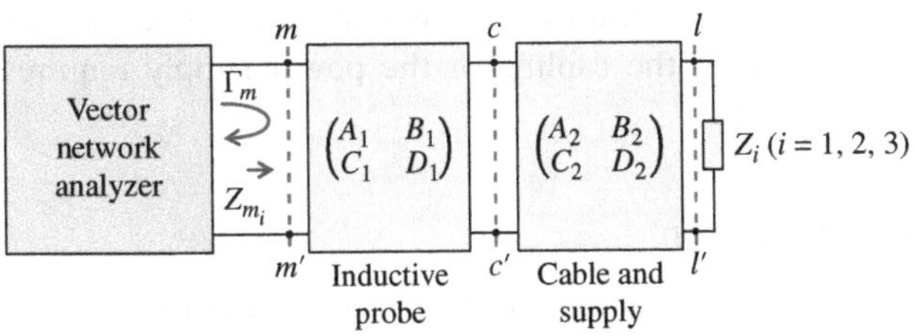

Figure 4.9 Pre-measurement characterization of the VNA-based FD-SPS at l–l'.

Similarly, combining Equations (4.55), (4.60) and (4.61) leads to

$$\left(\frac{Z_2 - Z_3}{Z_3 - Z_L}\right) = \left(\frac{\dfrac{h_1\Gamma_{m2} + h_2}{\Gamma_{m2} + h_3} - \dfrac{h_1\Gamma_{m3} + h_2}{\Gamma_{m3} + h_3}}{\dfrac{h_1\Gamma_{m3} + h_2}{\Gamma_{m3} + h_3} - \dfrac{h_1\Gamma_{m} + h_2}{\Gamma_{m} + h_3}}\right) \tag{4.63}$$

By multiplying Equations (4.62) and (4.63), the load impedance Z_L can be expressed as

$$Z_L = \frac{a + b\Gamma_m}{c + d\Gamma_m} \tag{4.64}$$

where

$$a = \Gamma_{m_1}\Gamma_{m_3}Z_2(Z_3 - Z_1) + \Gamma_{m_2}\Gamma_{m_3}Z_1(Z_2 - Z_3) + \Gamma_{m_1}\Gamma_{m_2}Z_3(Z_1 - Z_2) \tag{4.65}$$

$$b = \Gamma_{m_1}Z_1(Z_2 - Z_3) + \Gamma_{m_2}Z_2(Z_3 - Z_1) + \Gamma_{m_3}Z_3(Z_1 - Z_2) \tag{4.66}$$

$$c = \Gamma_{m_1}Z_1(\Gamma_{m_2} - \Gamma_{m_3}) + \Gamma_{m_2}Z_2(\Gamma_{m_3} - \Gamma_{m_1}) + \Gamma_{m_3}Z_3(\Gamma_{m_1} - \Gamma_{m_2}) \tag{4.67}$$

$$d = Z_1(\Gamma_{m_3} - \Gamma_{m_2}) + Z_2(\Gamma_{m_1} - \Gamma_{m_3}) + Z_3(\Gamma_{m_2} - \Gamma_{m_1}) \tag{4.68}$$

Comparing Equations (4.55) and (4.64), the coefficients h_1, h_2, and h_3 can be obtained as

$$h_1 = \frac{b}{d} \tag{4.69}$$

$$h_2 = \frac{a}{d} \tag{4.70}$$

$$h_3 = \frac{c}{d} \tag{4.71}$$

With the measured Γ_{m_1}, Γ_{m_2}, and Γ_{m_3} corresponding to the known impedances Z_1, Z_2, and Z_3, the coefficients h_1, h_2, and h_3 can be determined. The relationship between Z_L and Γ_m is thus fully established, enabling the extraction of an unknown load impedance Z_L from the measured reflection coefficient Γ_m at m–m'. It should be noted that both h_1 and h_2 have units of ohms (Ω), whereas h_3 is dimensionless. Any change in the cabling or the power supply requires re-characterization of h_1, h_2, and h_3.

4.1.4 Systematic Error Cancellation

With the pre-measurement characterization of the FD-SPS to determine k_1, k_2, and k_3, or h_1, h_2, and h_3, systematic errors inherent in the measurement instrument, whether an IA or a VNA, can be effectively canceled. This section briefly explains the underlying principle of such systematic error cancellation. For a linear error model, the systematic measurement error can be decomposed into two components: a proportional error and an offset error. Assuming that Γ_m measured by the VNA contains an offset error α and a proportional error β, the measured Γ_m and the true reflection coefficient Γ_T at the reference plane m–m' are then related by

$$\Gamma_m = (1+\beta)\Gamma_T + \alpha \tag{4.72}$$

Accordingly, the measurements obtained during the pre-measurement characterization of the FD-SPS can be expressed in terms of α and β for the purpose of illustrating error cancellation. From Equations (4.45) and (4.46), the following expressions can be formed:

$$\left(\frac{Z_c - Z_1}{Z_1 - Z_2}\right) = \left(\frac{\frac{k_1\Gamma_m + k_2}{\Gamma_m + k_3} - \frac{k_1\Gamma_{m_1} + k_2}{\Gamma_{m_1} + k_3}}{\frac{k_1\Gamma_{m_1} + k_2}{\Gamma_{m_1} + k_3} - \frac{k_1\Gamma_{m_2} + k_2}{\Gamma_{m_2} + k_3}}\right) = \left(\frac{\frac{k_1((1+\beta)\Gamma_T + \alpha) + k_2}{(1+\beta)\Gamma_T + \alpha + k_3} - \frac{k_1((1+\beta)\Gamma_{T_1} + \alpha) + k_2}{(1+\beta)\Gamma_{T_1} + \alpha + k_3}}{\frac{k_1((1+\beta)\Gamma_{T_1} + \alpha) + k_2}{(1+\beta)\Gamma_{T_1} + \alpha + k_3} - \frac{k_1((1+\beta)\Gamma_{T_2} + \alpha) + k_2}{(1+\beta)\Gamma_{T_2} + \alpha + k_3}}\right) \tag{4.73}$$

$$\left(\frac{Z_2 - Z_3}{Z_3 - Z_c}\right) = \left(\frac{\frac{k_1\Gamma_{m_2} + k_2}{\Gamma_{m_2} + k_3} - \frac{k_1\Gamma_{m_3} + k_2}{\Gamma_{m_3} + k_3}}{\frac{k_1\Gamma_{m_3} + k_2}{\Gamma_{m_3} + k_3} - \frac{k_1\Gamma_m + k_2}{\Gamma_m + k_3}}\right) = \left(\frac{\frac{k_1((1+\beta)\Gamma_{T_2} + \alpha) + k_2}{(1+\beta)\Gamma_{T_2} + \alpha + k_3} - \frac{k_1((1+\beta)\Gamma_{T_3} + \alpha) + k_2}{(1+\beta)\Gamma_{T_3} + \alpha + k_3}}{\frac{k_1((1+\beta)\Gamma_{T_3} + \alpha) + k_2}{(1+\beta)\Gamma_{T_3} + \alpha + k_3} - \frac{k_1((1+\beta)\Gamma_T + \alpha) + k_2}{(1+\beta)\Gamma_T + \alpha + k_3}}\right) \tag{4.74}$$

where Γ_{T_1}, Γ_{T_2}, and Γ_{T_3} denote the true reflection coefficients when Z_1, Z_2, and Z_3 are connected at the c–c' plane, respectively. The expressions in Equations (4.73) and (4.74) can be simplified as

$$\left(\frac{Z_c - Z_1}{Z_1 - Z_2}\right) = \left(\frac{\Gamma_m - \Gamma_{m_1}}{\Gamma_{m_1} - \Gamma_{m_2}}\right)\left(\frac{k_3 + \Gamma_{m_2}}{k_3 + \Gamma_m}\right) = \left(\frac{\Gamma_T - \Gamma_{T_1}}{\Gamma_{T_1} - \Gamma_{T_2}}\right)\left(\frac{\alpha + (1+\beta)\Gamma_{T_2} + k_3}{\alpha + (1+\beta)\Gamma_T + k_3}\right) \tag{4.75}$$

$$\left(\frac{Z_2 - Z_3}{Z_3 - Z_c}\right) = \left(\frac{\Gamma_{m_2} - \Gamma_{m_3}}{\Gamma_{m_3} - \Gamma_m}\right)\left(\frac{k_3 + \Gamma_m}{k_3 + \Gamma_{m_2}}\right) = \left(\frac{\Gamma_{T_2} - \Gamma_{T_3}}{\Gamma_{T_3} - \Gamma_T}\right)\left(\frac{\alpha + (1+\beta)\Gamma_T + k_3}{\alpha + (1+\beta)\Gamma_{T_2} + k_3}\right) \tag{4.76}$$

By multiplying Equations (4.75) and (4.76), the following relationship is obtained:

$$\left(\frac{Z_c - Z_1}{Z_1 - Z_2}\right)\left(\frac{Z_2 - Z_3}{Z_3 - Z_c}\right) = \left(\frac{\Gamma_m - \Gamma_{m_1}}{\Gamma_{m_1} - \Gamma_{m_2}}\right)\left(\frac{\Gamma_{m_2} - \Gamma_{m_3}}{\Gamma_{m_3} - \Gamma_m}\right) = \left(\frac{\Gamma_T - \Gamma_{T_1}}{\Gamma_{T_1} - \Gamma_{T_2}}\right)\left(\frac{\Gamma_{T_2} - \Gamma_{T_3}}{\Gamma_{T_3} - \Gamma_T}\right) \tag{4.77}$$

As shown in Equation (4.77), the two systematic error terms α and β are eliminated, demonstrating that these errors are canceled during the pre-measurement characterization process at the l–l' reference plane. Similarly, systematic errors in IA-based measurements are also canceled by the same pre-measurement characterization procedure, owing to the identical underlying calibration principle.

4.2 Setup for Time-domain Measurement

4.2.1 Basic Principle

The time-domain single-probe setup (TD-SPS) for online impedance measurement of an energized SUT is illustrated in Figure 4.10. The SUT consists of a load powered by either an AC or a DC power supply through connecting cables. The measurement plane at the load is denoted by l–l'. An inductive probe is clamped onto one of the power cables at the location denoted by c–c'. The inductive probe is connected to an arbitrary waveform generator (AWG) to provide excitation through a directional coupler. The coupled ports of the directional coupler are terminated by a two-channel analog-to-digital converter (ADC), whose channels are denoted as CH1 and CH2 for measuring the time-domain signals v_{CH1} and v_{CH2}, respectively. The corresponding frequency-domain quantities V_{CH1} and V_{CH2} are obtained from the measured time-domain signals using a digital signal processing (DSP) algorithm. Typically, the AWG and the ADC are integrated into a computer-controlled SGAS.

The equivalent network representation of the TD-SPS shown in Figure 4.10 is presented in Figure 4.11. The termination impedances of the two ADC channels are denoted by Z_{CH1} and Z_{CH2}. The AWG has an output impedance of Z_{AWG}. The reflection coefficients associated with the terminations of CH1, CH2, and the AWG

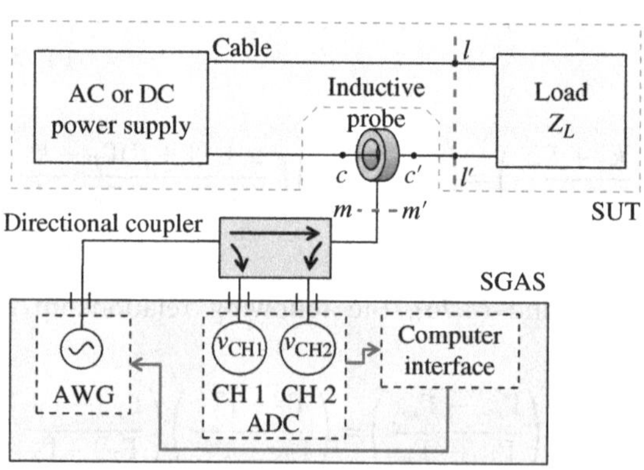

Figure 4.10 Schematic diagram of the TD-SPS for online impedance measurement.

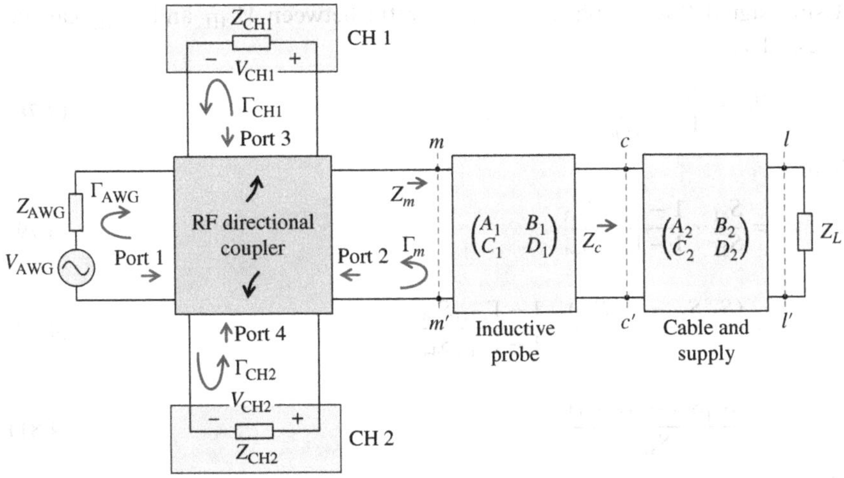

Figure 4.11 Equivalent network representation of the TD-SPS.

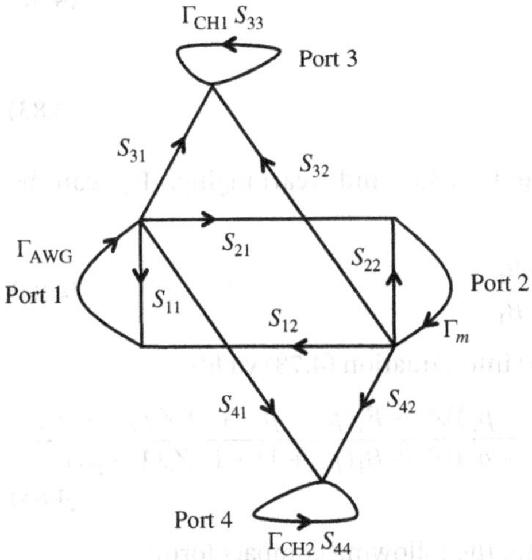

Figure 4.12 Signal flow graph of the directional coupler.

are denoted by Γ_{CH1}, Γ_{CH2}, and Γ_{AWG}, respectively. The directional coupler is modeled as a four-port network.

The signal flow graph of the directional coupler is shown in Figure 4.12, illustrating the dominant coupling paths [2]. The characteristics of the directional coupler are described using four-port S-parameters. The reflection coefficients observed at the individual ports are denoted by their respective reflection coefficients Γ_{CH1}, Γ_{CH2}, Γ_{AWG}, and Γ_m.

Using signal flow graph analysis, the ratio between V_{CH1} and V_{CH2} can be expressed as

$$\frac{V_{\mathrm{CH2}}}{V_{\mathrm{CH1}}} = \frac{p_1 + p_2\Gamma_m}{1 + p_3\Gamma_m} \tag{4.78}$$

where

$$p_1 = \frac{S_{41}}{S_{31}} \cdot \frac{1 - \Gamma_{\mathrm{CH1}} S_{33}}{1 - \Gamma_{\mathrm{CH2}} S_{44}} \tag{4.79}$$

$$p_2 = \frac{(S_{21}S_{42} - S_{41}S_{22})}{S_{31}} \cdot \frac{1 - \Gamma_{\mathrm{CH1}} S_{33}}{1 - \Gamma_{\mathrm{CH2}} S_{44}} \tag{4.80}$$

$$p_3 = \frac{(S_{21}S_{32} - S_{31}S_{22})}{S_{31}} \tag{4.81}$$

and

$$Z_m = Z_0 \frac{1 + \Gamma_m}{1 - \Gamma_m} \tag{4.82}$$

$$Z_c = -\frac{Z_m D_1 - B_1}{Z_m C_1 - A_1} \tag{4.83}$$

By combining Equations (4.82) and (4.83) and rearranging, Γ_m can be expressed as

$$\Gamma_m = -\frac{Z_c(C_1 Z_0 - A_1) + D_1 Z_0 - B_1}{Z_c(A_1 + C_1 Z_0) + D_1 Z_0 + B_1} \tag{4.84}$$

Substituting Γ_m from Equation (4.84) into Equation (4.78) yields

$$\frac{V_{\mathrm{CH2}}}{V_{\mathrm{CH1}}} = \frac{(A_1(p_1 + p_2) + C_1 Z_0(p_1 - p_2))Z_c + B_1(p_1 + p_2) + D_1 Z_0(p_1 - p_2)}{(A_1(p_3 + 1) + C_1 Z_0(1 - p_3))Z_c + B_1(p_3 + 1) + D_1 Z_0(1 - p_3)} \tag{4.85}$$

Equation (4.85) can be rearranged into the following compact form:

$$Z_c = \frac{k_1\left(\frac{V_{\mathrm{CH2}}}{V_{\mathrm{CH1}}}\right) + k_2}{\left(\frac{V_{\mathrm{CH2}}}{V_{\mathrm{CH1}}}\right) + k_3} \tag{4.86}$$

where

$$k_1 = \frac{D_1(1 - p_3)Z_0 + B_1(1 + p_3)}{C_1(1 - p_3)Z_0 + A_1(1 + p_3)} \tag{4.87}$$

$$k_2 = -\frac{D_1(p_1 - p_2)Z_0 + B_1(p_1 + p_2)}{C_1(1 - p_3)Z_0 + A_1(1 + p_3)} \tag{4.88}$$

$$k_3 = \frac{C_1(p_1 - p_2)Z_0 + A_1(p_1 + p_2)}{C_1(1 - p_3)Z_0 + A_1(1 + p_3)} \tag{4.89}$$

The vector voltage ratio (VVR) is defined as $V_{\text{CH2}}/V_{\text{CH1}}$. Once the coefficients k_1, k_2, and k_3 are determined, the impedance Z_c seen by the inductive probe can be extracted directly from the measured VVR. The coefficients p_1, p_2, and p_3 depend on the characteristics of the directional coupler, including its directivity, insertion loss, coupling factor, and isolation. In contrast, k_1, k_2, and k_3 depend on both the directional coupler and the inductive probe characteristics, as well as the selected reference impedance Z_0.

Since the effects of p_1, p_2, and p_3 are already embedded in k_1, k_2, and k_3, it is sufficient to determine k_1, k_2, and k_3 for online impedance measurement using the TD-SPS. Similar to the FD-SPS, pre-measurement characterization of the TD-SPS can be performed by connecting c–c' to three known and distinct impedances Z_1, Z_2, and Z_3, where $Z_1 \neq Z_2 \neq Z_3$. Following the same procedure described in Section 4.1, substituting Z_1, Z_2, and Z_3 into Equation (4.86) yields

$$Z_1 = \frac{k_1\left(\frac{V_{\text{CH2}}}{V_{\text{CH1}}}\right)_{Z1} + k_2}{\left(\frac{V_{\text{CH2}}}{V_{\text{CH1}}}\right)_{Z1} + k_3} \tag{4.90}$$

$$Z_2 = \frac{k_1\left(\frac{V_{\text{CH2}}}{V_{\text{CH1}}}\right)_{Z2} + k_2}{\left(\frac{V_{\text{CH2}}}{V_{\text{CH1}}}\right)_{Z2} + k_3} \tag{4.91}$$

$$Z_3 = \frac{k_1\left(\frac{V_{\text{CH2}}}{V_{\text{CH1}}}\right)_{Z3} + k_2}{\left(\frac{V_{\text{CH2}}}{V_{\text{CH1}}}\right)_{Z3} + k_3} \tag{4.92}$$

where $(V_{\text{CH2}}/V_{\text{CH1}})_{Z1}$, $(V_{\text{CH2}}/V_{\text{CH1}})_{Z2}$, and $(V_{\text{CH2}}/V_{\text{CH1}})_{Z3}$ are the measured VVRs corresponding to Z_1, Z_2, and Z_3 at c–c', respectively. Combining Equations (4.86), (4.90), and (4.91) leads to

$$\left(\frac{Z_c - Z_1}{Z_1 - Z_2}\right) = \left(\frac{\dfrac{k_1\left(\frac{V_{\text{CH2}}}{V_{\text{CH1}}}\right)+k_2}{\left(\frac{V_{\text{CH2}}}{V_{\text{CH1}}}\right)+k_3} - \dfrac{k_1\left(\frac{V_{\text{CH2}}}{V_{\text{CH1}}}\right)_{Z1}+k_2}{\left(\frac{V_{\text{CH2}}}{V_{\text{CH1}}}\right)_{Z1}+k_3}}{\dfrac{k_1\left(\frac{V_{\text{CH2}}}{V_{\text{CH1}}}\right)_{Z1}+k_2}{\left(\frac{V_{\text{CH2}}}{V_{\text{CH1}}}\right)_{Z1}+k_3} - \dfrac{k_1\left(\frac{V_{\text{CH2}}}{V_{\text{CH1}}}\right)_{Z2}+k_2}{\left(\frac{V_{\text{CH2}}}{V_{\text{CH1}}}\right)_{Z2}+k_3}}\right) \tag{4.93}$$

Similarly, combining Equations (4.86), (4.91), and (4.92) yields

$$\left(\frac{Z_2 - Z_3}{Z_3 - Z_c}\right) = \left(\frac{\dfrac{k_1\left(\frac{V_{\text{CH2}}}{V_{\text{CH1}}}\right)_{Z2}+k_2}{\left(\frac{V_{\text{CH2}}}{V_{\text{CH1}}}\right)_{Z2}+k_3} - \dfrac{k_1\left(\frac{V_{\text{CH2}}}{V_{\text{CH1}}}\right)_{Z3}+k_2}{\left(\frac{V_{\text{CH2}}}{V_{\text{CH1}}}\right)_{Z3}+k_3}}{\dfrac{k_1\left(\frac{V_{\text{CH2}}}{V_{\text{CH1}}}\right)_{Z3}+k_2}{\left(\frac{V_{\text{CH2}}}{V_{\text{CH1}}}\right)_{Z3}+k_3} - \dfrac{k_1\left(\frac{V_{\text{CH2}}}{V_{\text{CH1}}}\right)+k_2}{\left(\frac{V_{\text{CH2}}}{V_{\text{CH1}}}\right)+k_3}}\right) \tag{4.94}$$

By multiplying Equations (4.93) and (4.94), Z_c can be expressed as

$$Z_c = \frac{a + b\left(\frac{V_{\text{CH2}}}{V_{\text{CH1}}}\right)}{c + d\left(\frac{V_{\text{CH2}}}{V_{\text{CH1}}}\right)} \tag{4.95}$$

where

$$\begin{aligned} a = {} & Z_2(Z_3 - Z_1)\left(\frac{V_{\text{CH2}}}{V_{\text{CH1}}}\right)_{Z1}\left(\frac{V_{\text{CH2}}}{V_{\text{CH1}}}\right)_{Z3} + Z_1(Z_2 - Z_3)\left(\frac{V_{\text{CH2}}}{V_{\text{CH1}}}\right)_{Z2}\left(\frac{V_{\text{CH2}}}{V_{\text{CH1}}}\right)_{Z3} \\ & + Z_3(Z_1 - Z_2)\left(\frac{V_{\text{CH2}}}{V_{\text{CH1}}}\right)_{Z1}\left(\frac{V_{\text{CH2}}}{V_{\text{CH1}}}\right)_{Z2} \end{aligned} \tag{4.96}$$

$$b = Z_1(Z_2 - Z_3)\left(\frac{V_{\text{CH2}}}{V_{\text{CH1}}}\right)_{Z1} + Z_2(Z_3 - Z_1)\left(\frac{V_{\text{CH2}}}{V_{\text{CH1}}}\right)_{Z2} + Z_3(Z_1 - Z_2)\left(\frac{V_{\text{CH2}}}{V_{\text{CH1}}}\right)_{Z3} \tag{4.97}$$

$$\begin{aligned} c = {} & Z_1\left(\frac{V_{\text{CH2}}}{V_{\text{CH1}}}\right)_{Z1}\left(\left(\frac{V_{\text{CH2}}}{V_{\text{CH1}}}\right)_{Z2} - \left(\frac{V_{\text{CH2}}}{V_{\text{CH1}}}\right)_{Z3}\right) \\ & + Z_2\left(\frac{V_{\text{CH2}}}{V_{\text{CH1}}}\right)_{Z2}\left(\left(\frac{V_{\text{CH2}}}{V_{\text{CH1}}}\right)_{Z3} - \left(\frac{V_{\text{CH2}}}{V_{\text{CH1}}}\right)_{Z1}\right) \\ & + Z_3\left(\frac{V_{\text{CH2}}}{V_{\text{CH1}}}\right)_{Z3}\left(\left(\frac{V_{\text{CH2}}}{V_{\text{CH1}}}\right)_{Z1} - \left(\frac{V_{\text{CH2}}}{V_{\text{CH1}}}\right)_{Z2}\right) \end{aligned} \tag{4.98}$$

$$\begin{aligned} d = {} & Z_1\left(\left(\frac{V_{\text{CH2}}}{V_{\text{CH1}}}\right)_{Z3} - \left(\frac{V_{\text{CH2}}}{V_{\text{CH1}}}\right)_{Z2}\right) + Z_2\left(\left(\frac{V_{\text{CH2}}}{V_{\text{CH1}}}\right)_{Z1} - \left(\frac{V_{\text{CH2}}}{V_{\text{CH1}}}\right)_{Z3}\right) \\ & + Z_3\left(\left(\frac{V_{\text{CH2}}}{V_{\text{CH1}}}\right)_{Z2} - \left(\frac{V_{\text{CH2}}}{V_{\text{CH1}}}\right)_{Z1}\right) \end{aligned} \tag{4.99}$$

Comparing Equations (4.86) and (4.95), the coefficients k_1, k_2, and k_3 can be expressed as

$$k_1 = b/d \tag{4.100}$$

$$k_2 = a/d \tag{4.101}$$

$$k_3 = c/d \tag{4.102}$$

With the measured $(V_{CH2}/V_{CH1})_{Z1}$, $(V_{CH2}/V_{CH1})_{Z2}$, and $(V_{CH2}/V_{CH1})_{Z3}$ corresponding to the known impedances Z_1, Z_2, and Z_3, respectively, the coefficients k_1, k_2, and k_3 can be determined. Once these coefficients are obtained, the complete relationship between Z_c and the measured VVR is established, enabling any unknown Z_c to be extracted from a newly measured VVR. Note that both k_1 and k_2 have units of ohms (Ω), whereas k_3 is dimensionless. The above derivations can be extended in a similar manner to establish the relationship between Z_L and the measured VVR. Accordingly, Z_L can be expressed as

$$Z_L = \frac{h_1\left(\frac{V_{CH2}}{V_{CH1}}\right) + h_2}{\left(\frac{V_{CH2}}{V_{CH1}}\right) + h_3} \tag{4.103}$$

where

$$h_1 = b/d \tag{4.104}$$

$$h_2 = a/d \tag{4.105}$$

$$h_3 = c/d \tag{4.106}$$

and

$$\begin{aligned} \alpha = {} & Z_2(Z_3 - Z_1)\left(\frac{V_{CH2}}{V_{CH1}}\right)_{Z1}\left(\frac{V_{CH2}}{V_{CH1}}\right)_{Z3} + Z_1(Z_2 - Z_3)\left(\frac{V_{CH2}}{V_{CH1}}\right)_{Z2}\left(\frac{V_{CH2}}{V_{CH1}}\right)_{Z3} \\ & + Z_3(Z_1 - Z_2)\left(\frac{V_{CH2}}{V_{CH1}}\right)_{Z1}\left(\frac{V_{CH2}}{V_{CH1}}\right)_{Z2} \end{aligned} \tag{4.107}$$

$$\begin{aligned} b = {} & Z_1(Z_2 - Z_3)\left(\frac{V_{CH2}}{V_{CH1}}\right)_{Z1} + Z_2(Z_3 - Z_1)\left(\frac{V_{CH2}}{V_{CH1}}\right)_{Z2} \\ & + Z_3(Z_1 - Z_2)\left(\frac{V_{CH2}}{V_{CH1}}\right)_{Z3} \end{aligned} \tag{4.108}$$

$$\begin{aligned} c = {} & Z_1\left(\frac{V_{CH2}}{V_{CH1}}\right)_{Z1}\left(\left(\frac{V_{CH2}}{V_{CH1}}\right)_{Z2} - \left(\frac{V_{CH2}}{V_{CH1}}\right)_{Z3}\right) \\ & + Z_2\left(\frac{V_{CH2}}{V_{CH1}}\right)_{Z2}\left(\left(\frac{V_{CH2}}{V_{CH1}}\right)_{Z3} - \left(\frac{V_{CH2}}{V_{CH1}}\right)_{Z1}\right) \\ & + Z_3\left(\frac{V_{CH2}}{V_{CH1}}\right)_{Z3}\left(\left(\frac{V_{CH2}}{V_{CH1}}\right)_{Z1} - \left(\frac{V_{CH2}}{V_{CH1}}\right)_{Z2}\right) \end{aligned} \tag{4.109}$$

$$d = Z_1\left(\left(\frac{V_{\mathrm{CH2}}}{V_{\mathrm{CH1}}}\right)_{Z3} - \left(\frac{V_{\mathrm{CH2}}}{V_{\mathrm{CH1}}}\right)_{Z2}\right) + Z_2\left(\left(\frac{V_{\mathrm{CH2}}}{V_{\mathrm{CH1}}}\right)_{Z1} - \left(\frac{V_{\mathrm{CH2}}}{V_{\mathrm{CH1}}}\right)_{Z3}\right) + Z_3\left(\left(\frac{V_{\mathrm{CH2}}}{V_{\mathrm{CH1}}}\right)_{Z2} - \left(\frac{V_{\mathrm{CH2}}}{V_{\mathrm{CH1}}}\right)_{Z1}\right) \tag{4.110}$$

where $(V_{\mathrm{CH2}}/V_{\mathrm{CH1}})_{Z1}$, $(V_{\mathrm{CH2}}/V_{\mathrm{CH1}})_{Z2}$, and $(V_{\mathrm{CH2}}/V_{\mathrm{CH1}})_{Z3}$ are the measured VVRs corresponding to Z_1, Z_2 and Z_3 at l–l', respectively. Note that both h_1 and h_2 have units of ohms (Ω), whereas h_3 is dimensionless. Similar to the FD-SPS, systematic errors in the TD-SPS are also effectively canceled through the pre-measurement characterization process.

4.2.2 Multitone Excitation Signal Synthesis

The excitation signal can be either a single-tone sinusoid or a sweeping sinusoid within the frequency range of interest. However, the true potential of the TD-SPS lies in its flexibility to excite the SUT using a multitone signal. Multitone excitation can significantly improve measurement efficiency compared with other types of excitation signals, owing to reduced dead time and settling time [3–5].

A multitone signal can be synthesized with arbitrary frequency and power spectra according to the measurement requirements, such as the desired SNR. When no prior knowledge of the SUT is available, it is common to select either a linearly (equidistant) spaced or a logarithmically spaced frequency spectrum with equal power allocated to each tone [6, 7]. A logarithmic frequency distribution allows the excitation signal to span several decades of frequency with a limited number of spectral components [8]. In contrast, when a linear placement of excitation frequencies across a wide frequency range is required, the synthesis methods described in [9, 10] can be employed. However, this approach either results in reduced frequency resolution over a wide frequency span or necessitates a large number of excitation tones, which can lead to a reduced SNR due to the limited power allocated to each frequency component [11].

Figure 4.13 compares two excitation signals with linear and logarithmic spectral distributions in both the time domain and frequency domain. In this example, both signals consist of 31 spectral components spanning from 150 kHz to 30 MHz. The logarithmic distribution exhibits more pronounced spectral leakage during the DSP stage, resulting in a higher noise floor compared with the linear distribution. Therefore, logarithmic multitone signal synthesis with minimum spectral leakage is adopted and discussed in the following.

A logarithmic multitone signal with N frequency components follows the relationship

$$f_n = \alpha f_{n-1} \tag{4.111}$$

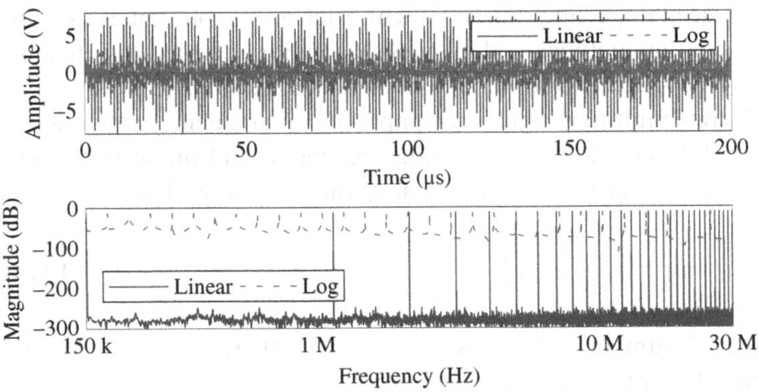

Figure 4.13 Comparison of excitation signals with linear and logarithmic frequency distributions in the time domain and frequency domain.

where $n = 1, 2, 3, \dots, N$ and $\alpha > 1$; f_n denotes the nth frequency component. Once N, f_1, and f_N are specified, α can be determined as

$$\alpha = 10^{\frac{\log\left(\frac{f_N}{f_1}\right)}{N-1}} \tag{4.112}$$

After α is obtained, the frequency components f_n can be generated for $n = 1, 2, 3, \dots, N$ using Equation (4.111). However, these frequencies are not guaranteed to align with inverse DFT bins during the synthesis stage or DFT bins during the DSP stage. This misalignment causes spectral leakage, which degrades the SNR. If the generated frequencies are instead aligned with the nearest DFT bins in the DSP stage, the excitation signal is referred to as having a quasi-logarithmic spectral distribution. Accordingly, Equation (4.111) can be approximated [12] as

$$f_n \approx \alpha f_{n-1} \tag{4.113}$$

under the constraint

$$f_n = H\frac{F_s}{K} \tag{4.114}$$

where F_s is the sampling rate, K is the number of samples, and $H \in \mathbb{N}$ [13].

In addition to frequency placement, multitone signal synthesis must consider the phase relationship among the sinusoidal components. The phase distribution should be selected to minimize the crest factor of the excitation signal. The crest factor characterizes the required maximum amplitude of the excitation signal for a given power distribution across the frequency components. Minimizing the crest factor allows the excitation signal power to be maximized under amplitude constraints [5]. This is advantageous during both the signal synthesis stage and the signal analysis stage of impedance measurement. Such excitation signals can outperform other excitation methods under low-SNR conditions for a given effective number of bits of the ADCs used in time-domain data acquisition [5–7].

Consider a discrete multitone signal x_n composed of K frequency components, where a_k, f_k, and ϕ_k denote the magnitude, frequency, and phase of the kth frequency component, respectively. The signal x_n can be expressed as [4]

$$x_n = \sum_{k=1}^{K} a_k \cos\left(2\pi f_k \frac{n}{N} + \phi_k\right) \tag{4.115}$$

where N is the total number of samples in x_n, and $n = 0, 1, 2, \ldots, (N-1)$. The discrete-time crest factor CF of x_n is defined as [4]

$$\mathrm{CF} = \frac{\max|x_n|}{\sqrt{\frac{1}{N}\sum_{n=0}^{N-1} x_n^2}} = \frac{\max|x_n|}{\sqrt{\sum_{k=1}^{K} a_k^2/2}} = \frac{\max|x_n|}{\sqrt{\sum_{k=1}^{K} p_k}} \tag{4.116}$$

where a_k and p_k denote the amplitude and power of the kth sinusoidal component, respectively. For a normalized multitone signal x_n with unit total power, the power spectrum satisfies

$$\sum_{k=1}^{K} p_k = 1 \tag{4.117}$$

When an equal power distribution among the frequency components is assumed, typically in the absence of prior knowledge of the SUT [6], the power allocated to each tone is given by

$$p_k = 1/K \tag{4.118}$$

Accordingly, a_k of each frequency component can be determined as

$$a_k = \sqrt{2p_k} \tag{4.119}$$

ϕ_k is initially determined using the Schroeder method and is subsequently optimized through an iterative procedure to further reduce the crest factor [4]. Figure 4.14 illustrates a quasi-logarithmic multitone excitation signal synthesized using the above method, consisting of 31 spectral components spanning from 150 kHz to 30 MHz. This signal can be generated by the AWG and injected into the SUT through the inductive probe.

4.2.3 Short-time Discrete Fourier Transform

The short-time discrete Fourier transform (STDFT) can be adopted for the time–frequency analysis of discrete signals. Accordingly, it is employed to convert the

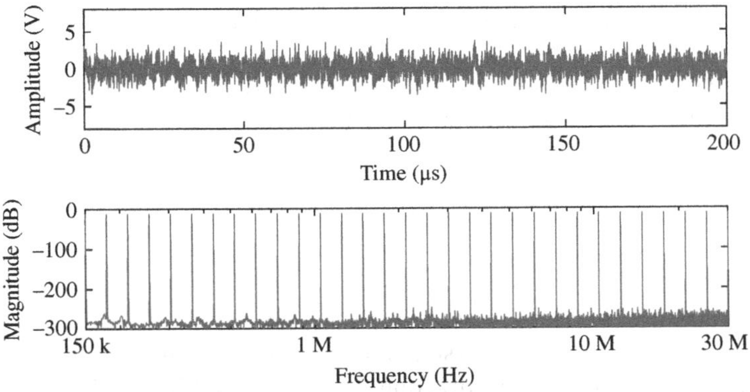

Figure 4.14 Quasi-logarithmic multitone test signal in the time domain and frequency domain.

time-domain signals v_{CH1} and v_{CH2} into their frequency-domain spectral components V_{CH1} and V_{CH2}. This enables subsequent analysis of time-variant online impedance measurement. Within the STDFT framework, an appropriate trade-off between time and frequency localization must be selected according to application requirements. Following the notation in [13] and Equation (4.86), the STDFT can be defined for real and time-limited v_{CH1} and v_{CH2} as follows:

$$V_{CH1}[\omega, q] = \sum_{n=1}^{K} w[n] v_{CH1}[n + Hq] e^{-j2\pi \frac{\omega n}{K}} \tag{4.120}$$

$$V_{CH2}[\omega, q] = \sum_{n=1}^{K} w[n] v_{CH2}[n + Hq] e^{-j2\pi \frac{\omega n}{K}} \tag{4.121}$$

where v_{CH1} and v_{CH2} are the time-domain signals measured at CH1 and CH2, respectively, and subsequently discretized and digitized by the ADC. The local time index relative to the analysis window is $n \in \{1, 2, 3, \ldots, K\}$, where $K \in \mathbb{N}$ denotes the window length. The analysis frame index is $q \in \mathbb{N}$, and $H \in \mathbb{N}$ represents the time advance from between adjacent analysis frames, which can alternatively be interpreted as the number of samples between consecutive frames (i.e., the hop size). $\mathbb{N}$ is the total number of samples in the acquired discrete signals v_{CH1} and v_{CH2}. A time-limited window function $w[n]$ is used to minimize spectral leakage. The frequency index is $\omega \in \{1, 2, 3, \ldots, K\}$, where ω corresponds to the discrete frequency localization of V_{CH1} and V_{CH2}, with the associated linear frequency given by $f = \omega f_s / K$. Temporal localization of V_{CH1} and V_{CH2} is represented by the analysis frame index q, and the corresponding continuous-time instant is $t = qH/F_s$, where F_s denotes the sampling rate. Consequently, the time resolution of the analysis is H/f_s, while the frequency resolution is f_s/K.

The STDFT outputs V_{CH1} and V_{CH2} are complex-valued sequences, whereas v_{CH1} and v_{CH2} are real-valued sequences. From Equations (4.120) and (4.121), the time-frequency localized VVR, defined as $V_{CH2}[\omega, q]/V_{CH1}[\omega, q]$, can be computed and used to determine the time-variant online impedance according to

$$Z_c[\omega, q] = \frac{k_1[\omega]\left(\frac{V_{CH2}[\omega,q]}{V_{CH1}[\omega,q]}\right) + k_2[\omega]}{\left(\frac{V_{CH2}[\omega,q]}{V_{CH1}[\omega,q]}\right) + k_3[\omega]} \tag{4.122}$$

The overall procedure, including multitone excitation signal synthesis, time-domain signal acquisition, signal processing, pre-measurement characterization, and time-variant online impedance extraction, is summarized in the flowchart shown in Figure 4.15.

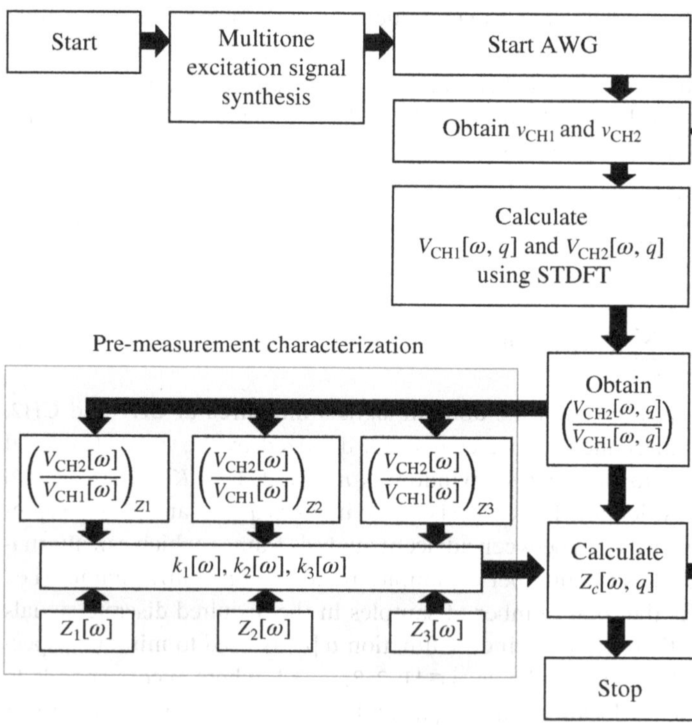

Figure 4.15 Flowchart illustrating multitone excitation signal synthesis, time-domain signal acquisition, signal processing, pre-measurement characterization, and time-variant online impedance measurement.

4.3 Practical Considerations

4.3.1 Signal-to-noise Ratio Requirements

To quantify the required signal-to-noise ratio (SNR) for achieving satisfactory measurement accuracy, a comprehensive analysis is conducted in this section. The FD-SPS is selected as a representative example, while the same practical considerations also apply to the TD-SPS. Figure 4.16 illustrates the experimental setup used to evaluate the impact of SNR on online impedance measurement accuracy using the FD-SPS. The FD-SPS consists of an inductive probe (Solar 9144-1N) and a VNA (Bode 100). The SUT comprises a pair of wires connected to a 100 Ω resistor. The power supply is intentionally excluded to ensure that the SNR analysis is not influenced by noise generated by the power supply itself.

To emulate background noise, a signal generator (National Instruments PXI-5412), together with an inductive probe (Tektronix CT1), is used to inject a noise signal with known magnitude and frequency into the setup. Another inductive probe (Tektronix CT1), connected to a spectrum analyzer (R&S FSH4), is employed to measure both the excitation signal generated by the VNA and the injected noise signal. The signal generator is configured to operate at three discrete frequencies, 1, 5, and 10 MHz, with adjustable signal levels. By separately measuring the excitation and noise signals using the inductive probe and spectrum analyzer, the corresponding SNR can be computed.

Figure 4.17 presents the measurement errors of the extracted online impedance in the presence of the emulated noise at 1, 5, and 10 MHz for three different SNR levels. It is clearly observed that a higher SNR leads to improved measurement accuracy. A closer examination of Figure 4.17 reveals that when the SNR reaches

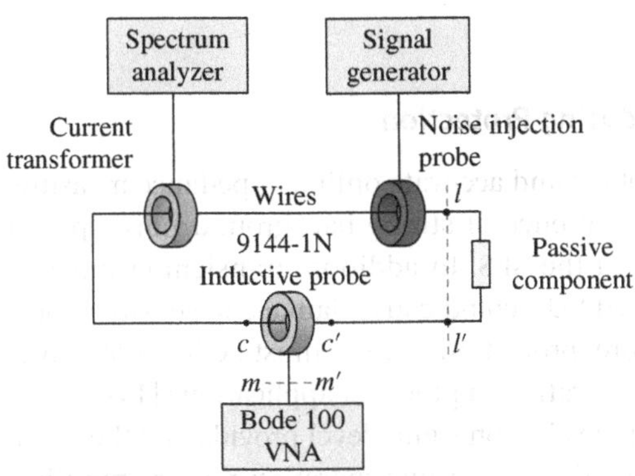

Figure 4.16 Experimental setup used to evaluate the impact of background noise on FD-SPS measurement accuracy.

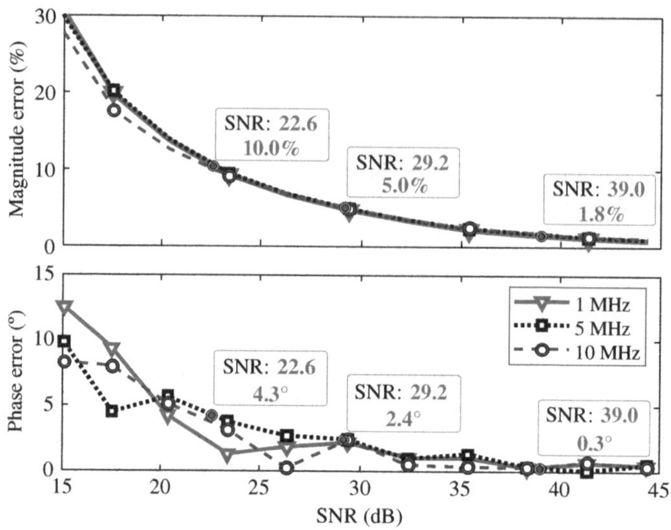

Figure 4.17 Relationship between measurement error and SNR for the FD-SPS shown in Figure 4.16.

approximately 30 dB or higher, the measurement accuracy achieves a magnitude error of less than 5% and a phase error of less than 2°.

In practical scenarios where strong background noise is present and the VNA's built-in signal source cannot provide sufficient excitation power to achieve an adequate SNR, an external power amplifier may be incorporated into the measurement setup to improve the SNR. In addition, protection devices can be added to enhance the robustness of the measurement system, as discussed in the following section.

4.3.2 Improving SNR and Introducing Protection

To maintain a sufficient SNR for robust and accurate online impedance measurement of an energized SUT in the presence of strong background noise, power amplification can be incorporated in the SPS. In addition, transient events and high excitation levels in an energized SUT could cause damage to sensitive measurement instrumentation. Therefore, protection devices must be integrated into the SPS to ensure safe and reliable operation in practical applications [14].

To address the issue of insufficient excitation signal level provided by the VNA, Figure 4.18 illustrates an FD-SPS configuration incorporating power amplification. To protect the measurement instrumentation from potential damage, surge protection devices and attenuators are also included in the setup. The excitation

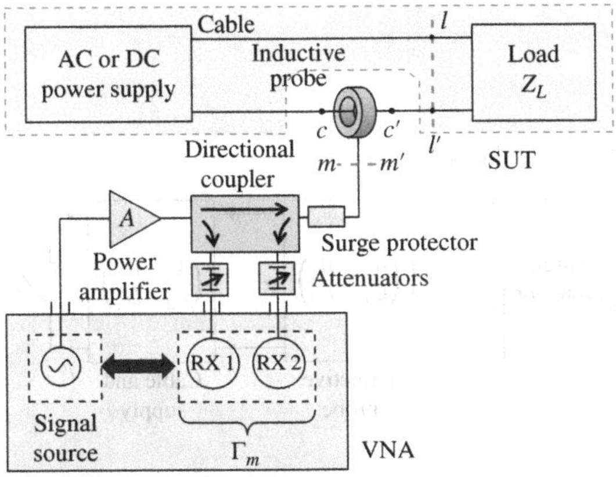

Figure 4.18 FD-SPS incorporating power amplification and protection devices.

signal generated by the VNA is amplified by a power amplifier with gain A. The required excitation level can be determined by first measuring the background noise level of the energized SUT prior to online impedance measurement. The VNA output power is then selected such that the amplified excitation signal maintains an adequate SNR, as established in the previous evaluation.

The inductive probe is connected to the directional coupler through a surge protector, which safeguards against any high voltages induced in the probe windings due to surge and transient events in the energized SUT. Two attenuators are placed between the directional coupler outputs and the VNA receivers (RX1 and RX2) to ensure that the measured signals remain within the permissible input range of the VNA receivers. The VNA and the directional coupler are configured in a reflectometer configuration, enabling measurement of the reflection coefficient at the m–m' reference plane, denoted by Γ_m. With this enhanced configuration, online impedance measurement using the FD-SPS can be reliably deployed in practical electrical systems where surge and transient disturbances are present.

The equivalent network representation of the FD-SPS incorporating power amplification and protection devices is shown in Figure 4.19. It comprises a directional coupler, two attenuators, and a surge protector, which together can be modeled as a four-port network with ports labeled from 1 to 4. The output reflection coefficient of the power amplifier is denoted by Γ_A at Port 1. The reflection coefficients associated with the VNA receiver terminations are denoted by Γ_{RX1} and Γ_{RX2} at Ports 3 and 4, respectively. Since the VNA measures Γ_m, the extraction of Z_c or Z_L follows the same procedure as that of the FD-SPS without power amplification and protection devices.

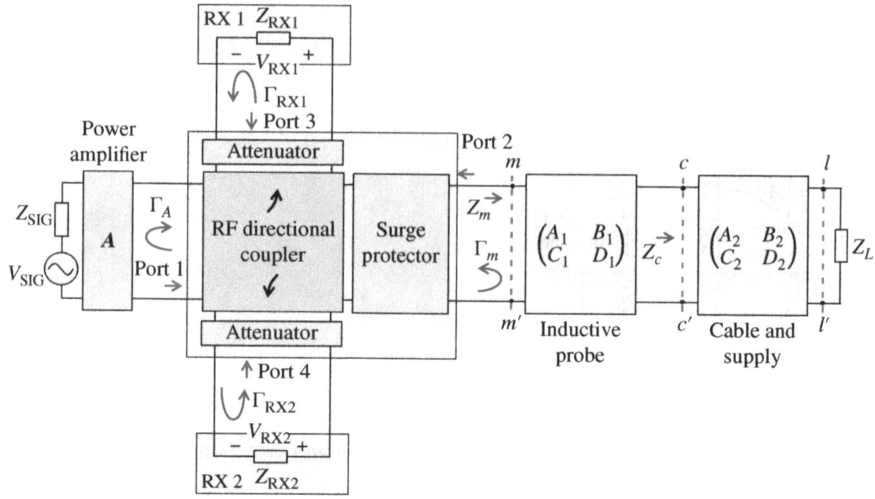

Figure 4.19 Network representation of the FD-SPS incorporating power amplification and protection devices.

This section has addressed the key practical considerations for reliable online impedance measurement using SPSs in realistic electrical environments. Through a systematic investigation of SNR requirements, it has been shown that sufficient measurement accuracy can be achieved when the excitation signal maintains an adequate SNR relative to background noise. Strategies for improving SNR, including power amplification, have been presented, together with the integration of protection devices such as surge protectors and attenuators to ensure safe operation under high-noise and transient conditions. Importantly, these enhancements do not alter the underlying impedance extraction framework, as their effects can be incorporated through pre-measurement calibration. Collectively, these considerations enable robust, accurate, and safe deployment of SPS-based online impedance measurement techniques in practical energized systems.

4.4 Chapter Summary

This chapter has presented a comprehensive treatment of the SPS for electrically contactless online impedance measurement. By establishing unified analytical frameworks for both frequency-domain and time-domain implementations, the SPS is shown to simplify measurement hardware while preserving good accuracy and flexibility for in-circuit applications. Systematic pre-measurement characterization and calibration procedures enable reliable impedance extraction and effective cancellation of systematic measurement errors. Multitone excitation and time–frequency analysis techniques further extend the SPS to multifrequency and time-variant online impedance monitoring under dynamic operating conditions. Practical considerations, including SNR requirements, power amplification, and protection against surges and transients, ensure robust and safe operation in realistic electrical environments. Collectively, the methodologies developed in this chapter establish the SPS as a scalable, efficient, and reliable solution for online impedance measurement in energized electrical systems.

References

1 A. Weerasinghe, Z. Zhao, N. B. Narampanawe et al., "Single-probe inductively coupled in-circuit impedance measurement," *IEEE Trans. Electromagn. Compat.*, vol. 64, no. 1, pp. 2–10, Aug. 2022.

2 G. H. Bryant, *Principles of microwave measurements*. London, UK: Institution of Engineering and Technology, 1993.

3 M. Friese, "Multitone signals with low crest factor," *IEEE Trans. Commun.*, vol. 45, no. 10, pp. 1338–1344, Oct. 1997.

4 Y. Yang, F. Zhang, K. Tao et al., "An improved crest factor minimization algorithm to synthesize multisines with arbitrary spectrum," *Physiol. Meas.*, vol. 36, no. 5, pp. 895, Apr. 2015.

5 F. M. Janeiro, Y. Hu, and P. M. Ramos, "Peak factor optimization of multi-harmonic signals using artificial bee colony algorithm," *Measurement*, vol. 150, p. 107040, Jan. 2020.

6 B. Sanchez, G. Vandersteen, R. Bragos et al., "Basics of broadband impedance spectroscopy measurements using periodic excitations," *Meas. Sci. Technol.*, vol. 23, no. 10, p. 105501, Aug. 2012.

7 B. Sanchez, G. Vandersteen, R. Bragos et al., "Optimal multisine excitation design for broadband electrical impedance spectroscopy," *Meas. Sci. Technol.*, vol. 22, no. 11, p. 115601, Sep. 2011.

8 P. Guillaume, J. Schoukens, R. Pintelon et al., "Crest-factor minimization using nonlinear Chebyshev approximation methods," *IEEE Trans. Instrum. Meas.*, vol. 40, no. 6, pp. 982–989, Dec. 1991.

9 M. Schroeder, "Synthesis of low-peak-factor signals and binary sequences with low autocorrelation (Corresp.)," *IEEE Trans. Inf. Theory*, vol. 16, no. 1, pp. 85–89, Jan. 1970.

10 L. Greenstein and P. Fitzgerald, "Phasing multitone signals to minimize peak factors," *IEEE Trans. Commun.*, vol. 29, no. 7, pp. 1072–1074, Jul. 1981.

11 O. I. Olayiwola and P. S. Barendse, "Characterization of silicon-based photovoltaic cells using broadband impedance spectroscopy," *IEEE Trans. Ind. Appl.*, vol. 54, no. 6, pp. 6309–6319, Jun. 2018.

12 E. Geerardyn, Y. Rolain, and J. Schoukens, “Design of quasi-logarithmic multisine excitations for robust broad frequency band measurements,” *IEEE Trans. Instrum. Meas.*, vol. 62, no. 5, pp. 1364–1372, Dec. 2013.

13 H. Zhivomirov, “A novel visual representation of the signals in the time-frequency domain,” *UPB Sci. Bull. Ser. C Electr. Eng. Comput. Sci.*, vol. 80, pp. 75–84, 2018.

14 F. Fan, K. Y. See, K. Li et al., “Efficient inductive coupled in-circuit impedance extraction with enhanced SNR and instrument protection,” in *Proc. Int. Symp. Electromagn. Compat. Asia-Pacific Symp. Electromagn. Compat.*, Singapore, May 2018.

5

Multi-probe Setups

This chapter introduces multi-probe measurement setups based on the inductive coupling approach, which extend online impedance measurement capabilities to electrical systems with multiple branches. While two-probe and single-probe configurations are effective for characterizing a single measurement point, some practical systems, such as power distribution networks and multi-load power converters, may require simultaneous impedance monitoring at multiple locations. To address this requirement, multi-probe setups enable concurrent extraction of online impedances from several branches powered by the same source. This chapter presents both frequency-domain and time-domain multi-probe measurement techniques, emphasizing their operating principles, calibration procedures, and applicability to complex multibranch systems.

5.1 Frequency-domain Measurement with Three-probe Setup

Multi-probe setups are introduced to extend the applicability of the inductive coupling approach for online impedance measurement, particularly in scenarios where multiple branch impedances must be evaluated simultaneously. In this section, a FD-THPS [1], consisting of three inductive probes and a three-port VNA, is introduced. The FD-THPS enables the simultaneous measurement of the online impedances of two SUTs located on two different branches.

The configuration of the FD-THPS is illustrated in Figure 5.1a. Port 1 of the VNA excites the circuit with a test signal through the injecting inductive probe (IIP), while the resulting signals are independently measured by Ports 2 and 3 of the VNA using RIP_1 and RIP_2, each clamped on a different branch. In contrast to the FD-TPS, this three-probe configuration allows the simultaneous extraction of impedances from two individual branches. Based on the equivalent circuit model

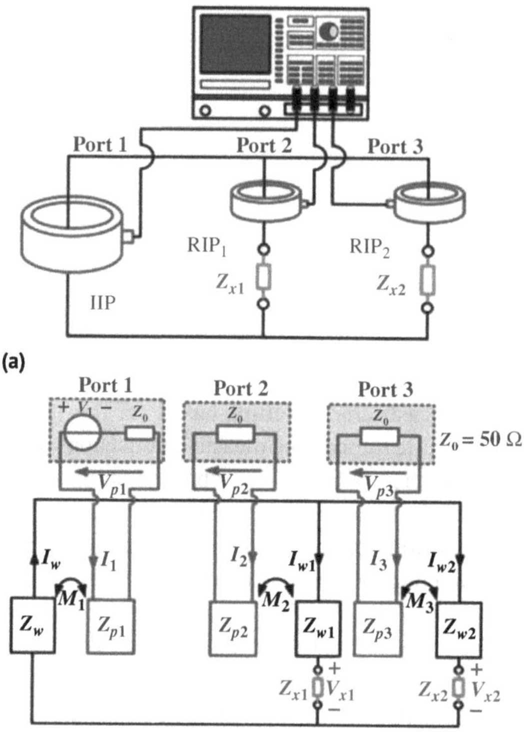

Figure 5.1 (a) Configuration of the FD-THPS. (b) Equivalent circuit model of the FD-THPS. *Source*: [1]/with permission of IEEE.

of the FD-THPS shown in Figure 5.1b, the corresponding impedance expressions can be derived as [1]

$$Z_{x1} = K_1\left(\frac{S_{11}+1}{S_{21}}\right) - Z_{\text{setup1}} - Q_1\left(\frac{S_{31}}{S_{21}}\right) \tag{5.1}$$

$$Z_{x2} = K_2\left(\frac{S_{11}+1}{S_{31}}\right) - Z_{\text{setup2}} - Q_2\left(\frac{S_{21}}{S_{31}}\right) \tag{5.2}$$

where K_1 denotes the coupling between the probes at Ports 1 and 2, including the associated wiring, and K_2 characterizes the corresponding coupling between the probes at Ports 1 and 3. Z_{setup1} and Z_{setup2} represent the setup impedances of the probe-wiring paths for Ports 1 and 2 and Ports 1 and 3, respectively. The parameters Q_1 and Q_2 account for the mutual coupling between RIP_1 and RIP_2, as well as the impedance contribution of the IIP and its wiring.

The six unknown parameters, $K_1, K_2, Z_{\text{setup1}}, Z_{\text{setup2}}, Q_1$, and Q_2, are determined through a calibration procedure. During calibration, Z_{x1} and Z_{x2} in Figure 5.1a are replaced with two precision standard resistors, R_1 and R_2, yielding the two equations shown in Equations (5.3) and (5.4). Since six parameters must be determined, the calibration procedure is repeated twice, and in each repetition, another pair of standard resistors with different values is used to generate an additional set of independent equations,

$$R_1 = K_1 \left(\frac{S_{11} + 1}{S_{21}} \right) \bigg|_{Z_{x1}=R_1} - Z_{\text{setup1}} - Q_1 \left(\frac{S_{31}}{S_{21}} \right) \bigg|_{Z_{x1}=R_1} \tag{5.3}$$

$$R_2 = K_2 \left(\frac{S_{11} + 1}{S_{31}} \right) \bigg|_{Z_{x2}=R_2} - Z_{\text{setup2}} - Q_2 \left(\frac{S_{21}}{S_{31}} \right) \bigg|_{Z_{x2}=R_2} \tag{5.4}$$

After the six unknown parameters (i.e., K_1, K_2, Z_{setup1}, Z_{setup2}, Q_1, and Q_2) are obtained through the above calibration procedure, the impedances Z_{x1} and Z_{x2} of unknown SUTs can be determined from the measured S-parameters using the FD-THPS according to Equations (5.1) and (5.2).

5.2 Time-domain Measurement with Multi-probe Setup

The abovementioned FD-THPS can measure online impedances at only two branches simultaneously. To overcome this limitation, a TD-MPS was proposed in [2]. The TD-MPS consists of one IIP, multiple RIPs depending on the number of SUTs to be measured, and a computer-controlled SGAS, as illustrated in Figure 5.2. The TD-MPS enables the simultaneous measurement of online impedances of multiple SUTs distributed across multiple branches, even when the number of branches exceeds two.

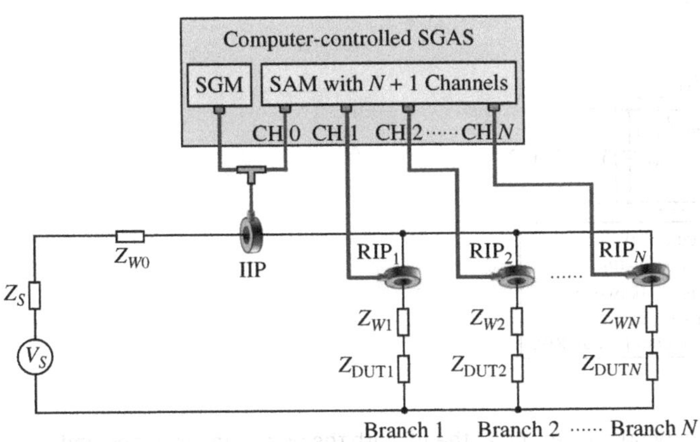

Figure 5.2 Configuration of the TD-MPS for online impedances measurement at multiple branches.

As shown in Figure 5.3, the IIP with the wire being clamped can be modeled as a two-port network $\boldsymbol{M}_I$ [3], where Port 1 corresponds to the input of the IIP and Port 2 corresponds to the two ends of the wire being clamped. Similarly, the RIP at the ith branch (denoted as RIP_i) with the wire being clamped can also be modeled as a two-port network $\boldsymbol{M}_{Ri}$, where Port 1 corresponds to the two ends of the wire being clamped and Port 2 corresponds to the output of RIP_i. Since the IIP is mounted on the main power feeder and each RIP is mounted on its respective branch, the inductive probes are typically well separated in space. As a result, probe-to-probe coupling among the inductive probes can be neglected [4].

With the above two-port network models defined, the configuration shown in Figure 5.2 can be represented by the equivalent circuit model shown in Figure 5.4.

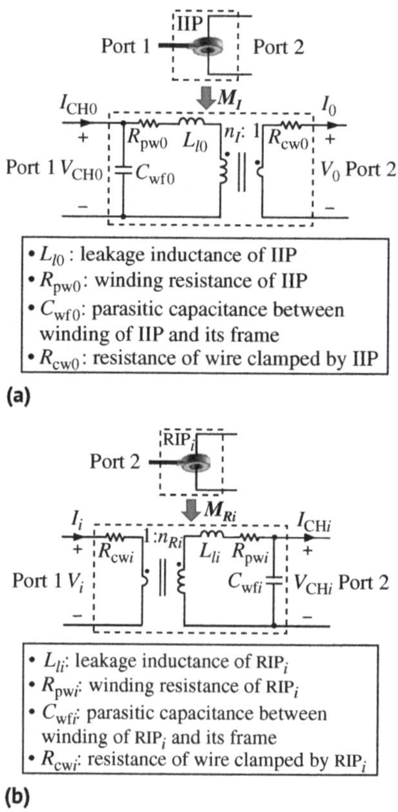

Figure 5.3 Two-port network models of (a) the IIP with the wire being clamped and (b) the RIP at the ith branch with the wire being clamped.

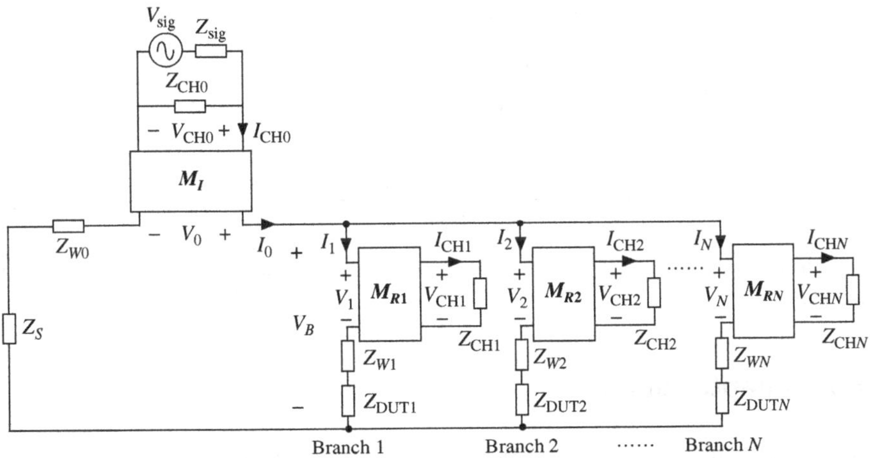

Figure 5.4 Equivalent circuit model of the TD-MPS measurement configuration.

Based on the derived formulations, the online impedance of the SUT at the ith branch, denoted as $Z_{\text{DUT}i}$, can be determined using Equation (5.5),

$$Z_{\text{DUT}i} = a_i + b_i \cdot (Z_S + Z_{W0}) - Z_{Wi} \tag{5.5}$$

where

$$a_i = \frac{\frac{V_{\text{CH0}}}{A_0} - \frac{B_0}{A_0}\sum_{i=1}^{N}\left[\left(C_i + \frac{D_i}{Z_{\text{CH}i}}\right)V_{\text{CH}i}\right]}{\left(C_i + \frac{D_i}{Z_{\text{CH}i}}\right)V_{\text{CH}i}} - \frac{A_i + \frac{B_i}{Z_{\text{CH}i}}}{C_i + \frac{D_i}{Z_{\text{CH}i}}} \tag{5.6}$$

$$b_i = -\frac{\sum_{i=1}^{N}\left[\left(C_i + \frac{D_i}{Z_{\text{CH}i}}\right)V_{\text{CH}i}\right]}{\left(C_i + \frac{D_i}{Z_{\text{CH}i}}\right)V_{\text{CH}i}} \tag{5.7}$$

To obtain the coefficients a_i and b_i, the two-port networks $\boldsymbol{M_I}$ and $\boldsymbol{M_{Ri}}$ are first pre-characterized using a test jig, as described in [3]. The voltages V_{CH0} and $V_{\text{CH}i}$ are then directly measured using the TD-MPS. Once a_i and b_i are known, the online impedance $Z_{\text{DUT}i}$ can be determined, provided that $(Z_s + Z_{w0})$ and Z_{wi} are known. To determine $(Z_s + Z_{w0})$ and Z_{wi}, the device under test (DUT) can be replaced by two known high-precision standard resistors R_{1i} and R_{2i} ($R_{1i} \neq R_{2i}$) prior to the actual measurement.

Thus, from Equation (5.5), the following expressions can be obtained:

$$R_{1i} = a_i|_{R_{1i}} + b_i|_{R_{1i}}(Z_S + Z_{W0}) - Z_{Wi} \tag{5.8}$$

$$R_{2i} = a_i|_{R_{2i}} + b_i|_{R_{2i}}(Z_S + Z_{W0}) - Z_{Wi} \tag{5.9}$$

From Equations (5.8) and (5.9), $(Z_S + Z_{W0})$ and Z_{Wi} can be solved as

$$Z_S + Z_{W0} = \frac{(R_{1i} - R_{2i}) - \left(a_i|_{R_{1i}} - a_i|_{R_{2i}}\right)}{b_i|_{R_{1i}} - b_i|_{R_{2i}}} \tag{5.10}$$

$$Z_{Wi} = \frac{R_{1i}\, b_i|_{R_{2i}} - R_{2i}\, b_i|_{R_{1i}} + a_i|_{R_{2i}}\, b_i|_{R_{1i}} - a_i|_{R_{1i}}\, b_i|_{R_{2i}}}{b_i|_{R_{1i}} - b_i|_{R_{2i}}} \tag{5.11}$$

Therefore, the online impedance Z_{DUTi} of any unknown DUT can be extracted accordingly.

5.3 Chapter Summary

This chapter has presented multi-probe online impedance measurement techniques that extend the inductive coupling approach to multibranch electrical systems powered by a common power source. The FD-THPS enables simultaneous impedance extraction from two branches, while the TD-MPS further generalizes this concept to support an arbitrary number of branches. By modeling each inductive probe as a two-port network and employing systematic pre-measurement characterization, both FD-THPS and TD-MPS achieve accurate impedance extraction without direct electrical contact with energized systems. These configurations provide a scalable and flexible solution for real-time impedance monitoring in complex power and energy infrastructures, complementing the single-probe and two-probe techniques discussed in earlier chapters and completing the framework of electrically contactless online impedance measurement setups.

References

1 K. Li, A. Videt, and N. Idir, "Multiprobe measurement method for voltage-dependent capacitances of power semiconductor devices in high voltage," *IEEE Trans. Power Electron.*, vol. 28, no. 11, pp. 5414–5422, Jan. 2013.

2 Z. Zhao and K. Y. See, "A multiprobe inductive coupling method for online impedance measurement of electrical devices distributed in multibranch cables," *IEEE Trans. Instrum. Meas.*, vol. 69, no. 9, pp. 5975–5977, Jul. 2020.

3 S. B. Rathnayaka, K. Y. See, and K. Li, "Online impedance monitoring of transformer based on inductive coupling approach," *IEEE Trans. Dielectr. Electr. Insul.*, vol. 24, no. 2, pp. 1273–1279, Apr. 2017.

4 A. S. Narayanan, Z. Zhao, E. K. Chua et al., "Study on the parameters affecting the impedance extraction accuracy by inductive coupling method," in *Proc. 2018 Int. Conf. Intell. Rail Transp. (ICIRT)*, 12–14 Dec. 2018.

6
Comparison of Different Contactless Setups

This chapter presents a systematic comparison of electrically contactless online impedance measurement setups based on the inductive coupling approach. Building on the detailed theoretical formulations and implementation methodologies developed in Chapters 3–5, this chapter focuses on evaluating the relative advantages, limitations, and practical characteristics of two-probe (TPS), single-probe (SPS), and multi-probe (MPS) configurations. Rather than introducing new measurement principles, emphasis is placed on a comparative assessment of these setups in terms of measurement capability, calibration requirements, installation complexity, robustness, scalability, and suitability for different application scenarios. Furthermore, explicit selection criteria are established to guide readers in choosing the most appropriate contactless impedance measurement configuration, thereby bridging the gap between measurement methodology and real-world applications.

6.1 Advantages and Limitations

Electrically contactless online impedance measurement based on inductive coupling can be realized through three representative configurations: the TPS, SPS, and MPS. Although all three configurations share the same underlying measurement principle, they exhibit distinct characteristics in terms of measurement flexibility, system complexity, calibration effort, and application scope.

The TPS, discussed in Chapter 3, represents the earliest and most intuitive form of inductively coupled online impedance measurement. By employing separate injecting and receiving probes, the TPS provides strong signal coupling and is well suited for online impedance measurements. Time-variant online impedance monitoring can also be achieved when the TD-TPS is combined with appropriate time–frequency analysis techniques. However, the presence of probe-to-probe coupling introduces additional calibration complexity, and the requirement to mount two probes on the same conductor may limit installation flexibility in space-constrained environments.

The SPS, presented in Chapter 4, simplifies the measurement hardware by using a single inductive probe for both signal injection and response acquisition. This configuration inherently eliminates probe-to-probe coupling effects and reduces installation complexity. Through pre-measurement characterization and systematic error cancellation, both FD-SPS and TD-SPS can achieve high measurement accuracy. Moreover, when combined with multi-sine excitation and STDFT techniques, the TD-SPS enables efficient multifrequency and time-variant impedance measurement. A potential limitation of the SPS is its reduced coupling strength compared with the TPS, which may impose higher SNR requirements in electrically noisy environments.

The MPS, introduced in Chapter 5, extends the inductive coupling approach to systems with multiple branches powered by the same source. In the frequency domain, an FD-THPS enables simultaneous impedance extraction from two branches. In the time domain, a TD-MPS supports concurrent impedance monitoring of multiple distributed loads. Although the MPS provides unmatched scalability and monitoring capability for complex systems, it also introduces increased system complexity, calibration effort, synchronization requirements, and data processing overhead.

Table 6.1 summarizes the key characteristics, advantages, and limitations of the contactless measurement setups reported in [1–6]. Rather than ranking these setups by absolute performance, the comparison highlights that each

Table 6.1 Comparison of contactless online impedance measurement setups.

Measurement Setup Features	FD-TPS [1]	TD-TPS [2]	FD-SPS [3]	TD-SPS [4]	FD-THPS [5]	TD-MPS [6]
Excitation signal	Swept-sine	Single-sine	Swept-sine	Multi-sine	Swept-sine	Single-sine
Time-variant in-circuit impedance measurement ability	No	Yes	No	Yes	No	Yes
Elimination of probe-to-probe coupling	No	No	Yes	Yes	No	No
Multifrequency simultaneous measurement ability	No	No	No	Yes	No	No
Measurement efficiency	Low	Low	Low	High	Moderate	High

configuration is optimized for a specific class of measurement objectives and application constraints.

6.2 Selection Criteria Based on Application Requirements

Although all contactless impedance measurement setups discussed in this book are based on the inductive coupling principle, their practical suitability strongly depends on specific application requirements and measurement configurations, particularly the choice between frequency-domain and time-domain implementations, as well as the selection among TPS, SPS, and MPS architectures. To facilitate rational and application-oriented selection, the following criteria are introduced.

6.2.1 Number of Measurement Points

If impedance measurement is required at a single location, either TPS or SPS configurations may be employed, using frequency-domain or time-domain implementations depending on the measurement objective. For systems with multiple branches or distributed loads, MPS configurations are preferred, as they enable simultaneous impedance extraction across multiple branches. In such scenarios, FD-THPS is suitable for steady-state characterization of two branches, whereas TD-MPS is more appropriate when concurrent monitoring of time-varying impedances across multiple branches is required.

6.2.2 Measurement Domain and Dynamics

For applications primarily concerned with steady-state impedance characteristics, frequency-domain configurations (FD-TPS, FD-SPS, and FD-THPS) are generally sufficient. In contrast, applications involving dynamic operating conditions, such as fault evolution, transient events, or load variations, benefit from time-domain configurations (TD-TPS, TD-SPS, and TD-MPS), which enable time-resolved impedance tracking.

6.2.3 Installation Constraints and Accessibility

When installation space is limited or rapid deployment is required, SPS configurations generally provide the greatest flexibility, as both FD-SPS and TD-SPS require only one clamp-on inductive probe. TPS configurations may be constrained by probe spacing and conductor geometry, while MPS configurations require additional physical space to accommodate multiple probes, which may restrict their applicability in compact or densely packed environments.

6.2.4 Calibration Complexity and Robustness

SPS configurations benefit from inherently simpler calibration procedures in both frequency-domain and time-domain implementations. TPS configurations require explicit handling of probe-to-probe coupling effects, and recalibration is generally necessary if the relative spacing between probes changes. MPS configurations involve more extensive calibration procedures due to the increased number of probes and signal paths, with TD-MPS typically requiring more elaborate synchronization and signal-processing schemes than FD-THPS.

6.2.5 Measurement Accuracy and SNR Requirements

Applications characterized by high background noise or weak coupling conditions may favor TPS configurations due to their stronger signal injection and reception capability in both frequency-domain and time-domain implementations. SPS configurations may require power amplification or advanced signal-processing techniques to maintain adequate SNR. MPS configurations balance coupling strength against scalability, trading higher system complexity for multibranch measurement capability.

6.2.6 System Complexity and Cost

From a system-level perspective, SPS configurations generally offer the lowest hardware complexity and cost for both frequency-domain and time-domain implementations. TPS configurations introduce moderate system complexity, while MPS configurations incur higher cost and complexity due to the need for multiple probes, additional acquisition channels, and more sophisticated calibration procedures. These trade-offs are justified in applications where simultaneous multibranch impedance monitoring is required.

6.3 Chapter Summary

This chapter has presented a systematic comparison of electrically contactless online impedance measurement setups based on inductive coupling. By examining the advantages and limitations of two-probe, single-probe, and multi-probe configurations, the chapter clarified how measurement capability, installation flexibility, calibration complexity, scalability, and cost differ across these setups. Both frequency-domain and time-domain implementations were considered, highlighting their respective suitability for steady-state characterization and time-variant impedance monitoring. Based on this comparative analysis, explicit selection criteria were established to guide application-driven configuration choices among TPS, SPS, and MPS architectures. This chapter, therefore, provides a practical framework for selecting appropriate contactless impedance measurement setups and establishes a clear link between these configurations and their preferred application scenarios.

References

1 K. Li, K. Y. See, and X. Li, "Inductive coupled in-circuit impedance monitoring of electrical system using two-port ABCD network approach," *IEEE Trans. Instrum. Meas.*, vol. 64, no. 9, pp. 2489–2495, Sep. 2015.

2 Z. Zhao, K. Y. See, E. K. Chua et al., "Time-variant in-circuit impedance monitoring based on the inductive coupling method," *IEEE Trans. Instrum. Meas.*, vol. 68, no. 1, pp. 169–176, 2019.

3 A. Weerasinghe, Z. Zhao, N. B. Narampanawe et al., "Single-probe inductively coupled in-circuit impedance measurement," *IEEE Trans. Electromagn. Compat.*, vol. 64, no. 1, pp. 2–10, Feb. 2022.

4 A. Weerasinghe, Z. Zhao, Q. Sun et al., "A novel single-probe setup for multifrequency simultaneous measurement of in-circuit impedance," *IEEE Trans. Ind. Electron.*, vol. 70, no. 9, pp. 9538–9549, Sep. 2023.

5 K. Li, A. Videt, and N. Idir, "Multiprobe measurement method for voltage-dependent capacitances of power semiconductor devices in high voltage," *IEEE Trans. Power Electron.*, vol. 28, no. 11, pp. 5414–5422, Nov. 2013.

6 Z. Zhao and K. Y. See, "A multiprobe inductive coupling method for online impedance measurement of electrical devices distributed in multibranch cables," *IEEE Trans. Instrum. Meas.*, vol. 69, no. 9, pp. 5975–5977, Sep. 2020.

7

Applications in Power Electronics

This chapter explores the application of electrically contactless online impedance measurement techniques within power electronics systems. Power electronic converters incorporate fast-switching semiconductor devices, passive filtering components, and complex control loops, resulting in impedance characteristics that are highly frequency-dependent and strongly influenced by operating conditions. Accurate online impedance data is crucial for ensuring reliable design, performance optimization, and health diagnostics. Building on the inductive-coupling-based measurement setups discussed in Chapters 3–6, this chapter demonstrates practical applications such as EMI (electromagnetic interference) filter design for switched-mode power supplies (SMPSs), health diagnosis of power semiconductor devices, and their voltage-dependent capacitances extraction. These examples show how online impedance measurement provides deeper insights into both component-level behavior and system-level performance under real-world operating conditions.

7.1 EMI Filter Design for Switched-mode Power Supplies

Built-in power line EMI filters are parts of the SMPS designs to limit conducted emissions, typically over the frequency range up to 30 MHz, in order to comply with international EMI regulatory requirements [1, 2]. Unlike filters used in communication and microwave applications, where both source and termination impedances are usually well defined as 50 Ω, the actual noise source and termination impedances of an EMI filter in an SMPS are generally far from 50 Ω [3]. In standard conducted emission measurement setups, the noise termination of an SMPS is typically provided by a line impedance stabilization network (LISN), whose impedance characteristics are well defined and standardized [4]. In contrast, the noise source impedance of an SMPS is highly variable and depends on multiple factors, such as converter topology, power rating, component parasitics, and board layout [5].

For example, the DM noise source impedance is strongly influenced by the reverse recovery behavior of the diode rectifier [6], as well as by the equivalent series resistance (ESR) and equivalent series inductance (ESL) of the bulk capacitor [7]. For the CM noise source impedance, the dominant contributors are the parasitic capacitance between the switching device and its heat sink, and the parasitic capacitance between the circuit board and the chassis [1]. Consequently, designing an EMI filter for an SMPS under the assumption of 50 Ω noise source and termination impedances generally lead to suboptimal EMI suppression performance. Therefore, obtaining accurate information on the noise source impedance of an SMPS is essential for effective EMI filter design [8–10]. In this section, the FD-TPS was selected for such an application. It should be noted that in addition to the content of the FD-TPS discussed in Chapter 3, the number of turns of the probe coil was increased in [11] to mitigate signal attenuation, particularly at low frequencies in FD-TPS measurements. In addition, an alternative method was reported [12], where an additional capacitance is connected in parallel with the SUT to compensate for low-frequency attenuation.

7.1.1 Conducted Emission Measurement with LISN

Compared with linear power supplies, which continuously dissipate power due to the linear operating region of the regulating transistor, SMPSs achieve significantly higher efficiency because the metal–oxide–semiconductor field-effect transistor (MOSFET) operates as a high-frequency switching element and dissipates minimal power [13]. In addition, SMPSs are substantially lighter than linear power supplies, as they employ much smaller magnetic cores operating at switching frequencies of tens or hundreds of kilohertz rather than at line frequency. Owing to these advantages in efficiency and power density, SMPSs have been widely adopted in modern electronic products. However, the high-frequency switching action of the MOSFET generates harmonics extending into the megahertz range, which become sources of conducted emissions. These unwanted noise currents originate within the SMPS and propagate through the connecting cables to the AC mains or nearby sensitive equipment [14].

In general, conducted noise currents can be categorized into two modes. The first is CM noise current. During switching operation, the pulsed voltage at the switching device Q_1 causes repetitive charging and discharging of the parasitic capacitance C_{para} between Q_1 and the heat sink. As a result, a CM current is generated, with the charging and discharging of C_{para} acting as the dominant CM noise mechanism. Figure 7.1 illustrates a typical CM current path in an SMPS.

As shown in Figure 7.1, the CM current (I_{CM}) flows along the dashed path, entering the AC mains through the live and neutral conductors and returning through

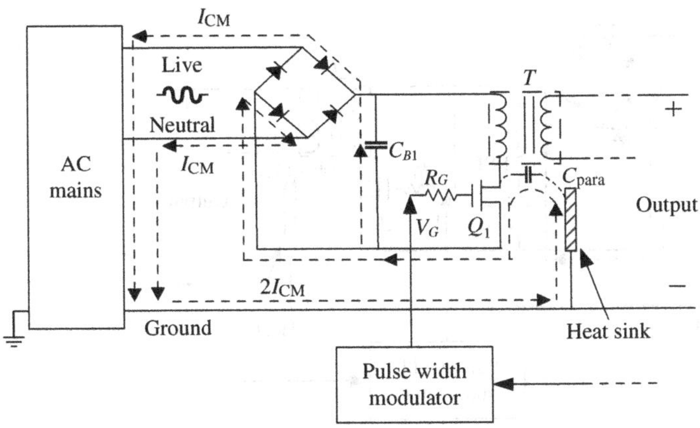

Figure 7.1 CM current path in an SMPS.

the protective ground conductor. Accordingly, the live and neutral conductors can be treated as electrically parallel and modeled as a single equivalent conductor between the SMPS and the AC mains for CM noise analysis. The CM current flows through Q_1, C_{B1}, two forward-biased diodes of the rectifier, the live and neutral conductors, and the AC mains, before returning to C_{para} via the ground conductor and the grounded heat sink.

The second mode of conducted noise is DM noise. When Q_1 switches, a pulsed current is drawn from the bulk capacitor C_{B1} into the primary winding. Since C_{B1} is nonideal and exhibits ESR and ESL, these parasitic elements dominate its impedance at the switching frequency. Consequently, the current flowing through C_{B1} produces a ripple voltage across the capacitor, which acts as a DM noise voltage source and drives DM noise current into the AC mains. Figure 7.2 illustrates a representative DM current path in an SMPS. As seen, the DM current I_{DM} flows along the dashed path, entering the AC mains through the live conductor and returning through the neutral conductor. Specifically, I_{DM} flows from the bulk capacitor C_{B1}, through one of the turn-on diodes of the rectifier, the live conductor, and the AC mains, and finally returns to C_{B1} via the neutral conductor and another turn-on diode of the rectifier.

With the CM current I_{CM} and the DM current I_{DM}, the noise currents on the live line (I_L) and the neutral line (I_N) can be expressed as the phasor sum and phasor difference, respectively, as follows:

$$I_L = I_{\text{CM}} + I_{\text{DM}} \tag{7.1}$$

$$I_N = I_{\text{CM}} - I_{\text{DM}} \tag{7.2}$$

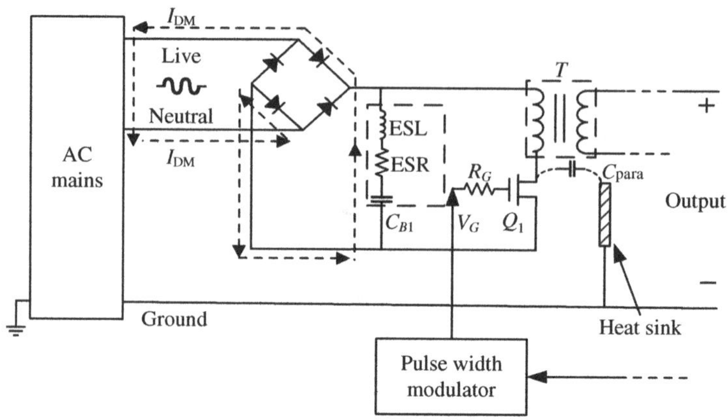

Figure 7.2 DM current path in an SMPS.

To design an EMI filter for suppressing conducted emissions, the magnitude of the conducted emission must first be quantified through measurement. Since different AC mains connections to the same SMPS can result in different conducted emission levels, directly using the AC mains as the noise-terminating load may lead to ambiguous measurement results. Therefore, international EMI standards require that the AC mains supply the SMPS through a LISN, which provides a well-defined and repeatable mains impedance. In addition to stabilizing the line impedance, the LISN prevents external conducted noise present on the AC mains from contaminating the measurement, while minimally disturbing the fundamental 50 or 60 Hz power delivered to the SMPS [15–17]. The schematic of a typical LISN and the corresponding measurement setup for I_L are shown in Figure 7.3.

As shown in Figure 7.3, the LISN, enclosed by the dashed box, is inserted between the AC mains and the SMPS. The SMPS together with a DC load constitutes the equipment under test (EUT), whose conducted emission current I_L is measured using a spectrum analyzer (SA). The LISN consists of two 1 μF capacitors, two 0.1 μF capacitors, two 50 μH inductors, and two 1 kΩ resistors. The 1 μF capacitors connected between the live and ground lines and between the neutral and ground lines are used to bypass external noise from the AC mains to ground, ensuring that the measured conducted emission originates solely from the SMPS. Similarly, the 50 μH inductors on the live and neutral conductors block high-frequency noise propagating toward the AC mains while minimally affecting the fundamental 50 or 60 Hz power delivery. The 0.1 μF capacitors connected between the live and ground lines and between the neutral and ground lines provide low-impedance paths to the I_L and I_N test ports, which are designed for connection to the SA. The 1 kΩ resistors connected between the live and ground lines and

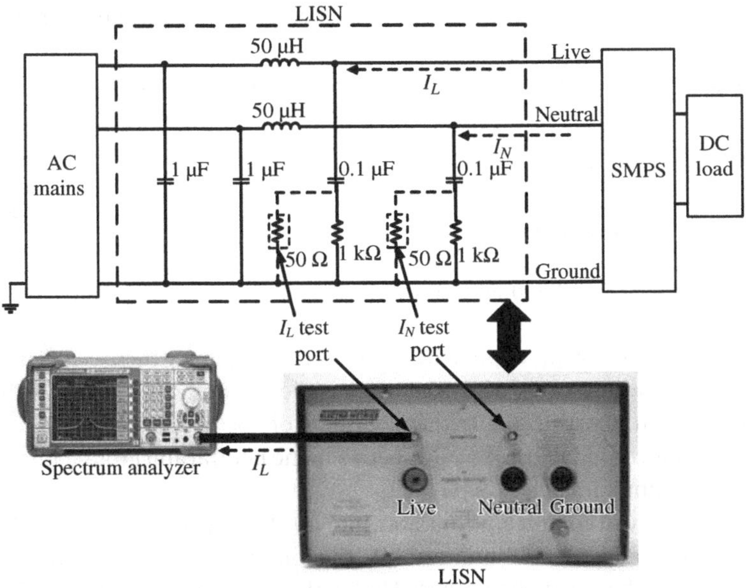

Figure 7.3 Schematic circuit of the LISN and measurement setup for the conducted emission current I_L.

between the neutral and ground lines provide discharge paths for the 0.1 μF capacitors when the test ports are not connected to the SA.

The image in Figure 7.3 shows the front panel of the LISN, where the live, neutral, and ground ports are connected to the input terminals of the SMPS, while the rear side interfaces with the AC mains. I_L is measured by connecting the SA, which has a 50 Ω input impedance, to the I_L test port using a coaxial cable, while the I_N test port is terminated with a 50 Ω dummy load. Consequently, the 50 Ω input impedance of the SA is placed in parallel with the 1 kΩ resistor between the live and ground lines, while the 50 Ω dummy load at the I_N test port is placed in parallel with the 1 kΩ resistor between the neutral and ground lines. Similarly, I_N is measured by connecting the SA to the I_N test port and terminating the I_L test port with a 50 Ω dummy load. Both I_L and I_N are measured over the frequency range from 150 kHz to 30 MHz in accordance with commonly used international conducted EMI standards, such as Federal Communications Commission (FCC) Part 15 and Comité International Spécial des Perturbations Radioélectriques (CISPR) 22.

Because the 1 kΩ resistors are connected in parallel with the equivalent 50 Ω terminations during the measurements of I_L or I_N, their influence can be neglected, as 1 kΩ ≫ 50 Ω. Therefore, the LISN circuit between the live and ground lines or between the neutral and ground lines can be simplified to the equivalent circuit shown in Figure 7.4.

Based on Figure 7.4, the LISN line impedance Z_{LISN}, as seen from the SMPS, is given by

$$Z_{\text{LISN}} = \frac{\left(R_1 + \frac{1}{j\omega C_1}\right)\left(j\omega L_1 + \frac{1}{j\omega C_2}\right)}{R_1 + \frac{1}{j\omega C_1} + j\omega L_1 + \frac{1}{j\omega C_2}} \tag{7.3}$$

Figure 7.5 shows the extracted magnitude of Z_{LISN} over the frequency range from 150 kHz to 30 MHz.

As shown in Figure 7.5, Z_{LISN} approaches 50 Ω as the frequency increases. This behavior indicates that the equivalent 50 Ω resistor dominates Z_{LISN} at high frequencies. Consequently, within the frequency range of 150 kHz–30 MHz, Z_{LISN} between the live and ground lines, as well as between the neutral and ground lines, can be reasonably approximated as 50 Ω.

From the perspective of conducted emission analysis, the capacitors in the LISN can therefore be treated as short circuits, while the inductors can be regarded as open circuits within this frequency range. Under these assumptions, I_L and I_N flow only through the equivalent parallel 50 Ω resistors and return to the SMPS through the ground line during measurement. The simplified equivalent circuit of the LISN is shown in Figure 7.4. Accordingly, I_L and I_N can be expressed in terms of the voltages measured by the SA, V_L, and V_N, as

$$V_L = 50 I_L \tag{7.4}$$

$$V_N = 50 I_N \tag{7.5}$$

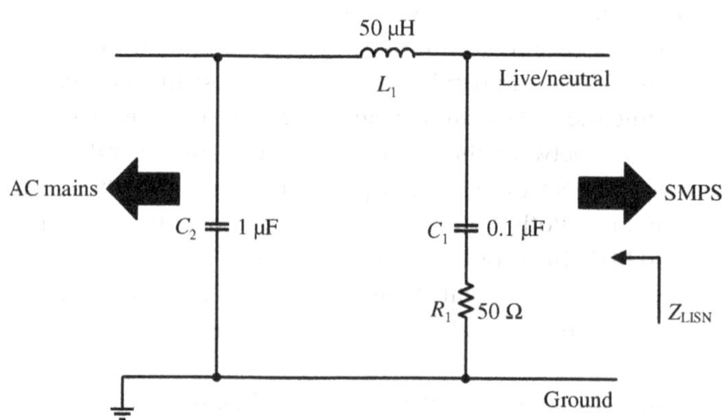

Figure 7.4 Simplified LISN circuit between the live and ground lines or between the neutral and ground lines.

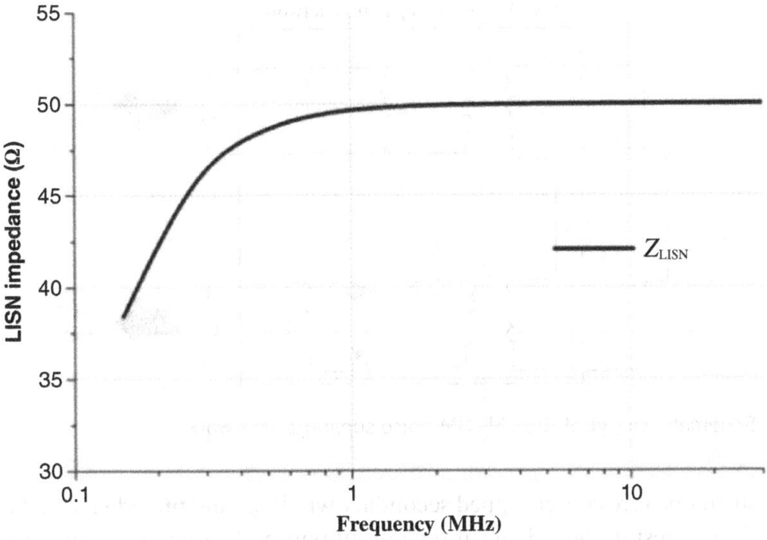

Figure 7.5 Magnitude of the extracted Z_{LISN}.

Substituting Equations (7.1) and (7.2) into Equations (7.4) and (7.5), respectively, yields

$$V_L = (I_{CM} + I_{DM})50 = V_{CM} + V_{DM} \tag{7.6}$$

$$V_N = (I_{CM} - I_{DM})50 = V_{CM} - V_{DM} \tag{7.7}$$

where $V_{CM} = 50I_{CM}$ and $V_{DM} = 50I_{DM}$ represent the CM and DM noise voltages measured by the SA, respectively. In practice, the magnitudes of V_L and V_N are typically expressed in units of dBμV (i.e., decibels referenced to 1 μV) and compared against the conducted EMI limits specified in FCC Part 15 and CISPR 22 standards.

To separately measure V_{CM} and V_{DM}, a CM–DM noise separation network is required, as shown in Figure 7.6. This network is inserted between the LISN and the SA. As illustrated in Figure 7.6, the separation network is enclosed within the dashed box, and the 50 Ω resistor represents the input termination of the SA. Two ports on one side of the separation network are connected to the LISN's I_L and I_N test ports, respectively, while the output port on the other side is connected to the SA.

The switch on the V_N side is used to change the polarity of V_N, such that either V_N or $-V_N$ appears on the secondary side of the transformer. The separation network employs two 1:1 wideband RF transformers (model: LF428, VARI-L Co.), which exhibit an essentially flat frequency response from 10 kHz to 50 MHz.

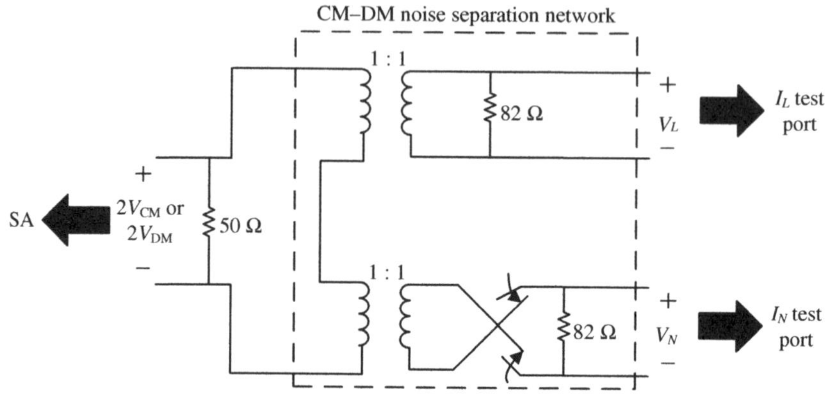

Figure 7.6 Schematic circuit of the CM–DM noise separation network.

These transformers have center-tapped secondary windings and provide DC isolation [18, 19]. As a result, the voltage at the output port of the separation network connected to the SA is either $V_L + V_N$ or $V_L - V_N$, corresponding to $2V_{CM}$ or $2V_{DM}$, respectively. Therefore, the CM- and DM-conducted emissions measured using the separation network and the SA are equal to $2V_{CM}$ and $2V_{DM}$, respectively.

By obtaining the magnitudes of the conducted emissions through measurement, the results can be directly compared with the conducted EMI limits specified in FCC Part 15 or CISPR 22. Through this comparison, the frequency ranges and magnitudes of excessive conducted emissions requiring suppression can be clearly identified. The conducted EMI limits specified in CISPR 22 are shown in Figure 7.7.

With the excessive conducted emission identified through comparison with the selected conducted EMI standards, an EMI filter can be designed based on online impedance measurement to attenuate the emission to a level below the specified conducted EMI limits.

7.1.2 CM and DM Impedance Extraction

Besides measuring conducted emissions to determine the required filter attenuation, accurate characterization of both the noise source impedance and the noise termination impedance is essential for optimal EMI filter design. Figure 7.8 compares the measured DM and CM output impedances of a LISN (Electro-Metrics MIL 7-25/2). The DM and CM impedances are measured using both the FD-TPS and a commercial impedance analyzer (HP4396B, 100 kHz–1.8 GHz). As shown, the measured DM impedance $Z_{LISN,DM}$ and CM impedance $Z_{LISN,CM}$ obtained using the two methods are in close agreement over the entire frequency range of interest.

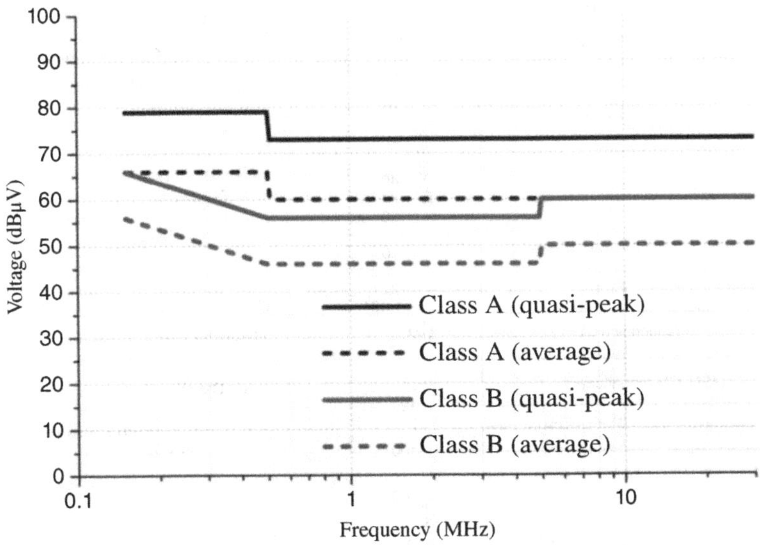

Figure 7.7 Conducted EMI limits specified in CISPR 22.

It should be noted that, when using the FD-TPS, the LISN impedance can be measured with the AC power applied. In contrast, measurements performed using the impedance analyzer must be conducted without AC power applied to the LISN in order to prevent damage to the measuring equipment.

For the FD-TPS, AC power is applied to the input of the LISN, and one or two 1 μF X-class capacitors are connected at the output of the LISN to establish an AC short circuit. Specifically, a 1 μF capacitor is connected between the line and neutral terminals for DM impedance measurement. For CM impedance measurement, two 1 μF capacitors are required: one connected between the line and ground, and the other between the neutral and ground. For DM output impedance measurement, the line conductor is treated as the outgoing conductor, while the neutral conductor serves as the return path. For CM measurement, the line and neutral conductors are treated as a single outgoing conductor, and the safety ground conductor is treated as the return path. The connecting wires between the LISN and the capacitors are kept as short as possible to minimize the parasitic inductance introduced by the connecting wires.

Figure 7.9 illustrates the measurement setups used to extract the DM noise source impedance $Z_{SMPS,DM}$ and the CM noise source impedance $Z_{SMPS,CM}$ of an SMPS using the FD-TPS. The DM and CM configurations are shown in Figure 7.9a,b, respectively. The SMPS under test is a VTM22WB module (15 W, +12 V_{dc}/0.75 A, −12 V_{dc}/0.5 A). The SMPS is powered through the MIL 7-25/2 LISN to ensure a stable and repeatable AC mains impedance, and a resistive load

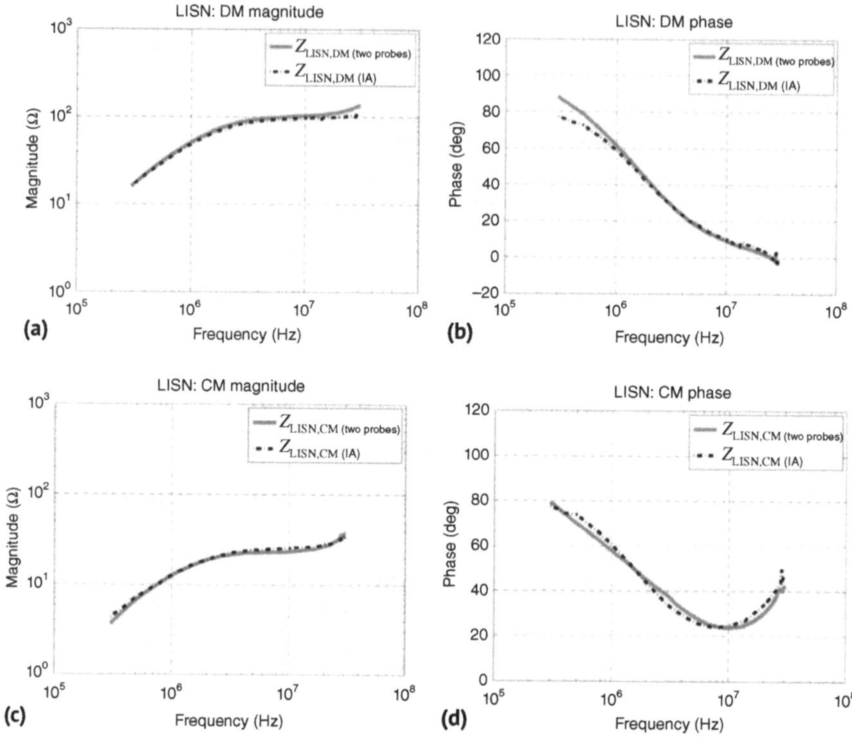

Figure 7.8 Comparison of measured LISN impedances. (a) DM magnitude. (b) DM phase. (c) CM magnitude. (d) CM phase.

is connected at the output of the SMPS to provide a well-defined loading condition. The DM and CM output impedances of the LISN, $Z_{\mathrm{LISN,DM}}$ and $Z_{\mathrm{LISN,CM}}$, have been measured in advance, as shown in Figure 7.8.

To extract the DM and CM noise source impedances of the SMPS, the FD-TPS is employed. The transmission-line effects of the interconnecting wires can be neglected because the total wire length between the LISN and the SMPS is approximately 70 cm, which is much shorter than the wavelength corresponding to the highest frequency of interest in this study (30 MHz). The impedance measured by the FD-TPS corresponds to the total loop impedance Z_T of the circuit connecting the SMPS and the LISN under actual AC-powered operating conditions. This total impedance is given by

$$Z_T = Z_{\mathrm{LISN}} + Z_{\mathrm{SMPS}} + Z_{\mathrm{wire}} \tag{7.8}$$

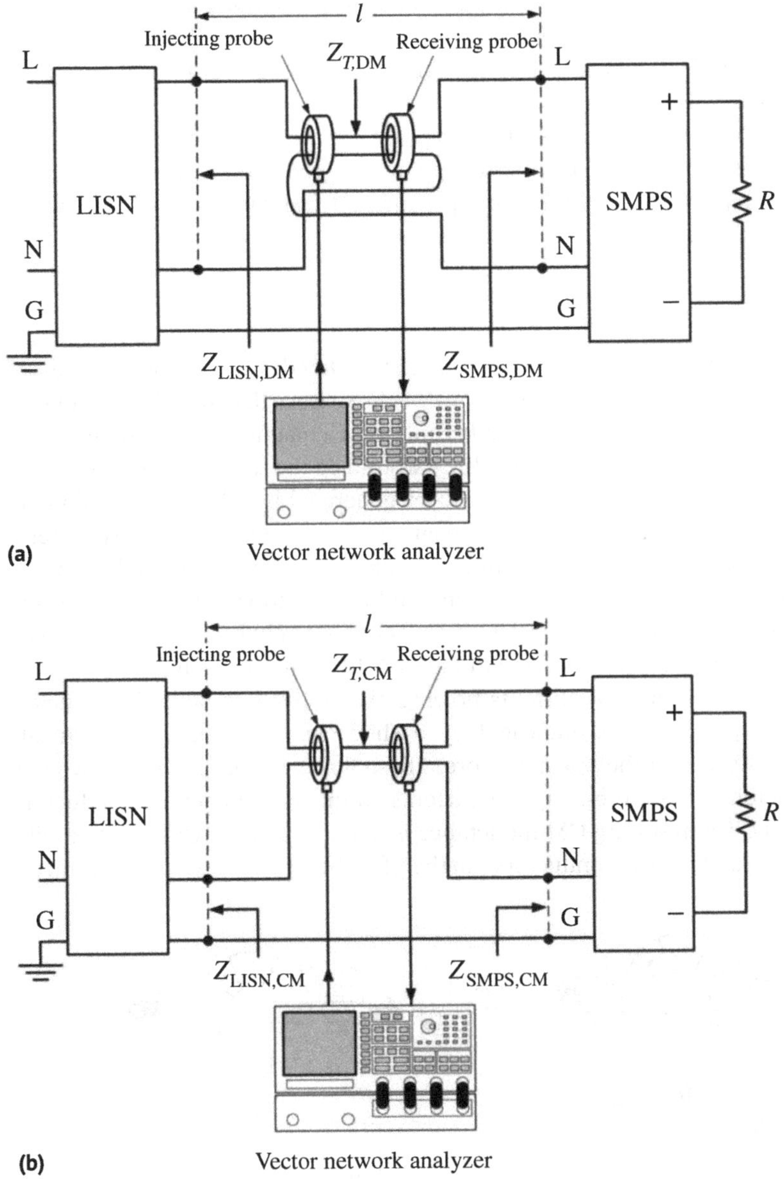

Figure 7.9 SMPS noise source impedance extraction using the FD-TPS. (a) DM configuration. (b) CM configuration.

where Z_{LISN} is the impedance of the LISN, Z_{SMPS} is the noise source impedance of the SMPS, and Z_{wire} represents the impedance contributed by the wiring connections. With Z_{LISN} and Z_{wire} known, Z_{SMPS} can be readily evaluated once Z_T is measured using Equation (7.8).

It should be noted that a two-turn loop probe was also recently reported for FD-SPS measurements aimed at simplifying both the test setup and the calibration procedure. Its effectiveness was also experimentally validated in EMI filter design applications [20].

7.1.3 EMI Filter Components

The passive components commonly used in EMI filter design include CM and DM chokes, as well as X and Y capacitors. A CM choke is particularly effective in suppressing CM noise currents. It typically consists of a magnetic core with high relative permeability and two sets of windings wound on the same core. The windings are arranged such that the magnetic fluxes generated by CM currents add constructively within the core, while the magnetic fluxes generated by DM currents cancel each other [21, 22]. As a result, a CM choke presents a high inductance to CM currents while introducing negligible inductance to DM currents. The configurations and simplified equivalent circuits of a CM choke under CM and DM excitation conditions are illustrated in Figure 7.10a,b, respectively.

In Figure 7.10a, the CM currents flowing on the live and neutral conductors generate magnetic fluxes, indicated by dashed clockwise arrows, which add constructively within the magnetic core. These fluxes are equivalent to two parallel inductors that suppress the CM currents on the live and neutral conductors, respectively. The resulting CM inductance is denoted as L_{CM}. Under CM excitation, the live and neutral conductors can therefore be treated as a single conductor.

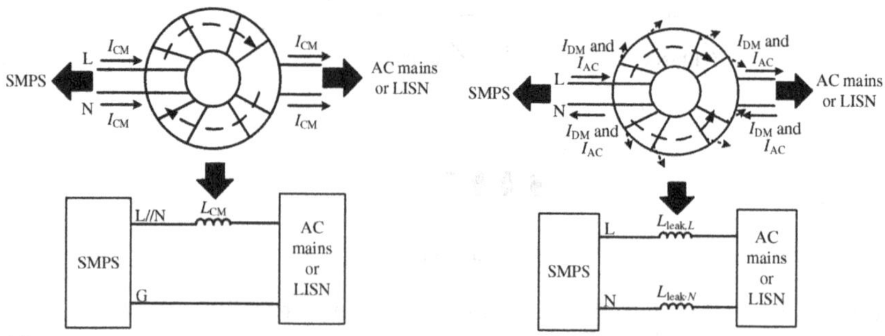

Figure 7.10 Configurations and simplified equivalent circuits of a CM choke under: (a) CM excitation and (b) DM excitation.

In Figure 7.10b, the high-frequency DM noise current and the 50/60 Hz AC current (I_{AC}) generate magnetic fluxes indicated by dashed clockwise and counterclockwise arrows, respectively, which largely cancel each other within the core. Since the magnitude of the 50/60 Hz AC current is typically much larger than that of the high-frequency DM noise current, the net magnetic flux remains small, thereby preventing core saturation. The high-frequency DM noise current, however, produces leakage fluxes, indicated by dashed straight arrows. These leakage fluxes result in incomplete flux cancellation and give rise to a finite DM inductance that suppresses high-frequency DM noise currents on the live and neutral conductors. Under DM excitation, the two DM inductances are connected in series, and the resultant leakage inductance is given by $L_{leak} = L_{leak,L} + L_{leak,N}$, which is typically much smaller than L_{CM}, as the magnitude of the leakage flux is significantly lower than that of the CM flux.

Since a single-phase CM choke has four terminals and exhibits two distinct inductance values, the measurement of L_{CM} and L_{leak} using an impedance analyzer differs from that of a conventional two-terminal inductor. The corresponding measurement configurations are illustrated in Figure 7.11.

In Figure 7.11, terminals T1, T2, T3, and T4 represent the four terminals of the CM choke. Figure 7.11a shows the measurement configuration for L_{CM}, where terminals T1 and T3 are shorted together and connected to Port 1 of the impedance analyzer, while terminals T2 and T4 are shorted together and connected to Port 2 [23]. Figure 7.11b shows the measurement configuration for L_{leak}, where terminals T1 and T2 are shorted together and terminals T3 and T4 are connected to Ports 1 and 2 of the impedance analyzer, respectively [23].

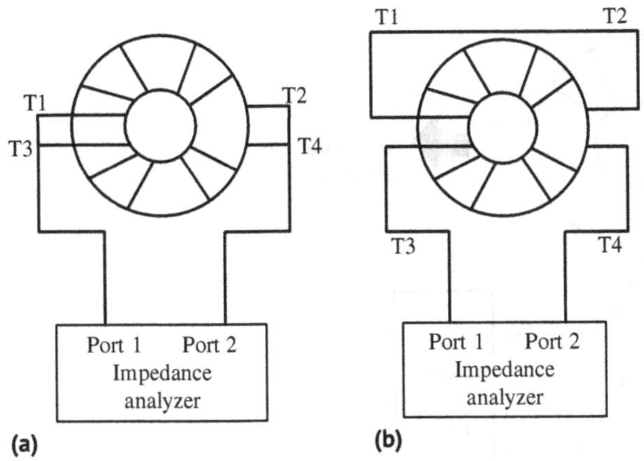

Figure 7.11 Measurement configurations of a single-phase CM choke. (a) Measurement of L_{CM}. (b) Measurement of L_{leak}.

In contrast, a DM choke operates based on the same principle as a conventional two-terminal inductor. It typically consists of an iron-powder magnetic core with a single winding wound around the core. Unlike a CM choke, in which magnetic fluxes cancel under DM excitation, a DM choke carries substantial magnetic flux generated by the AC current. Therefore, the magnetic core of a DM choke must be designed to withstand large flux densities without saturation [19]. The configuration and simplified equivalent circuit of a DM choke are illustrated in Figure 7.12.

As shown in Figure 7.12, two identical DM chokes, each with an inductance of L_{DM}, are placed on the live and neutral conductors, respectively. When CM currents flow through the two DM chokes, the chokes are effectively connected in parallel. Consequently, the resultant inductance seen by the CM current is $L_{DM}/2$.

In contrast, when DM currents flow, the two DM chokes are effectively connected in series, yielding a resultant inductance of $2L_{DM}$. Therefore, DM chokes provide strong attenuation for high-frequency DM noise while presenting negligible impedance to the low-frequency AC current, since the AC mains frequency is sufficiently low. In practice, DM chokes are not always required in an EMI filter. If the leakage inductance of the CM choke is sufficiently large, it can also provide effective DM noise suppression, allowing the CM choke to serve a dual role.

A line-to-line capacitor, commonly referred to as an X capacitor, is connected across the live and neutral conductors and is used to suppress DM noise currents. A line-to-ground capacitor, known as a Y capacitor, is used for CM noise suppression. Two Y capacitors are typically required: one connected between the live conductor and ground, and the other between the neutral conductor and ground. The capacitance of an X capacitor, denoted as C_X, is usually relatively large and typically falls in the microfarad range. In contrast, the capacitance of a Y capacitor,

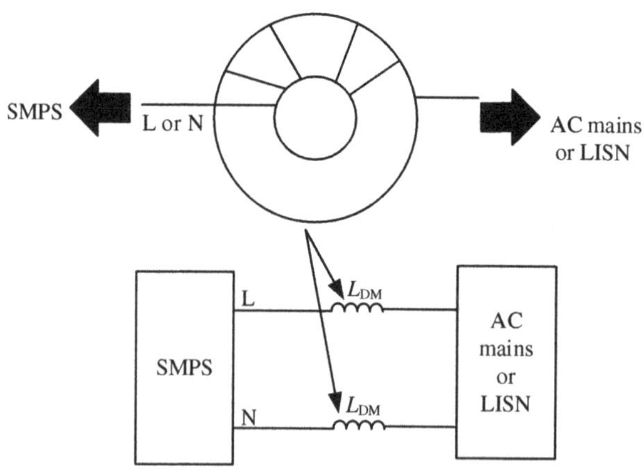

Figure 7.12 Configuration and simplified equivalent circuit of a DM choke.

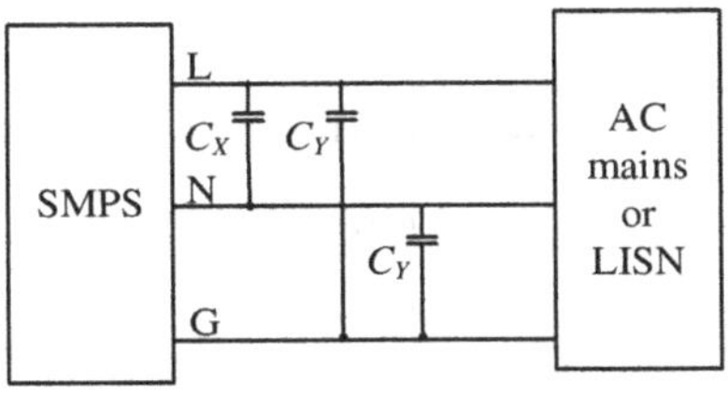

Figure 7.13 Configuration of X and Y capacitors used in an EMI filter.

denoted as C_Y, must be kept small to limit earth leakage current and satisfy electrical safety requirements. As a result, C_Y is usually limited to several thousand picofarads.

As shown in Figure 7.13, only the two Y capacitors contribute to CM noise suppression, whereas all three capacitors participate in DM noise suppression. For CM noise, the two Y capacitors are effectively connected in parallel, resulting in an equivalent capacitance of $2C_Y$. For DM noise suppression, the X capacitor is connected in parallel with the series combination of the two Y capacitors. Consequently, the resultant equivalent capacitance for DM noise suppression is $C_X + C_Y/2$.

7.1.4 EMI Filter Design

There are two main approaches to EMI filter design [24, 25]. The first approach [25] is relatively simple but neglects the influence of filter configurations as well as the source and termination impedances of the SMPS and the LISN. By ignoring these impedances, the resulting EMI filter is often overdesigned, as reported in [24]. In contrast, the second approach [24] provides a systematic EMI filter design procedure that explicitly accounts for both circuit impedances and filter configurations, thereby avoiding overdesign. In this work, the latter approach is adopted. The EMI filter is designed based on measurements performed with the LISN and is subsequently evaluated to demonstrate its impact on EMI performance under practical operating conditions without the LISN.

The first step in EMI filter design is to determine the required CM and DM attenuations. These attenuations are obtained by comparing the measured CM and DM conducted emissions with the applicable conducted EMI limits shown in Figure 7.7. However, since the conducted EMI limits are defined in terms of voltage (in units of dBμV) measured using a LISN, they do not directly apply to subsequent demonstrations where conducted emissions are measured in terms of current (in units of dBμA), including cases without the LISN. To address this issue, the CISPR 22 Class A quasi-peak voltage limit is converted into a current limit (dBμA) by subtracting 34 dBΩ, corresponding to the logarithmic representation

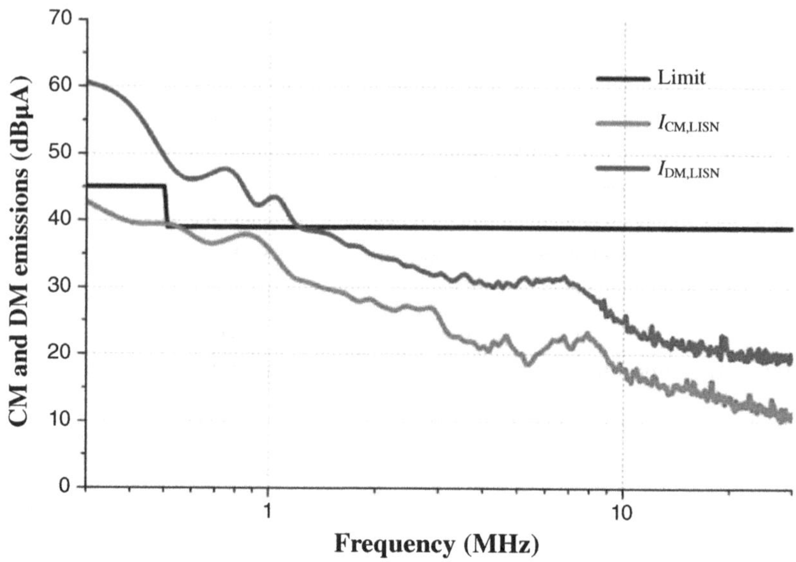

Figure 7.14 CM and DM emissions measured with the LISN and the corresponding converted current limit.

of the 50 Ω LISN impedance. The converted limit, together with the CM and DM emissions measured using the LISN, is shown in Figure 7.14.

As illustrated in Figure 7.14, $I_{CM,LISN}$ and $I_{DM,LISN}$ denote the CM and DM conducted emissions (in dBμA) measured with the LISN, respectively. The CM emission is generally below the limit line from 300 kHz to 30 MHz and can be further attenuated by a few decibels below 1 MHz. In contrast, the DM emission exceeds the limit below approximately 1.5 MHz and requires a maximum attenuation of about 15 dB at 300 kHz. Based on these observations, the required CM and DM attenuations are determined.

After defining the required attenuations, the filter configuration must be selected. Since the SMPS impedance is larger than the LISN impedance for both CM and DM over the measurement frequency range, a CL filter configuration [18, 26–28] is chosen. In this configuration, the shunt capacitors are placed on the high-impedance side, and the series inductors are placed on the low-impedance side. Accordingly, an X capacitor, two Y capacitors, and a CM choke are selected to form the CL filter. A DM choke is introduced only if the required DM attenuation cannot be achieved. The X and Y capacitors are connected in parallel with the SMPS, while the CM choke is placed in series with the LISN. The equivalent circuits of the resulting CL filter for CM and DM are shown in Figure 7.15.

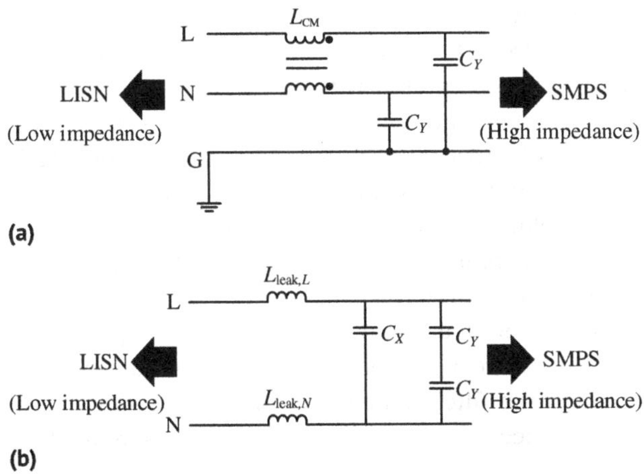

Figure 7.15 Equivalent CL-filter configuration for (a) CM and (b) DM.

As shown in Figure 7.15, the CM filter consists of the CM inductance L_{CM} and two parallel C_Y capacitors, whereas the DM filter consists of the leakage inductances $L_{leak,L}$ and $L_{leak,N}$, the X capacitor C_X, and two series C_Y capacitors. Based on these CM and DM configurations, the estimated CM attenuation ($A_{CM,est}$) and DM attenuation ($A_{DM,est}$) can be calculated using [3, 29, 30] as

$$A_{CM,est} = 20\log\left|s^2\cdot\left(\frac{L_{CM}C_YZ_{SMPS,CM}}{Z_{LISN,CM}+Z_{SMPS,CM}}\right) + s\cdot\left(\frac{L_{CM}+2C_YZ_{SMPS,CM}Z_{LISN,CM}}{Z_{LISN,CM}+Z_{SMPS,CM}}\right)+1\right| \tag{7.9}$$

$$A_{DM,est} = 20\log\left|s^2\cdot\left(\frac{L_{leak}\left(C_X+\frac{C_Y}{2}\right)Z_{SMPS,DM}}{Z_{LISN,DM}+Z_{SMPS,DM}}\right) + s\cdot\left(\frac{L_{leak}+\left(C_X+\frac{C_Y}{2}\right)Z_{SMPS,DM}Z_{LISN,DM}}{Z_{LISN,DM}+Z_{SMPS,DM}}\right)+1\right| \tag{7.10}$$

where $s = j\omega$. In Equations (7.9) and (7.10), all filter components are assumed to be ideal, and L_{leak}, L_{CM}, C_X, and C_Y are treated as frequency-invariant over the range from 300 kHz to 30 MHz.

Since L_{leak} depends on the selected CM choke, L_{CM} and C_Y are determined first. For CM attenuation, L_{CM} and C_Y are selected such that $A_{CM,est}$ satisfies the required CM attenuation. Given that the required CM attenuation is relatively small, $A_{CM,est}$ is chosen to be just sufficient to avoid overdesign. By selecting

$C_Y = 1000$ pF, L_{CM} is determined by substituting C_Y, $Z_{SMPS,CM}$, and $Z_{LISN,CM}$ into Equation (7.9), resulting in $L_{CM} = 0.8$ mH. A CM choke with $L_{CM} = 0.8$ mH is then selected, and its leakage inductance is measured as $L_{leak} = 3.7$ μH using the setup shown in Figure 7.11b. For DM attenuation, L_{leak}, C_Y, and C_X must satisfy the required DM attenuation, which is 16 dB at 300 kHz. Since L_{leak} and C_Y are already determined, C_X is obtained by substituting these values together with $Z_{SMPS,DM}$ and $Z_{LISN,DM}$ into Equation (7.10), yielding $C_X = 0.6$ μF. The estimated CM and DM attenuations calculated using ideal component values are plotted in Figure 7.16 over the frequency range from 300 kHz to 30 MHz.

The actual impedances of the selected components, including $L_{leak}(Z_{L_{leak}})$, $L_{CM}(Z_{L_{CM}})$, $C_X(Z_{C_X})$, and $C_Y(Z_{C_Y})$, are measured over the same frequency range and shown in Figure 7.17. With the filter components and configuration thus determined, the EMI filter designed based on LISN-assisted measurements is finalized.

This section has demonstrated a systematic workflow for EMI filter design in SMPSs based on electrically contactless online impedance measurement. By first analyzing conducted EMI mechanisms and separating CM and DM emissions using LISN-based measurements, the required attenuation levels were quantitatively determined. The CM and DM source and termination impedances were then extracted under realistic operating conditions using inductive coupling

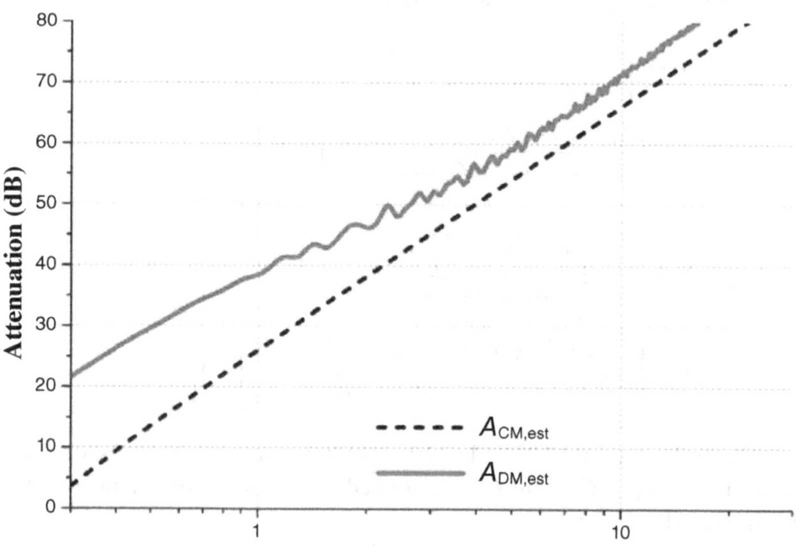

Figure 7.16 Estimated $A_{CM,est}$ and $A_{DM,est}$ determined using the selected ideal inductors and capacitors.

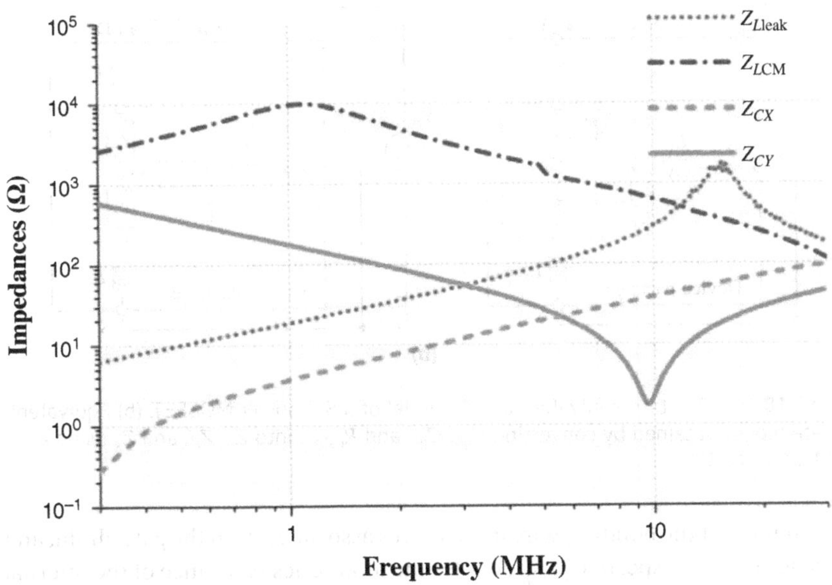

Figure 7.17 Measured impedances of $L_{\text{leak}}(Z_{L_{\text{leak}}})$, $L_{\text{CM}}(Z_{L_{\text{CM}}})$, $C_X(Z_{C_X})$, and $C_Y(Z_{C_Y})$.

techniques, enabling impedance-aware filter configuration selection. Based on these measured impedances, a CL-filter topology was designed, and its CM and DM attenuations were analytically estimated and experimentally validated. This impedance-informed design approach avoids overdesign and provides a practical and measurement-driven methodology for EMI filter development in real power-electronic systems.

7.2 Health Diagnosis of Power Semiconductor Devices

Understanding the mechanisms underlying defects in power semiconductor devices is essential for failure analysis and reliability assessment and therefore motivates the development of effective health diagnosis methodologies. Health diagnosis typically involves monitoring predefined electrical parameters of a power device under normal operating conditions and analyzing deviations from their nominal values to identify performance degradation or the onset of incipient faults.

7.2.1 On-state Equivalent Circuit Model of SiC Power MOSFETs

The equivalent circuit model of the drain-to-source on-state impedance of a Silicon Carbide (SiC) power MOSFET is illustrated in Figure 7.18a. In this model, L_g, L_d,

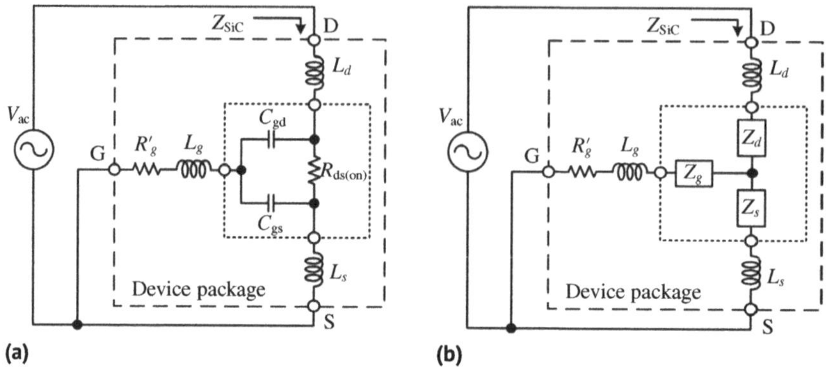

Figure 7.18 (a) On-state equivalent circuit model of a SiC power MOSFET. (b) Equivalent on-state model obtained by converting C_{gd}, C_{gs}, and $R_{ds(on)}$ into Z_g, Z_d, and Z_s using a Δ–Y transformation.

and L_s represent the bonding-wire inductances associated with the gate, drain, and source terminals, respectively. R_g' denotes the total series resistance of the internal and external gate resistances. The gate-to-drain and gate-to-source capacitances are denoted by C_{gd} and C_{gs}, respectively, and $R_{ds(on)}$ represents the drain-to-source on-state resistance.

By applying a Δ–Y transformation, the elements C_{gd}, C_{gs}, and $R_{ds(on)}$ can be converted into the equivalent impedances Z_g, Z_d, and Z_s, as shown in Figure 7.18b. For a sinusoidal steady-state excitation with angular frequency ω, the on-state impedance of the SiC power MOSFET can be expressed as

$$Z_{\text{SiC}} = (j\omega L_d + Z_d) + \left(R'_g + j\omega L_g + Z_g\right) \| (j\omega L_s + Z_s) \tag{7.11}$$

The impedances Z_g, Z_d, and Z_s obtained from the Δ–Y transformation are given by

$$Z_g = \frac{Z_{gd} Z_{gs}}{Z_{\text{total}}} \quad Z_d = \frac{Z_{gd} Z_{ds}}{Z_{\text{total}}} \quad Z_s = \frac{Z_{gs} Z_{ds}}{Z_{\text{total}}} \tag{7.12}$$

where $Z_{gd} = 1/(j\omega C_{gd})$, $Z_{gs} = 1/(j\omega C_{gs})$, $Z_{ds} = R_{ds(on)}$, and $Z_{\text{total}} = Z_{gd} + Z_{gs} + Z_{ds}$.

When the MOSFET operates in the fully on-state ($V_{ds} \approx 0$), Z_g is typically much larger than Z_s over the frequency range of interest, and the ratio $Z_g/Z_s \approx 1/(j\omega R_{ds(on)} C_{gd})$. According to the datasheet of the selected SiC power MOSFET (C2M0080120D, Cree Inc.) used for proof of concept in this study, $C_{gd} \approx 412$ pF and $R_{ds(on)} \approx 80$ mΩ at a rated current of 20 A. This yields $|Z_g/Z_s| \approx 2.41 \times 10^2$ at 20 MHz, allowing Equation (7.11) to be

$$Z_{\text{SiC}} \approx Z_d + Z_s + j\omega(L_d + L_s) \tag{7.13}$$

Substituting Z_d and Z_s from Equation (7.12) into Equation (7.13) gives

$$Z_{\text{SiC}} = \frac{R_{\text{ds(on)}}}{1 + j\omega R_{\text{ds(on)}}\left(C_{\text{gd}} \| C_{\text{gs}}\right)} + j\omega(L_d + L_s) \quad (7.14)$$

Equation (7.14) indicates that $R_{\text{ds(on)}}$ dominates Z_{SiC} at low frequencies, while the inductive reactance associated with the bonding wires dominates at high frequencies. Using the manufacturer-provided packaged SPICE model of the C2M0080120D, the simulated frequency response of the fully turned-on SiC power MOSFET with a nominal gate-source voltage $V_{\text{GS}} = 20$ V is shown in Figure 7.19. The simulated results exhibit the expected frequency-dependent behavior predicted by Equation (7.14).

From the simulated frequency response, Z_{SiC} can be accurately approximated by a series RL model with $R = 69.2$ mΩ and $L = 15.5$ nH, as illustrated in Figure 7.19. It should be noted that, within the frequency range of interest, the capacitive effects of C_{gd} and C_{gs} are negligible in this case study.

7.2.2 Measurement Setup

Figure 7.20 illustrates the measurement configuration used to extract the unknown impedance Z_x of a DUT powered by a DC voltage source using the FD-TPS. Two inductive probes are clamped onto the wiring connection between the DC source and the DUT. A coupling capacitor C_c is inserted to block the DC power while allowing high-frequency excitation and response signals to pass.

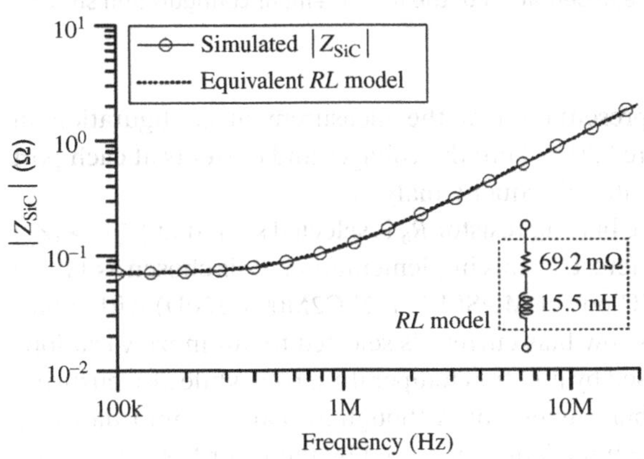

Figure 7.19 Simulated frequency response of the on-state impedance of a SiC power MOSFET (C2M0080120D) and its equivalent series RL model.

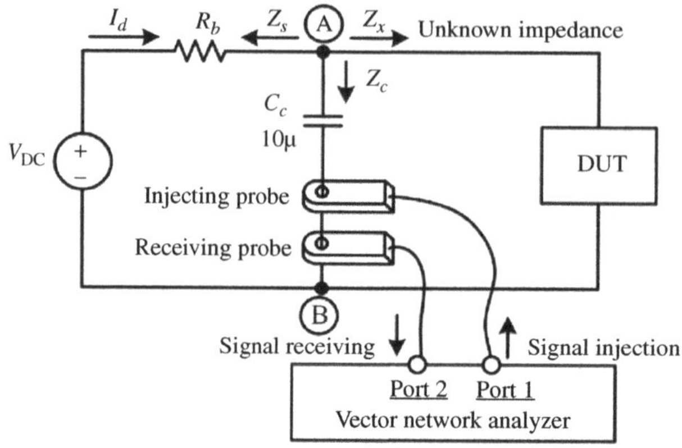

Figure 7.20 Measurement configuration for extracting the unknown impedance Z_x using the FD-TPS.

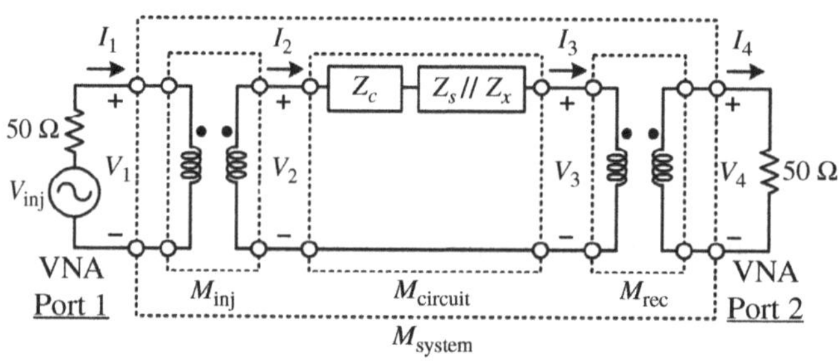

Figure 7.21 Two-port network representation of the measurement configuration shown in Figure 7.20.

The two-port network representation of the measurement configuration in Figure 7.20 is shown in Figure 7.21, where the voltages and currents at each port are explicitly labeled to facilitate subsequent analysis.

As shown in Figure 7.20, the biasing resistor R_b is selected such that $|Z_s| \gg |Z_x|$ over the frequency range of interest. In this implementation, R_b is chosen as 11.2 Ω to provide the DUT (i.e., a SiC power MOSFET, P/N C2M0080120D) with a bias current of 1 A. This relatively low bias current is selected to minimize variations in the on-state resistance caused by junction temperature rise, while also ensuring low power dissipation in the biasing resistor. Although the parasitic impedance Z_c can be de-embedded from the impedance measured by the IA or VNA, it is good practice to minimize Z_c so that the DUT impedance Z_x dominates the measurement. The FD-TPS employed for experimental validation consists of two identical inductive coupling probes (Tektronix CT6) and a VNA (Rohde & Schwarz ZNB20).

Prior to measuring the DUT impedance, the measurement setup is validated using several known precision surface-mounted device (SMD) resistors with negligible parasitic effects. These resistors are treated as unknown impedances Z_x for validation purposes. The total impedance Z_{total} is first measured, after which the DUT is replaced by a short circuit to obtain the parasitic impedance Z_c. The measured magnitude of Z_c remains well below 200 mΩ over the frequency range of interest, as shown in Figure 7.22.

It should be noted that Z_c includes the inductive reactance of the wiring connections along the path "node A → DUT → node B," which is much smaller than that contributed by the path "node A → C_c → node B," owing to the significantly shorter PCB trace length. The impedance Z_s is measured with the DUT removed. Finally, the unknown DUT impedance Z_x is obtained by subtracting Z_c from the measured Z_{total}.

Figure 7.23 presents the measured results for three low-resistance precision resistors used for validation. Standard 1 Ω resistors with ±1% tolerance are selected. Two and three 1 Ω resistors are connected in parallel to form equivalent resistances of 0.5 and 0.33 Ω, respectively. The frequency response of $|Z_s|$ is also plotted in Figure 7.23 to confirm that $|Z_s| \gg |Z_x|$ over the measurement bandwidth. The extracted impedance values using the FD-TPS agree well with their nominal resistance values.

To further validate the method introduced, a parallel *RC* circuit in series with an inductor is selected. This configuration emulates the typical on-state impedance behavior of a SiC power MOSFET, where the on-state resistance $R_{ds(on)}$ dominates at low frequencies and the inductive reactance becomes significant at higher frequencies. As shown in Figure 7.24, the measured and simulated impedance responses exhibit close agreement across the frequency range of interest.

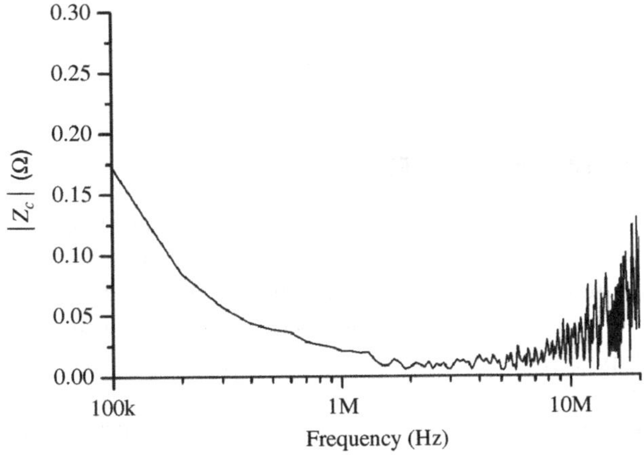

Figure 7.22 Measured frequency response of the parasitic impedance $|Z_c|$.

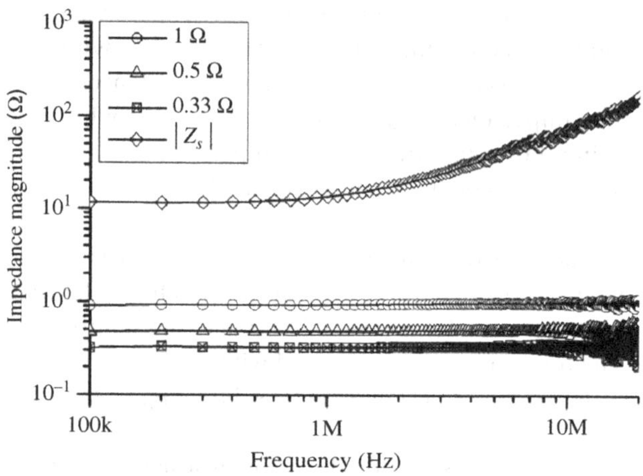

Figure 7.23 Measured frequency responses of precision resistors and $|Z_s|$.

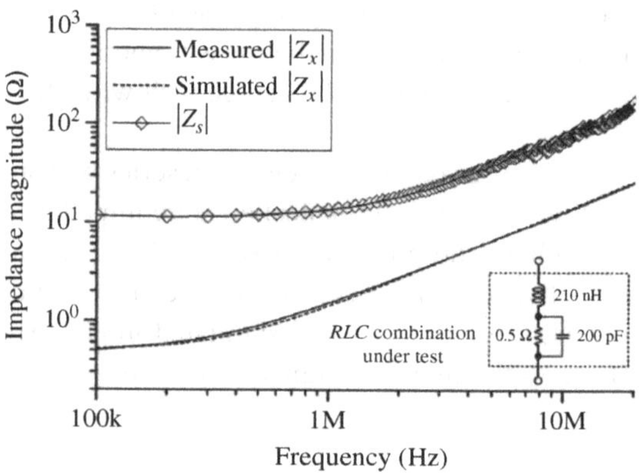

Figure 7.24 Measured impedance frequency response of a parallel *RC* circuit in series with an inductor.

7.2.3 Health Diagnosis of SiC Power MOSFETs

Based on the above-validated method, the on-state impedance frequency response of the DUT is measured. Unless specified otherwise, the DUT is fully turned on with a nominal gate bias voltage of 20 V. Figure 7.25 shows the actual test fixture fabricated on a printed circuit board (PCB). Terminal blocks are mounted on the PCB to facilitate DUT replacement between measurements while maintaining similar environmental and electrical conditions.

Figure 7.26a shows four similar DUTs used in the experiments. DUT-A and DUT-D are non-degraded power MOSFETs with different pin lengths. DUT-B

Figure 7.25 PCB-based test fixture for health diagnosis of the SiC power MOSFET.

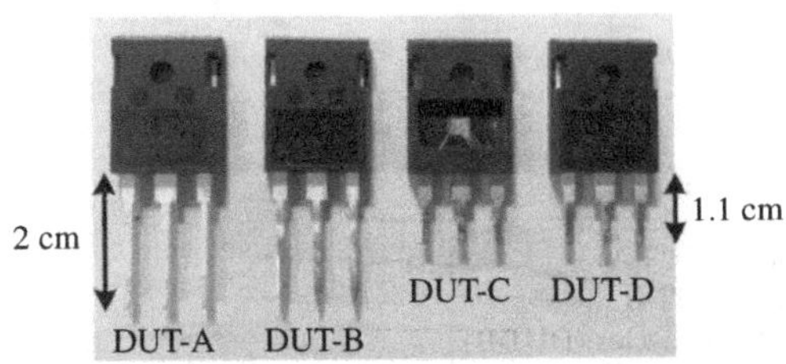

(a)

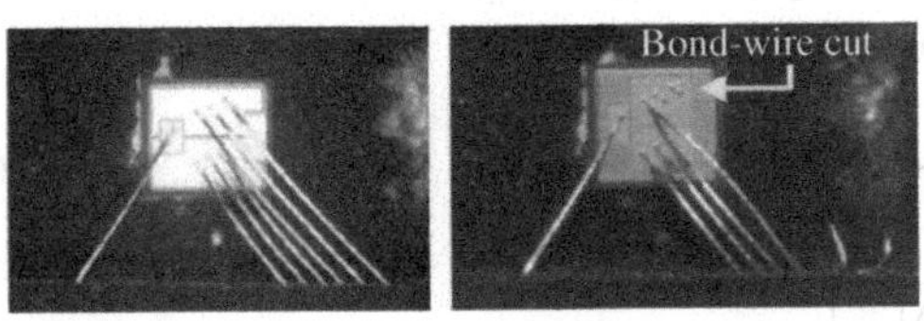

(b)

Figure 7.26 (a) Four power MOSFETs under test. DUT-A: non-degraded, pin length 2 cm; DUT-B: degraded after 20,000 temperature cycles; DUT-C: non-degraded, de-capsulated with damaged bonding wires; DUT-D: non-degraded, pin length 1.1 cm. (b) Images of DUT-C before (left) and after (right) cutting of bonding wires.

is a degraded power MOSFET subjected to 20,000 temperature cycles. The temperature cycling is implemented by repetitively changing the operating current between 24 and 0.5 A to emulate thermal stress, corresponding to case temperatures of 35 and 27 °C, respectively. DUT-C is a non-degraded and de-capsulated power MOSFET in which two bonding wires connected to the source terminal are intentionally cut to emulate a bonding-wire defect. The impedance frequency responses of DUT-A and DUT-B are compared to evaluate device degradation induced by accelerated thermal stress. Since the pin leads of DUT-C are cut during the de-capsulation process, its impedance frequency response is compared with that of DUT-D to diagnose damaged bonding wires. Images of DUT-C before and after cutting of bonding wires are shown in Figure 7.26b.

Figure 7.27 shows the I–V characteristics of DUT-A and DUT-B measured using an I–V curve tracer (Tektronix 371A). The results indicate that the $R_{ds(on)}$ of the degraded DUT-B deviates from its initial value by 29.1% at an operating current of 10 A. The on-state impedance measured using the FD-TPS at an operating current of 1 A is shown in Figure 7.28. The results clearly reveal that the on-state impedance of the degraded DUT-B is higher than that of non-degraded DUT-A.

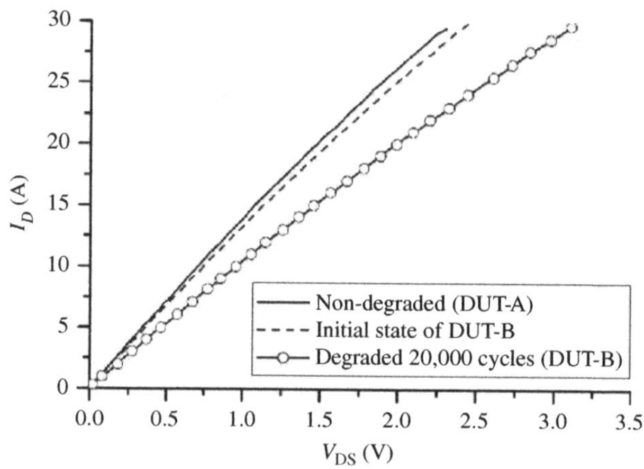

Figure 7.27 $I-V$ characteristics of DUT-A and DUT-B.

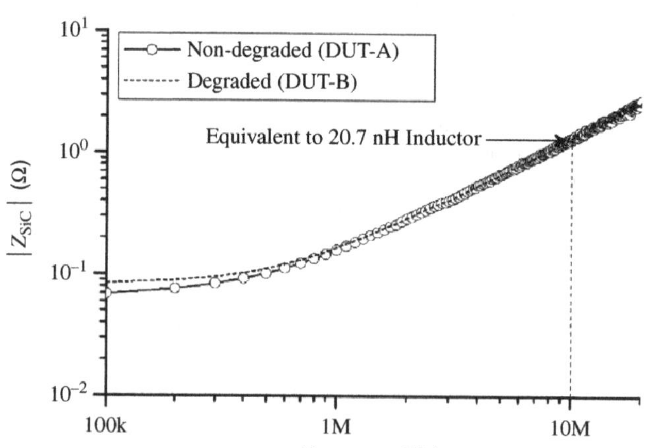

Figure 7.28 On-state impedance frequency responses of DUT-A and DUT-B using the FD-TPS.

Table 7.1 compares the $R_{ds(on)}$ values measured using the I–V curve tracer and the FD-TPS. It can be observed that the $R_{ds(on)}$ values obtained using both methods agree well.

Proper gate bias voltage is essential for the optimal operation of a power MOSFET in power conversion systems. Insufficient gate bias voltage results in an increased $R_{ds(on)}$, leading to higher power dissipation and an elevated risk of thermal runaway. Figure 7.29 demonstrates that the introduced method can effectively detect insufficient gate bias voltage for DUT-A. The measured $|Z_{SiC}|$ at 100 kHz using this method is dominated by $R_{ds(on)}$, which agrees well with the $R_{ds(on)}$ values obtained using the conventional V–I method (based on V_{ds} and I_D), as summarized in Table 7.2.

Bonding wires contribute significantly to the parasitic inductance within a power MOSFET package. At high frequencies, $|Z_{SiC}|$ is dominated by inductive reactance; therefore, a power MOSFET with damaged bonding wires is expected to exhibit a higher inductive reactance. Figure 7.30 compares the on-state impedance

Table 7.1 Comparison of measured $R_{ds(on)}$ using different methods.

Measurement Method	$R_{ds(on)}$(mΩ)@1 A, Room Temperature	
	DUT-A	**DUT-B (Degraded)**
I–V curve tracer	64.0	80.0
FD-TPS	68.8	84.4

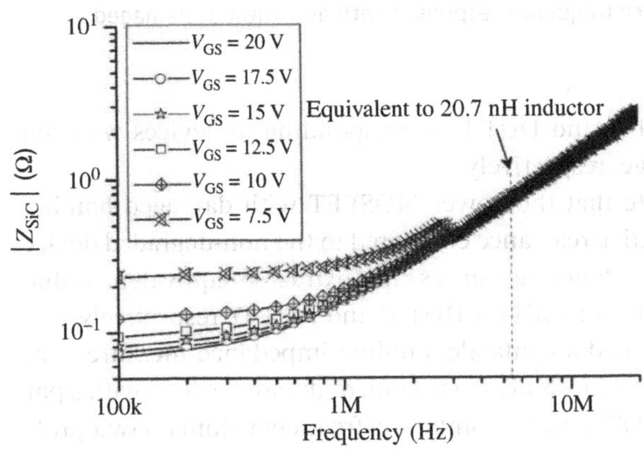

Figure 7.29 On-state impedance frequency responses of DUT-A under different V_{GS} biasing conditions.

Table 7.2 Measured $R_{ds(on)}$ under different V_{GS} biasing conditions.

V_{GS}(V)	$R_{ds(on)}$(mΩ)@1 A, Room Temperature	
	V–I Method	FD-TPS
20	67.3	68.8
17.5	72.9	73.8
15	80.2	80.7
12.5	92.4	94.4
10	120.5	123.2
7.5	225.3	229.7

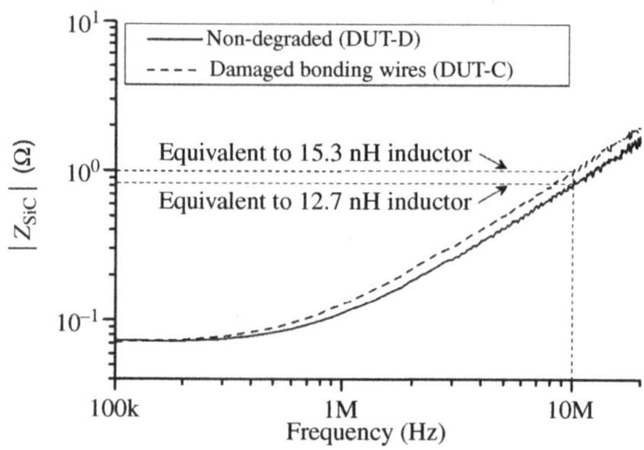

Figure 7.30 On-state impedance frequency responses with and without damaged bonding wires.

frequency responses of DUT-C and DUT-D, corresponding to devices with and without bonding-wire damage, respectively.

The results clearly indicate that the power MOSFET with damaged bonding wires exhibits a higher inductive reactance compared to the non-degraded device. Based on the measured impedance responses, the extracted equivalent inductances are found to be 15.3 and 12.7 nH for DUT-C and DUT-D, respectively.

This section has demonstrated a contactless online impedance measurement-based method for the health diagnosis of power semiconductor devices, with a particular focus on SiC power MOSFETs. By combining frequency-domain two-probe sensing with carefully designed measurement setups, the method enables accurate extraction of on-state impedance under normal operating conditions without intrusive electrical connections. Experimental validation confirms that key health indicators, including on-state resistance, gate-bias adequacy, and bonding-wire integrity, can be effectively identified through characteristic impedance signatures across frequency. Overall, this approach provides a practical, low-impact, and scalable solution for early fault diagnosis of power semiconductor devices in modern power electronic systems.

7.3 Voltage-dependent Capacitance Extraction of Power Semiconductor Devices

Compared with conventional power semiconductor devices, wide-bandgap (WBG) power semiconductor devices, such as SiC power MOSFETs, exhibit lower on-state resistance, higher switching frequency capability, and higher rated junction temperature. As a result, they are gaining increasing popularity in power electronic applications [31–34]. However, the higher switching speeds also give rise to undesirable voltage and current oscillations during switching transients, as well as resonances between circuit loop inductances and the interelectrode capacitances of SiC power MOSFETs [35, 36]. These phenomena can degrade system reliability and exacerbate EMI issues [37, 38]. To accurately evaluate and mitigate such adverse effects during the early design stage, reliable values of the interelectrode capacitances of SiC power MOSFETs are required [39–41]. Since these capacitances are inherently voltage-dependent [42], their characteristics over the entire operating voltage range must be obtained to enable accurate device modeling and robust circuit design.

Several methods have been reported in the literature for extracting voltage-dependent capacitances. One approach employs three-dimensional full-wave electromagnetic simulation tools, such as finite-element methods, to compute the capacitances [43]. This approach is feasible only when detailed device geometry and material properties are available, which is often not the case due to intellectual property constraints. Another method employs an impedance analyzer [44], but it requires numerous auxiliary components, including DC blocking capacitors, AC blocking inductors, various resistors, and protection diodes, to enable safe and effective online measurements. Time-domain reflectometry-based methods have also been reported [42]; however, proper electrical isolation is required to ensure that the measurement setup is fully disconnected from ground. Moreover, this approach is typically limited to extracting only the output capacitance of the power MOSFET. A frequency-domain multi-probe sensing approach has also been presented in [45], but parasitic components of the probes and probe-to-probe coupling are neglected during the measurement process, which may degrade the accuracy of the extracted capacitances. In addition to these methods, commercial power device analyzers (e.g., Keysight B1505A) are available and can extract voltage-dependent capacitances under high-voltage bias conditions using dedicated bias-tee configurations. However, because the bias-tee is physically connected to a high-voltage DC power supply, additional protection circuitry is required to satisfy safety compliance requirements. To extract the voltage-dependent capacitances of SiC power MOSFETs in a simple, safe, and accurate manner, this section introduces an alternative approach based on the TD-TPS.

7.3.1 Equivalent Circuit Model of SiC Power MOSFET

Figure 7.31 illustrates the small-signal equivalent circuit model of a SiC power MOSFET in the off-state, where d, g, and s represent the drain, gate, and source terminals, respectively. C_{gd}, C_{gs}, and C_{ds} denote the interelectrode capacitances. L_d, L_g, and L_s denote the terminal inductances. R_d, R_g, and R_s represent the terminal resistances. The terminal inductances are typically on the order of a few nanohenries [46–48] whereas the interelectrode capacitances are usually smaller than a few nanofarads [49–52]. Therefore, in the frequency range of $f_{\mathrm{sig}} \leq 1\,\mathrm{MHz}$, the impedance of the SiC power MOSFET in the off-state is dominated by its interelectrode capacitances, and the effects of terminal inductances can be neglected during the capacitance extraction process [44, 45]. Consequently, the equivalent circuit model shown in Figure 7.31 can be simplified to that shown in Figure 7.32. Figure 7.33 shows the star-connected equivalent circuit corresponding to Figure 7.32.

By applying a delta-to-star circuit transformation, the star-connected capacitances C_{dn}, C_{gn}, and C_{sn} can be expressed as

$$C_{\mathrm{dn}} = C_{\mathrm{gd}} + C_{\mathrm{ds}} + \frac{C_{\mathrm{gd}}C_{\mathrm{ds}}}{C_{\mathrm{gs}}} \tag{7.15}$$

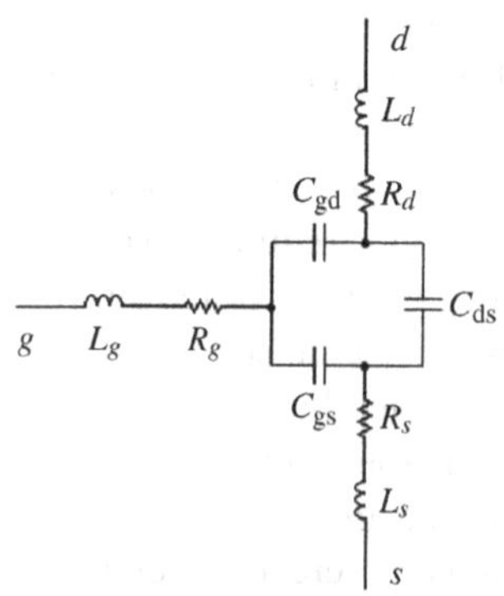

Figure 7.31 Small-signal equivalent circuit model of the SiC power MOSFET in the off-state.

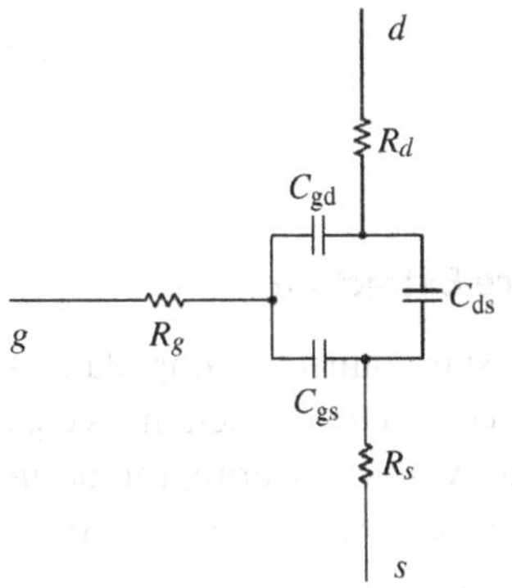

Figure 7.32 Simplified equivalent circuit model of Figure 7.31 when $f_{sig} \leq 1$ MHz.

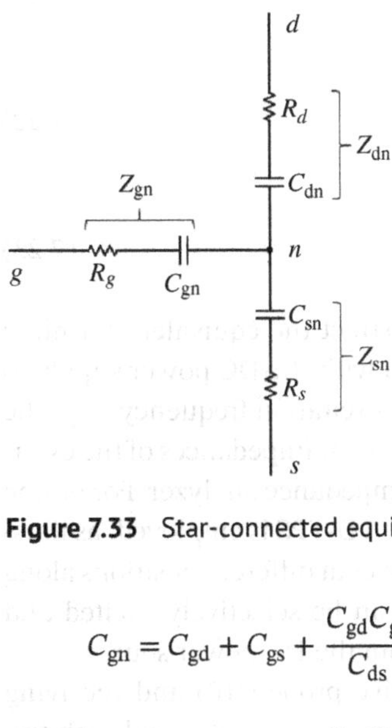

Figure 7.33 Star-connected equivalent circuit model derived from Figure 7.32.

$$C_{gn} = C_{gd} + C_{gs} + \frac{C_{gd}C_{gs}}{C_{ds}} \tag{7.16}$$

$$C_{sn} = C_{gs} + C_{ds} + \frac{C_{gs}C_{ds}}{C_{gd}} \tag{7.17}$$

Based on the star-connected capacitances, the corresponding star-connected impedances Z_{dn}, Z_{gn}, and Z_{sn} are obtained as

$$Z_{dn} = R_d + \frac{1}{j\omega C_{dn}} \tag{7.18}$$

$$Z_{gn} = R_g + \frac{1}{j\omega C_{gn}} \tag{7.19}$$

$$Z_{\mathrm{sn}} = R_s + \frac{1}{j\omega C_{\mathrm{sn}}} \tag{7.20}$$

where $\omega = 2\pi f_{\mathrm{sig}}$.

7.3.2 Principle of Voltage-dependent Capacitance Extraction

From Equations (7.18)–(7.20), once the equivalent star-connected impedances Z_{dn}, Z_{gn}, and Z_{sn} are obtained, the corresponding star-connected capacitances C_{dn}, C_{gn}, and C_{sn} can be determined at a static functionality. Furthermore, the original interelectrode capacitances C_{gd}, C_{gs}, and C_{ds} of the SiC power MOSFET can be derived from the extracted C_{dn}, C_{gn}, and C_{sn} according to

$$C_{\mathrm{gd}} = \frac{C_{\mathrm{gn}} C_{\mathrm{dn}}}{C_{\mathrm{dn}} + C_{\mathrm{gn}} + C_{\mathrm{sn}}} \tag{7.21}$$

$$C_{\mathrm{gs}} = \frac{C_{\mathrm{gn}} C_{\mathrm{sn}}}{C_{\mathrm{dn}} + C_{\mathrm{gn}} + C_{\mathrm{sn}}} \tag{7.22}$$

$$C_{\mathrm{ds}} = \frac{C_{\mathrm{dn}} C_{\mathrm{sn}}}{C_{\mathrm{dn}} + C_{\mathrm{gn}} + C_{\mathrm{sn}}} \tag{7.23}$$

Figure 7.34 illustrates the test circuits used to extract the equivalent terminal impedances. A capacitor C_1 is connected in parallel with the DC power supply to provide an AC short circuit for the test signal at the excitation frequency f_{sig}. The impedances Z_{ext1}, Z_{ext2}, and Z_{ext3} represent the equivalent impedances of the external wires, which can be measured offline using an impedance analyzer. For online impedance extraction under DC bias conditions, the TD-TPS is employed, as illustrated in Figure 7.35. By mounting the inductive probes at different positions along the signal path, individual terminal impedances can be selectively excited and measured while maintaining electrical isolation from the DC power source.

As shown in Figure 7.34a, the injecting inductive probe (IIP) and receiving inductive probe (RIP) are mounted between position a–a'. A test signal with frequency f_{sig} and current $I_{\mathrm{aa'}}$ is injected into the circuit from the SGC through the IIP. When the injected test signal current reaches node n, it splits into two paths: one flowing through the source terminal s, denoted as $I_{\mathrm{sn\_aa'}}$, and the other flowing through the gate terminal g, denoted as $I_{\mathrm{gn\_aa'}}$. According to Kirchhoff's current law (KCL), $I_{\mathrm{aa'}}$ is given by

$$I_{\mathrm{aa'}} = I_{\mathrm{gn\_aa'}} + I_{\mathrm{sn\_aa'}} \tag{7.24}$$

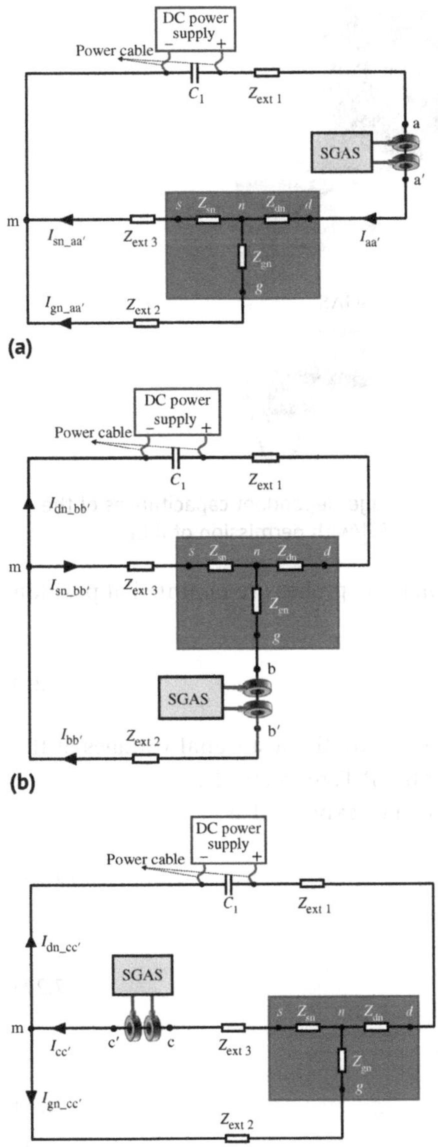

Figure 7.34 Test circuits to extract Z_{dn}, Z_{gn}, and Z_{sn} with the inductive probes of the TD-TPS mounted at different positions: (a) position $a-a'$, (b) position $b-b'$, and (c) position $c-c'$.

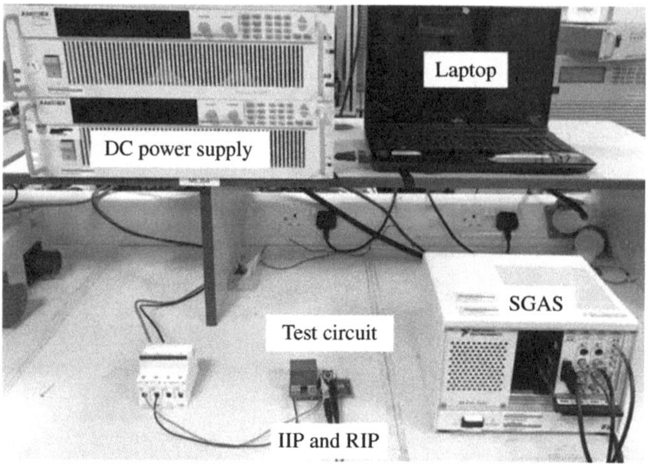

Figure 7.35 Experimental setup to extract the voltage-dependent capacitances of the Cree SiC power MOSFET using the TD-TPS. *Source*: [5]/with permission of IEEE.

The equivalent impedance measured when the probes are mounted at position a–a' is

$$Z_{aa'} = \frac{V_{aa'}}{I_{aa'}} \tag{7.25}$$

where $V_{aa'} = V_{aa'1} - V_{aa'2}$, and $V_{aa'1}$ and $V_{aa'2}$ are the test-signal voltages at the output port of the IIP and the input port of the RIP, respectively.

From Figure 7.34a, the branch currents can be expressed as

$$I_{gn\_aa'} = \frac{V_{nm\_aa'}}{Z_{ext2} + Z_{gn}} \tag{7.26}$$

$$I_{sn\_aa'} = \frac{V_{nm\_aa'}}{Z_{ext3} + Z_{sn}} \tag{7.27}$$

The measured voltage can be written as

$$V_{aa'} = I_{aa'}(Z_{ext1} + Z_{dn}) + V_{nm\_aa'} \tag{7.28}$$

where $V_{nm\_aa'}$ is the test-signal voltage between nodes n and m. Substituting Equation (7.28) into Equation (7.25) yields

$$Z_{aa'} = Z_{ext1} + Z_{dn} + \frac{V_{nm\_aa'}}{I_{aa'}} \tag{7.29}$$

By substituting Equations (7.26) and (7.27) into Equation (7.24), $I_{aa'}$ can be rewritten as

$$I_{aa'} = \frac{V_{nm\_aa'}}{Z_{ext2} + Z_{gn}} + \frac{V_{nm\_aa'}}{Z_{ext3} + Z_{sn}} \tag{7.30}$$

Substituting Equation (7.30) into Equation (7.29), the measured impedance $Z_{aa'}$ is finally obtained as

$$Z_{aa'} = Z_{ext1} + Z_{dn} + \frac{(Z_{ext2} + Z_{gn})(Z_{ext3} + Z_{sn})}{(Z_{ext2} + Z_{gn}) + (Z_{ext3} + Z_{sn})} \tag{7.31}$$

Similarly, Figure 7.34b,c illustrate the current paths when the probes are mounted between positions *b*–*b*′ and *c*–*c*′, respectively. Based on KCL, the injected currents in these configurations are given by

$$I_{bb'} = I_{dn\_bb'} + I_{sn\_bb'} \tag{7.32}$$

$$I_{cc'} = I_{dn\_cc'} + I_{gn\_cc'} \tag{7.33}$$

Accordingly, the measured impedances $Z_{bb'}$ and $Z_{cc'}$ can be expressed as

$$Z_{bb'} = Z_{ext2} + Z_{gn} + \frac{(Z_{ext1} + Z_{dn})(Z_{ext3} + Z_{sn})}{(Z_{ext1} + Z_{dn}) + (Z_{ext3} + Z_{sn})} \tag{7.34}$$

$$Z_{cc'} = Z_{ext3} + Z_{sn} + \frac{(Z_{ext1} + Z_{dn})(Z_{ext2} + Z_{gn})}{(Z_{ext1} + Z_{dn}) + (Z_{ext2} + Z_{gn})} \tag{7.35}$$

From Equations (7.31), (7.34), and (7.35), the unknown impedances Z_{dn}, Z_{gn}, and Z_{sn} can be solved simultaneously. Defining intermediate variables:

$$\begin{aligned}\Delta = Z_{aa'}^2 Z_{bb'}^2 - 2Z_{aa'}^2 Z_{bb'} Z_{cc'} + Z_{aa'}^2 Z_{cc'}^2 - 2Z_{aa'} Z_{bb'}^2 Z_{cc'} \\ -2Z_{aa'} Z_{bb'} Z_{cc'}^2 + Z_{bb'}^2 Z_{cc'}^2\end{aligned} \tag{7.36}$$

$$\Lambda = -2Z_{aa'} Z_{bb'} Z_{cc'} \tag{7.37}$$

The equivalent impedances are obtained as

$$Z_{dn} = \frac{\Lambda(Z_{aa'}Z_{bb'} + Z_{aa'}Z_{cc'} - Z_{bb'}Z_{cc'})}{\Delta} - Z_{ext1} \tag{7.38}$$

$$Z_{gn} = \frac{\Lambda(Z_{aa'}Z_{bb'} - Z_{aa'}Z_{cc'} + Z_{bb'}Z_{cc'})}{\Delta} - Z_{ext2} \tag{7.39}$$

$$Z_{\text{sn}} = \frac{\Lambda(-Z_{\text{aa}'}Z_{\text{bb}'} + Z_{\text{aa}'}Z_{\text{cc}'} + Z_{\text{bb}'}Z_{\text{cc}'})}{\Delta} - Z_{\text{ext3}} \tag{7.40}$$

As discussed earlier, $Z_{\text{aa}'}$, $Z_{\text{bb}'}$, and $Z_{\text{cc}'}$ can be extracted directly through online measurement, while Z_{ext1}, Z_{ext2}, and Z_{ext3} are measured offline using an impedance analyzer. Once Z_{dn}, Z_{gn}, and Z_{sn} are obtained, the corresponding star-connected capacitances C_{dn}, C_{gn}, and C_{sn} can be calculated using Equations (7.18)–(7.20). Subsequently, the interelectrode capacitances C_{gd}, C_{gs}, and C_{ds} can be derived using Equations (7.21)–(7.23).

Based on the extracted C_{dn}, C_{gn}, and C_{sn}, the input capacitance C_{iss}, output capacitance C_{oss}, and reverse transfer capacitance C_{rss} of the SiC power MOSFET can further be obtained as

$$C_{\text{iss}} = C_{\text{gs}} + C_{\text{gd}} \tag{7.41}$$

$$C_{\text{oss}} = C_{\text{ds}} + C_{\text{gd}} \tag{7.42}$$

$$C_{\text{rss}} = C_{\text{gd}} \tag{7.43}$$

7.3.3 Experimental Validation

Figure 7.35 shows the experimental measurement setup used for validation, in which the TD-TPS is employed for online impedance measurement. For the selected TD-TPS implementation, a computer-controlled SGAS is realized using a National Instruments (NI) PXI platform. The platform consists of a signal generation card (PXI-5412), a two-channel signal acquisition card (PXI-5122), a computer interface card (PXI-8360), and a back panel (PXI-1031). A laptop running programmable control software is connected to the SGAS for instrument control and data processing. Two Tektronix CT1 inductive probes are used as the IIP and RIP of the TD-TPS.

A Cree SiC power MOSFET (C2M0080120D, 1.2 kV) is selected as the DUT for extracting its voltage-dependent capacitances under varying V_{ds}. The DC bias is supplied by a Xantrex DC power supply (XDC 600-10, 0–600 V). In addition, an impedance analyzer (Agilent 4294A) is used to measure the external wire impedances Z_{ext1}, Z_{ext2}, and Z_{ext3} offline. For comparison, the capacitance values provided in the manufacturer's datasheet are used as reference values. Since the datasheet specifies C_{iss}, C_{oss}, and C_{rss} only at a frequency of 1 MHz, f_{sig} is set to 1 MHz for direct comparison. In addition, C_1 is selected as 1 μF (TDK B32656S7105K561, 1.25 kV). The measured values of Z_{ext1}, Z_{ext2}, and Z_{ext3} at 1 MHz are 0.4422 Ω $\angle 77.4°$, 0.2053 Ω $\angle 79.1°$, and 0.0493 Ω $\angle 54.8°$, respectively.

Figure 7.36 presents the extracted interelectrode capacitances of the Cree SiC power MOSFET as V_{ds} varies from 0 to 500 V. Since the datasheet only provides

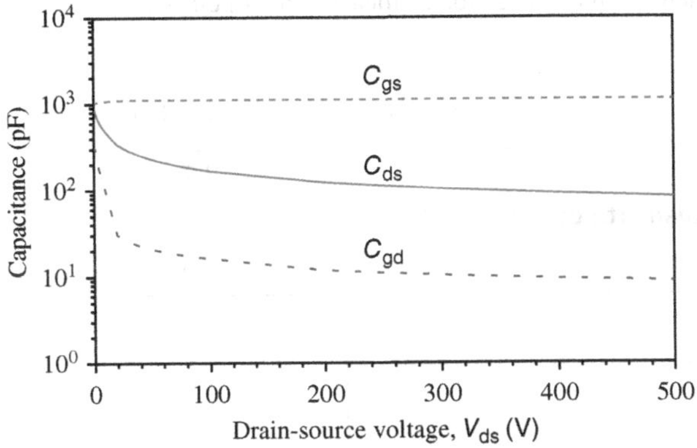

Figure 7.36 Extracted interelectrode capacitances of the Cree SiC power MOSFET using the TD-TPS at a frequency of 1 MHz.

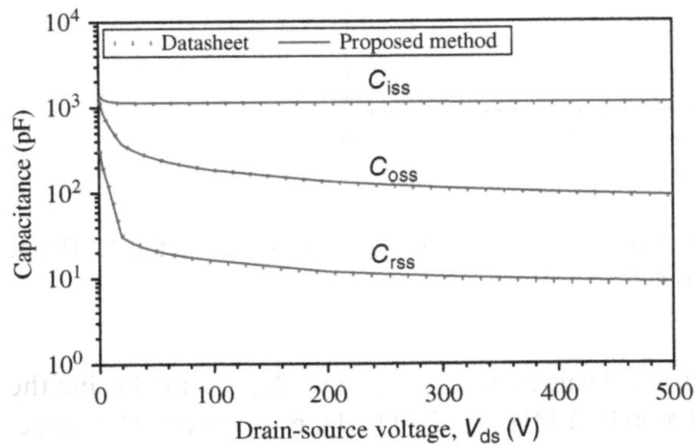

Figure 7.37 Comparison of the extracted C_{iss}, C_{oss}, and C_{rss} obtained using the TD-TPS (i.e., proposed method) with the values provided in the datasheet.

the values of C_{iss}, C_{oss}, and C_{rss}, these quantities are derived from the extracted interelectrode capacitances for comparison purposes. Figure 7.37 compares the values of C_{iss}, C_{oss}, and C_{rss} obtained using the TD-TPS with those provided in the datasheet. Based on 23 measurement points at different V_{ds}, the corresponding measurement errors are summarized in Table 7.3. As observed from Figure 7.37 and Table 7.3, the capacitances extracted using the TD-TPS show good agreement with the datasheet values, thereby validating the effectiveness of the presented method for accurately extracting voltage-dependent capacitances of SiC power MOSFETs over a wide voltage-biasing range.

Table 7.3 Errors of the extracted capacitances for 23 measurement points shown in Figure 7.37.

Capacitance	C_{iss} (%)	C_{oss} (%)	C_{rss} (%)
Mean absolute error	1.9	1.5	3.7
Standard deviation of absolute error	1.2	1.4	2.7
Maximum absolute error	4.3	4.4	9.8

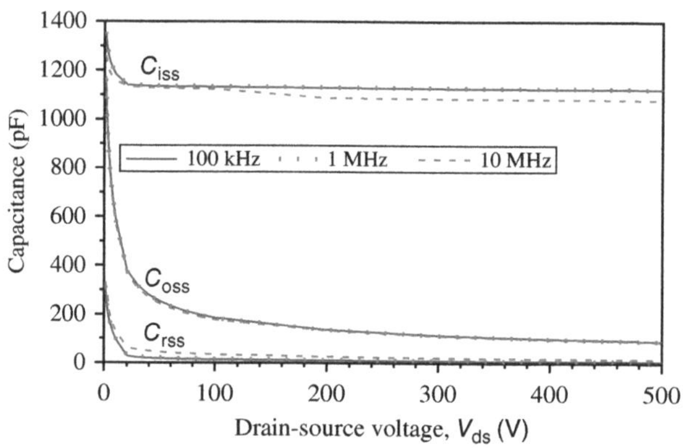

Figure 7.38 Comparison of the values of C_{iss}, C_{oss}, and C_{rss} obtained using the TD-TPS at frequencies of 100 kHz, 1 MHz, and 10 MHz.

Figure 7.38 shows the extracted values of C_{iss}, C_{oss}, and C_{rss} obtained using the TD-TPS when $f_{sig}$is set to 100 kHz, 1 MHz, and 10 MHz, respectively. The capacitance values extracted at 100 kHz are slightly different from those obtained at 1 MHz. In contrast, the capacitances extracted at 10 MHz exhibit significantly larger deviations compared with those obtained at 100 kHz and 1 MHz. This behavior is expected due to the increasing influence of terminal inductances (L_d, L_g, and L_s) at higher frequencies. When $f_{sig} \leq 1$ MHz, the impedances associated with the terminal inductances are much smaller than those of the interelectrode capacitances, and their effects can be neglected. However, when f_{sig} increases to 10 MHz, the terminal inductive impedances become nonnegligible, leading to increased extraction errors. Therefore, when applying the TD-TPS-based method to extract voltage-dependent capacitances of a SiC power MOSFET, f_{sig} should not exceed 1 MHz to ensure measurement accuracy.

In this section, an electrically contactless online impedance measurement method based on the TD-TPS has been presented and experimentally validated for extracting the voltage-dependent capacitances of SiC power MOSFETs.

This method enables capacitance extraction under actual high-voltage biasing conditions without any direct electrical contact with the energized device, making it both safe and relatively simple to implement. A 1.2 kV SiC power MOSFET was selected as a representative test case, and the accuracy of the extracted capacitances was verified through comparison with the manufacturer's datasheet. With the obtained voltage-dependent capacitances, an equivalent circuit model of the SiC power MOSFET can be constructed to support the evaluation and mitigation of potential EMI issues during the early design phase of power converters.

7.4 Chapter Summary

This chapter has demonstrated the effectiveness of electrically contactless online impedance measurement techniques in addressing several key challenges in power electronics applications. Through representative case studies, it has been shown that inductively coupled impedance measurement can support systematic EMI filter design for SMPSs, facilitate health diagnosis of power semiconductor devices, and enable the extraction of voltage-dependent capacitances under actual operating conditions. These examples highlight the capability of online impedance measurement to reveal critical device- and system-level characteristics that are difficult or impossible to obtain using conventional offline methods. By integrating measurement theory with practical power-electronic use cases, this chapter underscores the value of electrically contactless impedance measurement as a powerful diagnostic and design-enabling tool for modern power electronic systems, paving the way for its broader adoption in reliability-oriented and performance-critical applications.

References

1 H. Akagi and T. Shimizu, "Attenuation of conducted EMI emissions from an inverter-driven motor," *IEEE Trans. Power Electron.*, vol. 23, no. 1, pp. 282–290, Jan. 2008.

2 W. Shuo, F. C. Lee, J. D. van Wyk et al., "A study of integration of parasitic cancellation techniques for EMI filter design with discrete components," *IEEE Trans. Power Electron.*, vol. 23, no. 6, pp. 3094–3102, 2008.

3 B. Garry and R. Nelson, "Effect of impedance and frequency variation on insertion loss for a typical power line filter," in *Proc. IEEE EMC Symp., 1998. International Symposium on Electromagnetic Compatibility. Symposium Record (Cat. No.98CH36253)*, 24–28 Aug., 1998.

4 P. Code, and C. PRIX, *Specification for radio disturbance and immunity measuring apparatus and methods–Part 1-1: radio disturbance and immunity measuring apparatus–Measuring apparatus Spécifications des méthodes et des appareils de mesure des perturbations*, 2004.
5 Z. Zhao, K. Y. See, W. Wang et al., "Voltage-dependent capacitance extraction of SiC power MOSFETs using inductively coupled in-circuit impedance measurement technique," *IEEE Trans. Electromagn. Compat.*, vol. 61, no. 4, pp. 1322–1328, Aug. 2019.
6 A. Guerra, F. Maddaleno, and M. Soldano, "Effects of diode recovery characteristics on electromagnetic noise in PFCs," in *Proc. APEC' 98 Thirteenth Annual Applied Power Electronics Conf. Exposition*, USA, Feb. 1998.
7 Q. Liu, S. Wang, A. C. Baisden et al., "EMI suppression in voltage source converters by utilizing dc-link decoupling capacitors," *IEEE Trans. Power Electron.*, vol. 22, no. 4, pp. 1417–1428, Jul. 2007.
8 A. Nagel, and R. W. D. Doncker, "Systematic design of EMI-filters for power converters," in *Proc. Thirty-Fifth IAS Annual Meeting and World Conf. Industrial Applications of Electrical Energy*, Oct. 2000.
9 M. C. Caponet, F. Profumo, and A. Tenconi, "EMI filters design for power electronics," in *Proc. 33rd Annual IEEE Power Electronics Specialists Conf.*, Jun. 2002.
10 Y. Sheng, W. Eberle, and L. Yan-Fei, "A novel EMI filter design method for switching power supplies," *IEEE Trans. Power Electron.*, vol. 19, no. 6, pp. 1668–1678, Nov. 2004.
11 J. Rhee, H. Kim, K. Chu et al., "Methodology for extracting low-frequency input impedance of personal computer during operation using dual current probes method," *IEEE Trans. Electromagn. Compat.*, vol. 66, no. 2, pp. 584–598, Jan. 2024.
12 J. Rhee, C. Lee, S. Lee et al., "A dual current probe method with enhanced low-frequency sensitivity for high-voltage dynamic capacitance measurement," *IEEE Trans. Instrum. Meas.*, vol. 74, pp. 1–11, Jul. 2025.
13 S. Maniktala, *Switching power supplies A-Z*, Elsevier, 2012.
14 P. Degauque, and J. Hamelin, *Electromagnetic compatibility*, USA: Oxford University Press, 1993.
15 J. C. Fluke, *Controlling conducted emissions by design*, New York: Van Nostrand Reinhold, 1991.
16 F. Costa, C. Gautier, E. Labouré et al., *Electromagnetic compatibility in power electronics*, London: ISTE; Hoboken, NJ: Wiley, 2014.
17 L. Tihanyi, *Electromagnetic compatibility in power electronics*, New York, NY, USA: IEEE Press; Sarasota, FL, USA: J.K. Eckert; Oxford, UK: Butterworth Heinemann, 1995.
18 C. R. Paul, and K. B. Hardin, "Diagnosis and reduction of conducted noise emissions," *IEEE Trans. Electromagn. Compat.*, vol. 30, no. 4, pp. 553–560, 1988.

19 C. R. Paul, *Introduction to electromagnetic compatibility*, John Wiley & Sons, 2006.

20 L. Wan, S. Negri, G. Spadacini et al., "Enhanced impedance measurement to predict electromagnetic interference attenuation provided by EMI filters in systems with AC/DC converters," *Appl. Sci.*, vol. 12, no. 23, p. 12497, Dec. 2022.

21 R. L. Ozenbaugh, and T. M. Pullen, *EMI filter design*, CRC Press, 2011.

22 R. L. Ozenbaugh, *EMI Filter Design*, 2nd ed. New York, NY, USA: Marcel Dekker, 2001.

23 L. Dehong, and J. Xanguo, "High frequency model of common mode inductor for EMI analysis based on measurements," in *Proc. 3rd Int. Symp. Electromagn. Compat.*, 2002.

24 V. Tarateeraseth, K. Y. See, F. G. Canavero et al., "Systematic electromagnetic interference filter design based on information from in-circuit impedance measurements," *IEEE Trans. Electromagn. Compat.*, vol. 52, no. 3, pp. 588–598, Aug. 2010.

25 S. Fu-Yuan, D. Y. Chen, W. Yan-Pei et al., "A procedure for designing EMI filters for AC line applications," *IEEE Trans. Power Electron.*, vol. 11, no. 1, pp. 170–181, Jan. 1996.

26 J. J. Goedbloed, *Electromagnetic compatibility*, Prentice-Hall, 1992.

27 T. Williams, *EMC for product designers*, Newnes, 2011.

28 V. P. Kodali, *Engineering Electromagnetic Compatibility: Principles, Measurements, Technologies, and Computer Models*, 2nd ed. New York, NY, USA: IEEE Press/Wiley, 2001.

29 B. Audone, and L. Bolla, "Insertion loss of mismatched EMI suppressors," *IEEE Trans. Electromagn. Compat.*, vol. EMC-20, no. 3, pp. 384–389, Aug. 1978.

30 S. M. Vakil, "A technique for determination of filter insertion loss as a function of arbitrary generator and load impedances," *IEEE Trans. Electromagn. Compat.*, vol. EMC-20, no. 2, pp. 273–278, May 1978.

31 J. Biela, M. Schweizer, S. Waffler et al., "SiC versus Si – Evaluation of potentials for performance improvement of inverter and DC–DC converter systems by SiC power semiconductors," *IEEE Trans. Ind. Electron.*, vol. 58, no. 7, pp. 2872–2882, Jul. 2011.

32 X. Gong, and J. A. Ferreira, "Investigation of conducted EMI in SiC JFET inverters using separated heat sinks," *IEEE Trans. Ind. Electron.*, vol. 61, no. 1, pp. 115–125, Jan. 2014.

33 Y. Liu, K. Y. See, S. Yin et al., "LCL filter design of 50 kW 60 kHz SiC inverter with size and thermal considerations for aerospace applications," *IEEE Trans. Ind. Electron.*, vol. 64, no. 10, pp. 8321–8333, Oct. 2017.

34 S. Yin, K. J. Tseng, R. Simanjorang et al., "A 50-kW high-frequency and high-efficiency SiC voltage source inverter for more electric aircraft," *IEEE Trans. Ind. Electron.*, vol. 64, no. 11, pp. 9124–9134, Nov. 2017.

35 R. Khanna, A. Amrhein, W. Stanchina et al., "An analytical model for evaluating the influence of device parasitics on Cdv/dt induced false turn-on in SiC MOSFETs," in *Proc. IEEE Appl. Power Electron. Conf. Expo.*, Long Beach, CA, USA, 2013.

36 A. Lemmon, M. Mazzola, J. Gafford et al., "Stability considerations for silicon carbide field-effect transistors," *IEEE Trans. Power Electron.*, vol. 28, no. 10, pp. 4453–4459, Oct. 2013.

37 N. Oswald, P. Anthony, N. McNeill et al., "An experimental investigation of the tradeoff between switching losses and EMI generation with hard-switched all-Si, Si-SiC, and all-SiC device combinations," *IEEE Trans. Power Electron.*, vol. 29, no. 5, pp. 2393–2407, May 2014.

38 Z. Zhao, Y. Liu, K.-Y. See et al., "Extraction of loop inductances of SiC half-bridge power module using an improved two-port network method," in *Proc. Annu. Conf. IEEE Ind. Electron. Soc. (IECON)*, Washington, DC, USA, 2018, pp. 1204–1208.

39 Z. Chen, D. Boroyevich, R. Burgos et al., "Characterization and modeling of 1.2 kV, 20 A SiC MOSFETs," in *Proc. IEEE Energy Convers. Congr. Expo.*, San Jose, CA, USA, 2009.

40 P. Anthony, N. McNeill, and D. Holliday, "A first approach to a design method for resonant gate driver architectures," *IEEE Trans. Power Electron.*, vol. 27, no. 8, pp. 3855–3868, Aug. 2012.

41 Y. Liu, K. Y. See, R. Simanjorang et al., "Modeling and simulation of switching characteristics of half-bridge SiC power module in single leg T-type converter for EMI prediction," in *Proc. Joint IEEE Int. Symp. Electromagn. Compat. and Proc. Asia-Pacific Int. Symp. Electromagn. Compat.*, Singapore, 2018.

42 Z.-N. Ariga, K. Wada, and T. Shimizu, "TDR measurement method for voltage-dependent capacitance of power devices and components," *IEEE Trans. Power Electron.*, vol. 27, no. 7, pp. 3444–3451, Jul. 2012.

43 R. Elferich, T. López, and N. Koper, "Accurate behavioural modelling of power MOSFETs based on device measurements and FE-simulations," in *Proc. Eur. Conf. Power Electron. Appl.*, Dresden, Germany, 2005.

44 T. Funaki, N. Phankong, T. Kimoto et al., "Measuring terminal capacitance and its voltage dependency for high-voltage power devices," *IEEE Trans. Power Electron.*, vol. 24, no. 6, pp. 1486–1493, Jun. 2009.

45 K. Li, A. Videt, and N. Idir, "Multiprobe measurement method for voltage-dependent capacitances of power semiconductor devices in high voltage," *IEEE Trans. Power Electron.*, vol. 28, no. 11, pp. 5414–5422, 2013.

46 Z. Chen, "Characterization and modeling of high-switching-speed behavior of SiC active devices," M.S. thesis, Dept. Electrical Engineering, Virginia Polytechnic Institute and State University, Blacksburg, VA, USA, 2009.

47 M. Ando, and K. Wada, "Design of acceptable stray inductance based on scaling method for power electronics circuits," *IEEE J. Emerg. Sel. Top. Power Electron.*, vol. 5, no. 1, pp. 568–575, Mar. 2017.

48 Y. Mukunoki, Y. Nakamura, K. Konno et al., "Modeling of a silicon-carbide MOSFET with focus on internal stray capacitances and inductances, and its verification," *IEEE Trans. Ind. Appl.*, vol. 54, no. 3, pp. 2588–2597, May/Jun. 2018.

49 *Silicon carbide power MOSFET*, C2M0080120D Cree datasheet, 2015.

50 *Silicon carbide power MOSFET*, SCT2H12NZ ROHM datasheet, 2017.

51 *Silicon carbide power MOSFET*, UJ3C120150K3S UnitedSiC datasheet, 2018.

52 *Silicon carbide power MOSFET*, LSIC1MO120E0160 Littelfuse datasheet, 2019.

8

Applications in Transport Electrification

This chapter presents several representative application case studies of electrically contactless online impedance measurement techniques in transport electrification systems. As modern transportation platforms increasingly rely on high-power electronics, dense cabling, and complex EMI, ensuring electromagnetic compatibility (EMC), operational reliability, and overall system health has become a critical challenge. This chapter demonstrates how noncontact online impedance measurement enables systematic design, condition monitoring, and fault diagnosis without intrusive electrical connections, making it particularly attractive for safety-critical and in-service environments. Through applications spanning inverter-driven motor systems, current collectors of LRT trains, and MRT railway track systems, this chapter illustrates how impedance-based sensing provides quantitative insight into electromagnetic behavior, fault evolution, and electromechanical interactions across diverse transportation domains.

8.1 EMI Filter Design for Motor Drive Systems

To suppress CM noise in inverter-driven motor drive systems, a CM EMI filter is typically installed between the inverter and the motor. In conventional practice, filter design often relies on iterative trial-and-error: designers test different component selections and configurations until the measured CM noise meets the required limits. Even when off-the-shelf filters are available, repeated tuning is frequently necessary to achieve the desired attenuation, which can be time-consuming and may lead to conservative designs. To address this limitation, this section introduces a systematic EMI filter design approach enabled by electrically contactless online impedance measurement, which allows the CM noise source and its propagation paths to be quantitatively characterized under realistic operating conditions. With the extracted noise source and load impedances, the filter can be designed in a model-based and physically grounded manner, improving predictability and reducing unnecessary overdesign. This section further examines typical CM filter configurations, including a first-order single-choke CM filter and a second-order LC CM filter. Particular attention is paid to the DM-to-CM mode conversion that can occur in higher-order filters. As will be shown, mitigating this mode conversion may require an additional DM inductor, which can partially offset the expected size advantage of higher-order CM filters.

8.1.1 Determination of Filter Attenuation

CM EMI filters are installed between the inverter and the motor to reduce CM noise currents at the inverter output and the resulting circulating bearing currents, etc. Since the bearing impedance varies dynamically with operating conditions and is difficult to characterize accurately in real time, it is challenging to directly define an allowable CM current limit based on bearing degradation considerations alone. Therefore, in this study, a standard specified in DO-160, as shown in Figure 8.1, is adopted as a reference to determine the required filter attenuation.

Most EMI standards and conducted emission limits for electrical motor drive systems are traditionally specified for the incoming power mains. However, with the increasing adoption of sophisticated power conversion systems, such as those used in more-electric aircraft and other safety-critical applications, EMI requirements are now applied to both the power mains side and the load side. Among

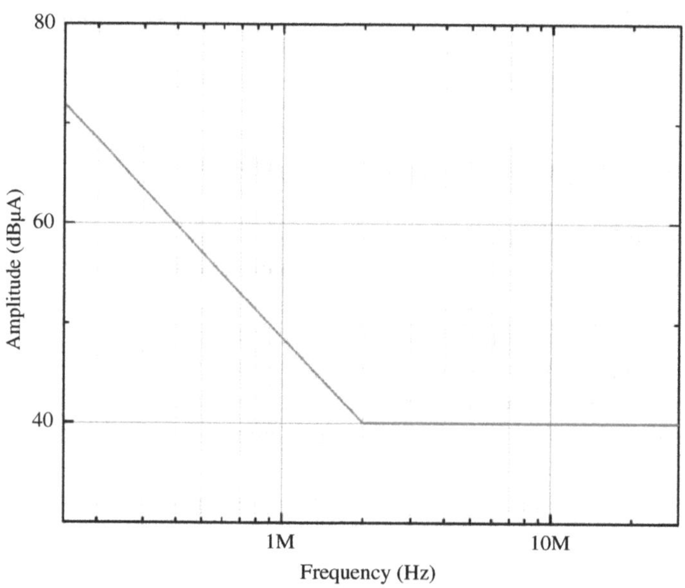

Figure 8.1 Conducted emissions current limits specified in the DO-160 standard.

these standards, DO-160 provides one of the most comprehensive conducted emission limits, with requirements defined according to installation location. Specifically, DO-160 specifies limits of 72 dBμA at 150 kHz, decreasing linearly to 40 dBμA at 2 MHz, and remaining at 40 dBμA from 2 to 30 MHz. These limits apply to both the power mains side (Category B) and the load side (Categories L, M, and H).

Although the full conducted emission frequency range defined in DO-160 extends from 150 kHz to 152 MHz, the impedance measurement bandwidth in this study is limited to 30 MHz due to the bandwidth constraints of the selected inductive probes. Consequently, the results presented here focus on the frequency range from 150 kHz to 30 MHz. Nevertheless, the impedance-based methodology introduced in this section remains applicable at higher frequencies, provided that suitable sensing hardware is employed. In addition, while DO-160 is primarily intended for avionics applications, it is adopted here as a conservative and representative reference, since the performance of a designed EMI filter is independent of the specific regulatory limit used. More importantly, the underlying design methodology based on extracted CM impedance characteristics is generally applicable to motor-side CM filter design across a wide range of motor drive systems.

Figure 8.2 shows the experimental setup of a 5.5 kW induction motor driven by a pulse-width modulation (PWM) inverter, connected via a 5 m unshielded cable.

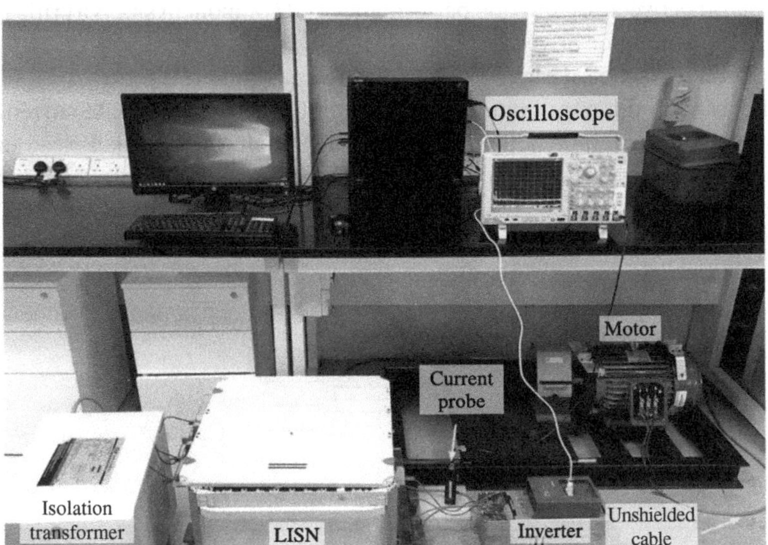

Figure 8.2 Experimental setup of a 5.5 kW induction motor driven by a PWM inverter.

The CM current path is illustrated in Figure 8.3, which depicts the system configuration excluding an additional isolation transformer and a three-phase LISN. The LISN provides a well-defined mains impedance and prevents externally conducted emissions from contaminating the measurement results. A mixed-signal oscilloscope with a built-in spectrum analyzer (Tektronix MDO4104B-3) and a current probe (Solar Electronics 9134-1N, 100 MHz bandwidth) are used to measure the CM current at the inverter output.

For the given inverter-driven motor system, compliance with the DO-160 conducted emission limits is required. Figure 8.4 shows the measured CM current without any EMI filter, from which it is evident that the emission level exceeds the specified DO-160 limit over a wide frequency range. Accordingly, the design objective of the CM filter is to suppress the CM current below the prescribed limit while minimizing the size and inductance value of the choke.

8.1.2 Design of First-order Single-choke Filter

As a cost-effective mitigation approach, three-phase CM chokes are widely employed in the AC power cable bundle between the inverter and the motor to suppress CM noise generated by high-frequency switching devices and to meet regulatory emission limits [1–5]. However, the physical size of CM magnetic components can become a major constraint, as they may occupy up to 25% of the total system volume in practical motor-drive installations [6]. Consequently, optimizing the trade-off between attenuation performance and component size is a critical design challenge in motor-drive systems.

In conventional practice, designers often rely on a trial-and-error process by testing CM chokes with different inductance values until the measured CM current falls below the required limit. Although such an approach can eventually lead to a compliant CM filter design, it is time-consuming and prone to overdesign. This

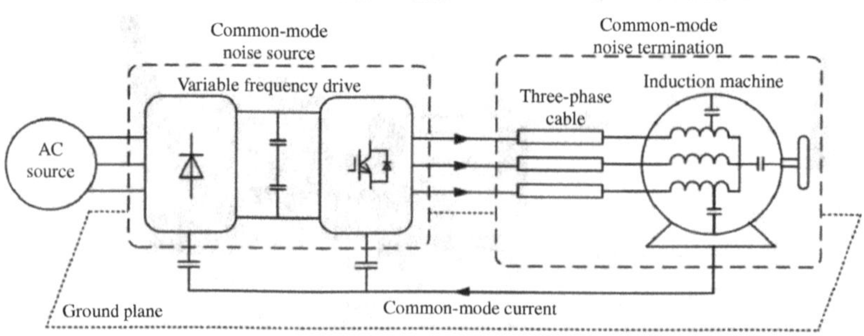

Figure 8.3 CM current paths in an inverter-driven motor drive system.

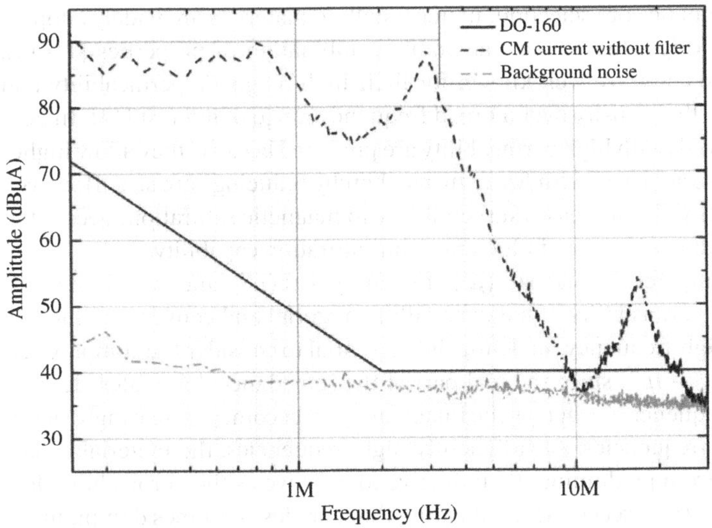

Figure 8.4 Measured CM conducted emission current of the inverter-driven motor system without an EMI filter.

issue becomes particularly significant in high-power inverter-driven motor systems rated at several hundred kilowatts or above, where CM chokes must withstand large currents and suitable off-the-shelf components may not be readily available. In such cases, multiple CM chokes with different specifications may need to be pre-designed and fabricated for iterative testing, resulting in increased cost and limited practicality.

Several analytical approaches have been proposed to improve the CM choke design process. Normalized attenuation has been calculated in [7] for choke optimization; however, the resulting estimation may exhibit considerable error, as it assumes a frequency-independent inductance over both low- and high-frequency ranges. In the latest research, a FD-TPS with VNA is applied to the in-circuit admittance parameters measurement of a single-phase motor drive system in [8]. References [9, 10] simplify the motor-drive system as a series *RLC* circuit to facilitate CM choke design, but this simplification prevents accurate prediction of the frequency-dependent behavior of the choke. In general, reliable prediction of CM choke performance requires prior knowledge of both the CM noise source impedance and the load termination impedance. Therefore, based on the inverter CM source impedance extracted using contactless online impedance measurement, this section presents a step-by-step CM choke design methodology that enables the determination of an appropriate choke specification at an early design stage, prior to fabrication.

First, for magnetic core selection, nanocrystalline materials are widely adopted in CM choke design due to their favorable combination of properties [11–13]. Nanocrystalline cores are well known for their high magnetic permeability and high saturation flux density over a broad frequency range of interest [14]. In general, core materials with high permeability are preferred because they allow higher inductance to be achieved with fewer turns, thereby reducing core size. However, high permeability also increases susceptibility to magnetic saturation, necessitating a careful balance between inductance and saturation capability.

Figure 8.5 compares the real part (μ'), imaginary part (μ''), and magnitude ($|\mu|$) of the complex permeability of nanocrystalline material and conventional ferrite material. For high-frequency modeling, it is essential to consider the complex permeability $\mu = \mu' - j\mu''$, since the real part of the impedance dominates attenuation at lower frequencies, whereas the imaginary part becomes increasingly significant at higher frequencies. At sufficiently high frequencies, the material behavior transitions from predominantly inductive to resistive as the permeability loss angle arctan (μ''/μ') exceeds 45°, indicating that core loss becomes dominant.

Compared with ferrite materials, nanocrystalline cores exhibit substantially higher permeability in the low-frequency range and maintain superior permeability even at elevated frequencies. This characteristic enables a significant reduction in EMI filter volume. Although high permeability facilitates compact designs, it also increases the risk of core saturation. Nanocrystalline materials offer a favorable compromise by combining high permeability with a relatively high saturation flux density. In addition, toroidal core geometries are preferred due to their lower cost and higher effective permeability compared with other core structures.

Consequently, off-the-shelf toroidal nanocrystalline cores (VITROPERM from VAC) are selected for the CM choke design, and their key specifications are summarized in Table 8.1.

Two critical frequencies are selected to enable a rapid analytical estimation of the core size and the required number of turns before proceeding to full modeling using the finite element method (FEM). The first frequency corresponds to the lowest frequency specified by the DO-160 standard, at which the required attenuation is primarily determined by the common-mode inductance L_{CM}. The second frequency is a higher frequency at which the maximum attenuation is required. At this frequency, the real part of the choke impedance R_{CM} contributes dominantly to the attenuation, as the imaginary part μ'' of the complex permeability becomes significant.

Assuming symmetrical three-phase windings and homogeneous core permeability, a lumped parameter L_{CM} can be used to represent the CM inductance of the choke, which is strongly frequency dependent [14],

$$L_{\mathrm{CM}}(f) = {N_L}^2 A_L \cdot \frac{\mu'(f)}{|\mu(f)|} \tag{8.1}$$

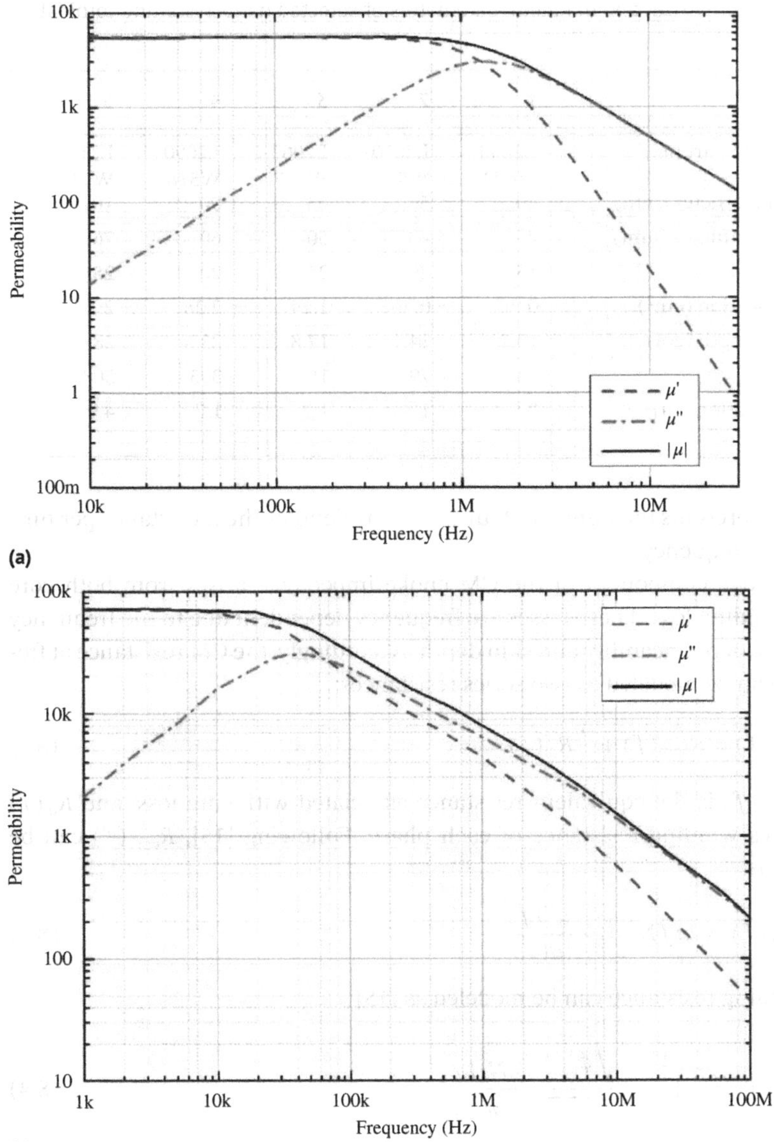

Figure 8.5 Frequency-dependent complex permeability of magnetic core materials. (a) Ferrite core. (b) Nanocrystalline core.

Table 8.1 Geometrical and magnetic parameters of selected nanocrystalline toroidal cores.

Core Number	1	2	3	4	5
Magnetic core part no.	L2040-W424	L2050-W516	L2063-W517	L2090-W518	L2102-W947
Core outer diameter (mm)	40	50	63	90	102
Core inner diameter (mm)	25	40	50	60	76
Core height (mm)	15	20	25	20	25
Iron cross-section (cm^2)	0.86	0.76	1.24	2.28	2.47
Mean path length (cm)	10.2	14.1	17.8	23.6	28
Weight (g)	64	79	161	395	508
Typical saturation current 10 kHz (A)	0.5	1	1.2	1.7	4.9

where N_L represents the number of turns, and A_L denotes the inductance per turn at a specific frequency f.

The resistive component of the CM choke impedance arises from both core loss and winding loss. These losses are frequency dependent due to the frequency dependence of permeability and skin depth. Accordingly, the CM resistance at frequency f can be modeled as two series resistances:

$$R_{\text{CM}}(f) = R_{\text{core}}(f) + \frac{1}{3}R_w(f) \tag{8.2}$$

where $R_{\text{core}}(f)$ is the equivalent resistance associated with core loss, and $R_w(f)$ is the series winding resistance of each phase. Following [15], $R_{\text{core}}(f)$ can be expressed as

$$R_{\text{core}}(f) = 2\pi f N_L^2 A_L \cdot \frac{\mu''(f)}{|\mu(f)|} \tag{8.3}$$

The winding resistance can be modeled as [15]

$$R_w(f) = \left(\frac{\pi}{4}\right)^{\frac{3}{4}} \cdot \frac{\sqrt{\phi_w^3}}{\delta\sqrt{d_{\text{wires}}}} \cdot \frac{4N_L l_t \rho_w}{\pi \phi_w^2} \tag{8.4}$$

with the skin depth given by

$$\delta = \sqrt{\frac{\rho_w}{\pi \mu_0 f}} \tag{8.5}$$

where ϕ_w is the wire diameter, d_{wires} is the estimated distance between two adjacent conductors, l_t is the average length of one turn of the winding, ρ_w is the resistivity of the wire material, and δ is the skin depth. This formulation is valid under the assumption that the CM choke is wound in a single layer, which is typically preferred to minimize parasitic winding capacitance and reduce proximity effects.

Neglecting stray capacitance, the total CM impedance of the choke at frequency f can be expressed as

$$Z_{\mathrm{CM}}(f) = R_{\mathrm{core}}(f) + \frac{1}{3}R_w(f) + j2\pi f L_{\mathrm{CM}}(f) \tag{8.6}$$

Since $R_{\mathrm{core}}(f)$, $R_w(f)$, and $L_{\mathrm{CM}}(f)$ are all frequency dependent, the total CM impedance can be rewritten in a compact form as

$$Z_{\mathrm{CM}}(f) = A \cdot N_L^2 + B \cdot N_L \tag{8.7}$$

where

$$A = 2\pi f A_L \cdot \frac{\mu''(f)}{|\mu(f)|} + j2\pi f A_L \cdot \frac{\mu'(f)}{|\mu(f)|} \tag{8.8}$$

$$B = \left(\frac{\pi}{4}\right)^{\frac{3}{4}} \cdot \frac{\sqrt{\phi_w^3}}{\delta\sqrt{d_{\mathrm{wires}}}} \cdot \frac{4l_t\rho_w}{\pi\phi_w^2} \tag{8.9}$$

With the measured CM noise source impedance $Z_S(f)$, CM choke impedance $Z_{\mathrm{CM}}(f)$, and CM load termination impedance $Z_L(f)$, the CM filter attenuation can be calculated as

$$\mathrm{ATT}(f) = 20\log\left|\frac{Z_S(f) + Z_L(f) + Z_{\mathrm{CM}}(f)}{Z_S(f) + Z_L(f)}\right| \tag{8.10}$$

Representative examples of the measured CM noise source impedance $Z_S(f)$ and CM load termination impedance $Z_L(f)$ using a FD-TPS in this case study are illustrated in Figure 8.6. The simulated and measured CM impedances of the designed CM choke, $Z_{\mathrm{CM}}(f)$, will be described later.

To achieve the required attenuation, the CM choke performance at the two selected frequencies must satisfy

$$20\log\left|\frac{Z_S(f) + Z_L(f) + Z_{\mathrm{CM}}(f)}{Z_S(f) + Z_L(f)}\right| \geq \mathrm{ATT\_}r(f) \tag{8.11}$$

where ATT_r(f) is the required attenuation obtained by subtracting the DO-160 limit from the measured CM emission, as shown in Figure 8.4. Substituting Equation (8.7) into Equation (8.11), the number of turns must satisfy

$$\left|Z_S(f) + Z_L(f) + A \cdot N_L^2 + B \cdot N_L\right| \geq 10^{\frac{\mathrm{ATT\_}r(f)}{20}} \cdot |Z_S(f) + Z_L(f)| \tag{8.12}$$

The toroid core window area imposes an additional design constraint. The maximum number of turns is limited by the inner diameter of the core according to [16]

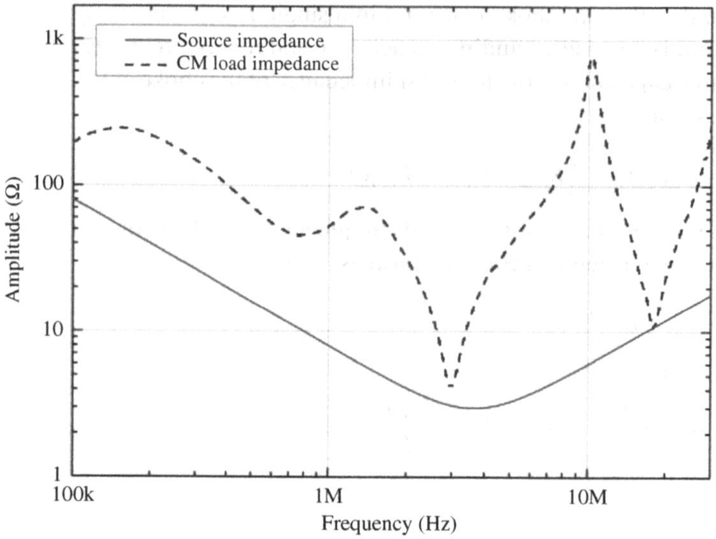

Figure 8.6 Measured CM noise source impedance and CM load termination impedance.

$$N_L \leq \frac{(\mathrm{ID} - D_w)\theta_{\max}}{2D_w} \tag{8.13}$$

where D_w represents the wire diameter, ID represents the inner diameter of the core, and $\theta_{\max}$ denotes the maximum angle that one phase covers. If the value of N_L obtained from Equation (8.12) fails to fit the window area, a larger core should be selected until N_L satisfies both Equations (8.12) and (8.13). In some extreme cases where there is no commercially available magnetic core with a suitable window or large inductance is needed without core saturation, two or more CM chokes can be added but at the expense of overall size.

Parasitic capacitances of the CM choke, particularly the winding-to-core capacitance, degrade choke performance at higher frequencies. Therefore, accurate prediction of parasitic parameters is essential for evaluating high-frequency attenuation. High-frequency CM choke behavior modeling has been reported in [17, 18]; however, the choke parameters rely on prior measurement results. In [16, 19], the total parasitic capacitance is modeled using a distributed network of turn-to-turn and turn-to-core capacitances. While this approach captures winding distribution effects, it relies on simplifying assumptions and is generally valid only when the number of turns exceeds 10. Consequently, a more accurate design requires FEM-based modeling, such as that implemented in Ansys Maxwell.

Core saturation is another critical consideration in CM choke design. The maximum magnetic flux density produced by the combined CM and DM currents should remain below the saturation flux density of the core material. While the peak DM current can be directly measured or estimated from converter ratings, the CM current magnitude depends on both system operation and CM filter performance. Therefore, the maximum flux density must be evaluated through the FEM simulation using the measured CM and DM currents during CM filter testing. During the design stage, only the DM current at some critical frequencies is considered, where the magnetic flux due to DM currents does not fully cancel [20].

As a summary, the overall CM choke design procedure is shown in Figure 8.7. Once the core size and number of turns are finalized, the CM choke can be fabricated accordingly. Based on this procedure, a CM choke using a VAC VITROPERM L2090-W518 nanocrystalline core with 10 turns per phase was designed for

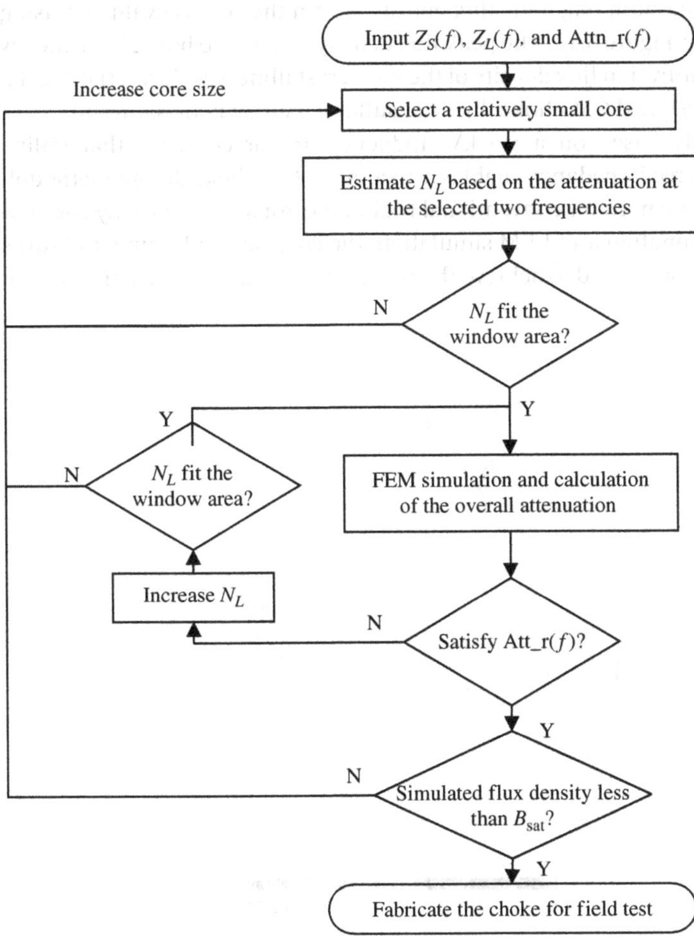

Figure 8.7 Systematic design procedure for the CM choke.

the motor drive system. The measured inductance of the choke at 10 kHz is 7 mH. Figure 8.8 presents the FEM model of the choke, including the refined mesh after iterative optimization.

Figure 8.9 compares the simulated CM impedance of the designed CM choke using FEM and the measurement result with an impedance analyzer, showing good agreement with each other.

Based on the FEM-derived choke impedance, the attenuation versus frequency is calculated and shown in Figure 8.10, where the simulated attenuation exceeds the required attenuation across the frequency range of interest.

The effectiveness of the designed choke is further validated by the measured CM current shown in Figure 8.11, where the CM current with the choke installed is significantly reduced and remains within the DO-160 limit.

Finally, the maximum magnetic flux density within the core is evaluated using FEM, as shown in Figure 8.12. The peak flux density is found to be 0.27 T, which is well below the saturation flux density of the nanocrystalline 500 F material (1.2 T). Therefore, the designed CM choke does not suffer from magnetic saturation.

This case study based on a 5.5 kW induction motor confirms that online extracted CM source impedance enables a systematic CM choke design methodology, eliminating the need for conventional trial-and-error approaches. By combining analytical estimation and FEM simulation, the core size and number of turns can be efficiently optimized to achieve the required attenuation with a minimized choke volume.

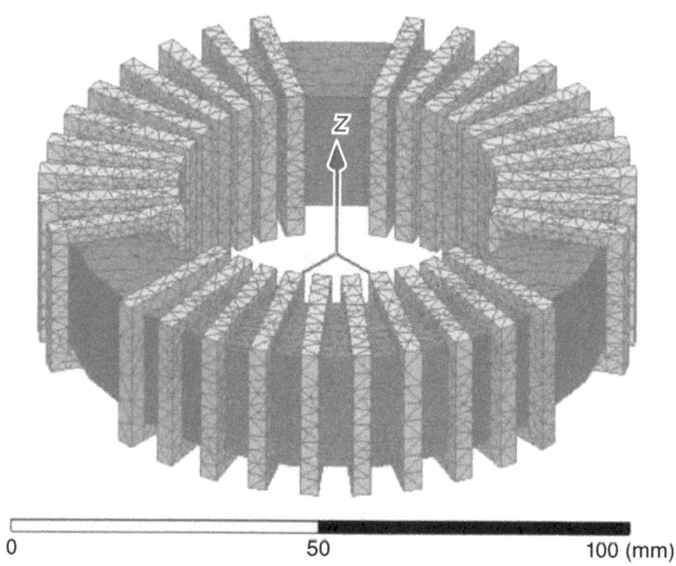

Figure 8.8 FEM model of the designed CM choke.

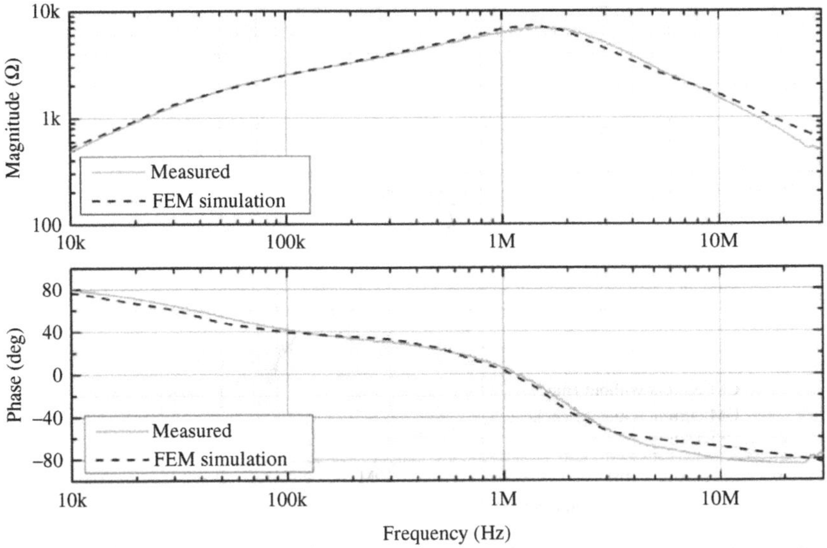

Figure 8.9 Comparison of measured and simulated CM impedance of the designed choke.

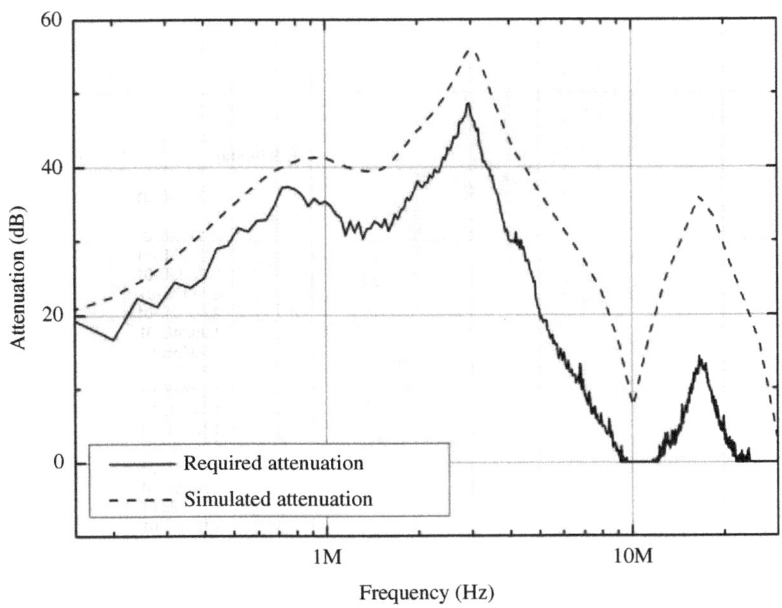

Figure 8.10 Simulated and required CM attenuations.

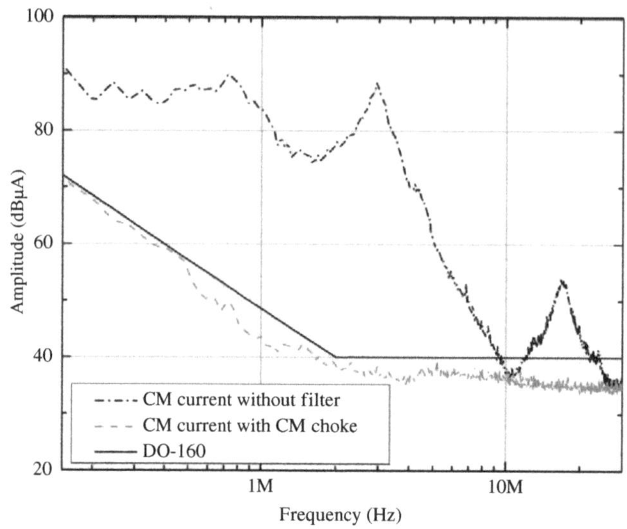

Figure 8.11 Measured CM current with and without the CM choke as well as DO-160 limits.

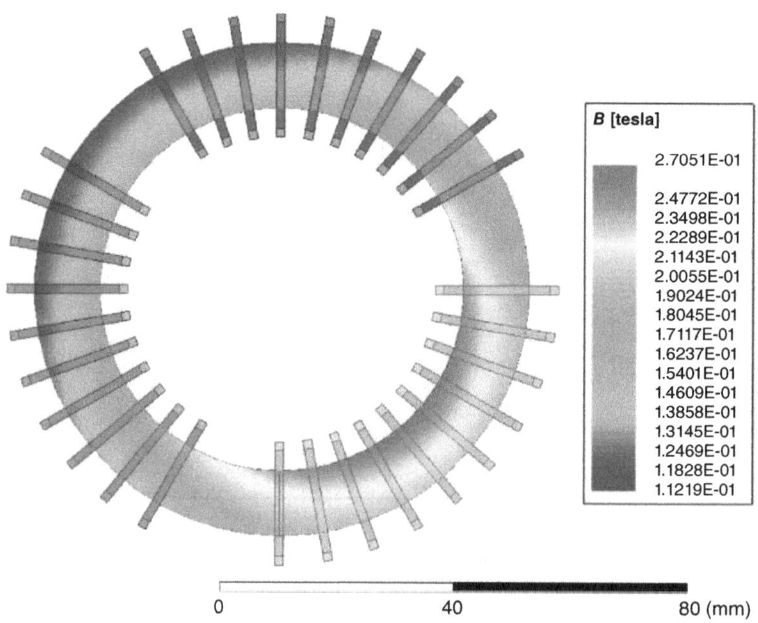

Figure 8.12 Resultant magnetic flux density distribution in the CM choke under combined CM and DM peak currents.

8.1.3 Design of Second-order LC Filter

Although a first-order CM filter employing only a CM choke is capable of reducing conducted emissions below the specified limit, the design of a second-order LC filter is further investigated with the objective of reducing the required CM choke size and achieving a more compact filter structure. However, it is found that second-order CM filters are inherently more susceptible to DM-to-CM conversion.

A CM choke of 1.5 mH is selected, with three *Y* capacitors connected to ground through a 22 Ω damping resistor [16], as illustrated in Figure 8.13. The CM emissions of the second-order LC filter with two different *Y* capacitor values (1 and 4.7 nF) are measured and presented in Figure 8.14. Ideally, a higher attenuation is expected from the second-order LC filter compared with the 1.5 mH CM choke alone. Interestingly, the results show that the second-order LC filter exhibits higher CM emissions at certain frequencies.

To further investigate this behavior, the corresponding DM emissions are also measured and shown in Figure 8.14. The frequencies at which elevated CM emissions are observed coincide with the resonant frequencies of the DM emissions, as indicated by the vertical dotted lines in Figure 8.14. This phenomenon confirms the presence of strong DM-to-CM conversion at these resonant frequencies, which is caused by the interaction between the leakage inductance of the CM choke and the CM capacitors in the second-order LC filter [21–23]. Such conversion is not observed when only the CM choke is employed.

Unbalanced parameters within an EMI filter, as well as asymmetries in the source and load terminations, can lead to mode conversion, as reported in [24–27]. To provide further insight into this mechanism, the simulation model shown in Figure 8.15 demonstrates how impedance imbalance in an induction motor converts DM excitation into CM current.

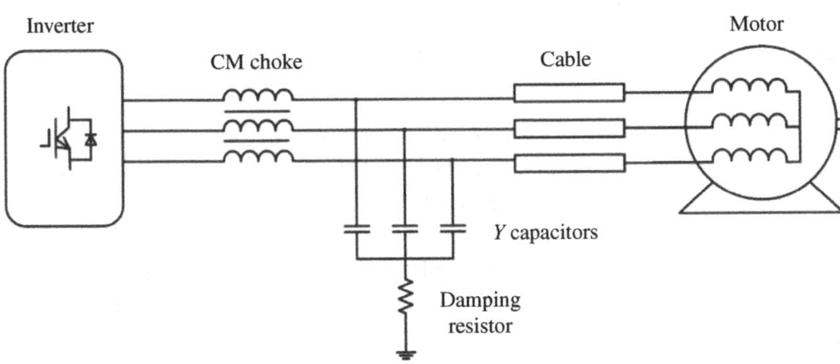

Figure 8.13 Setup for the output-side second-order LC filter.

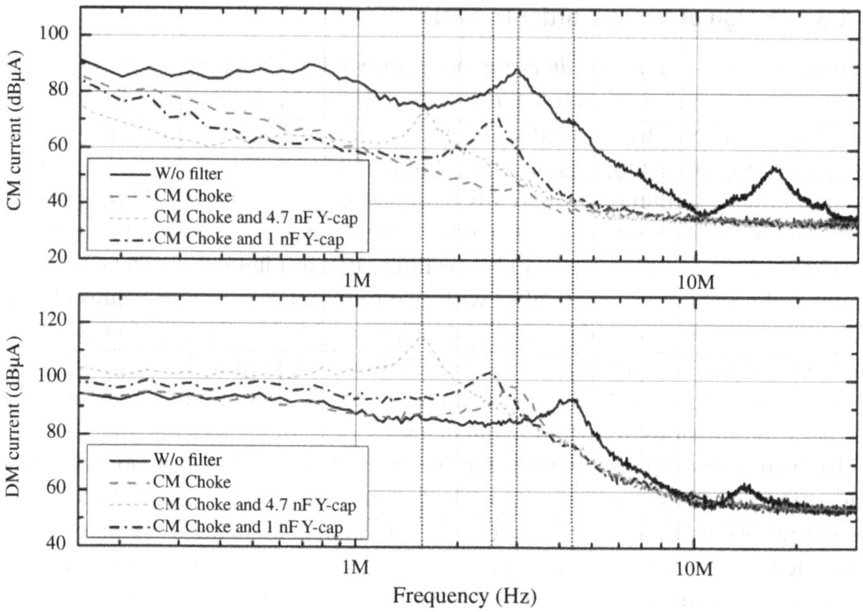

Figure 8.14 CM and DM emissions for different filter configurations.

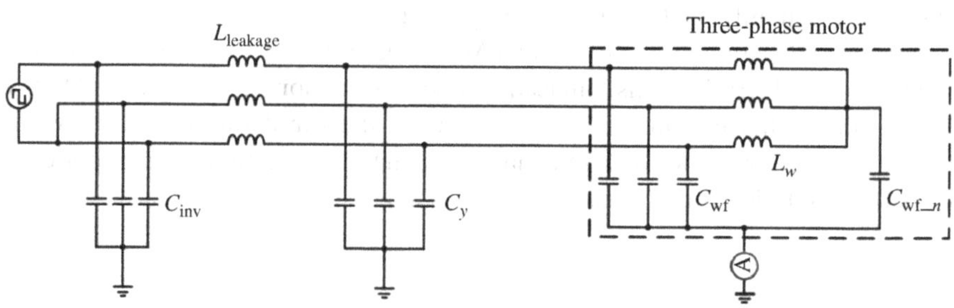

Figure 8.15 Demonstration of DM-to-CM noise conversion through simulation.

In the model, C_{inv} represents the stray capacitance of the switching devices, $L_{leakage}$ denotes the leakage inductance of the CM choke, and C_y represents the CM capacitors between each phase and ground. A detailed electrical model of a three-phase induction motor [28] is also included for simulation purposes. The CM current is simulated by introducing small imbalances of 2.5 and 5% in the winding stray capacitance C_{wf} and winding inductance L_w. The simulation results, shown in Figure 8.16, clearly illustrate DM-to-CM noise conversion caused by impedance imbalance in the induction motor.

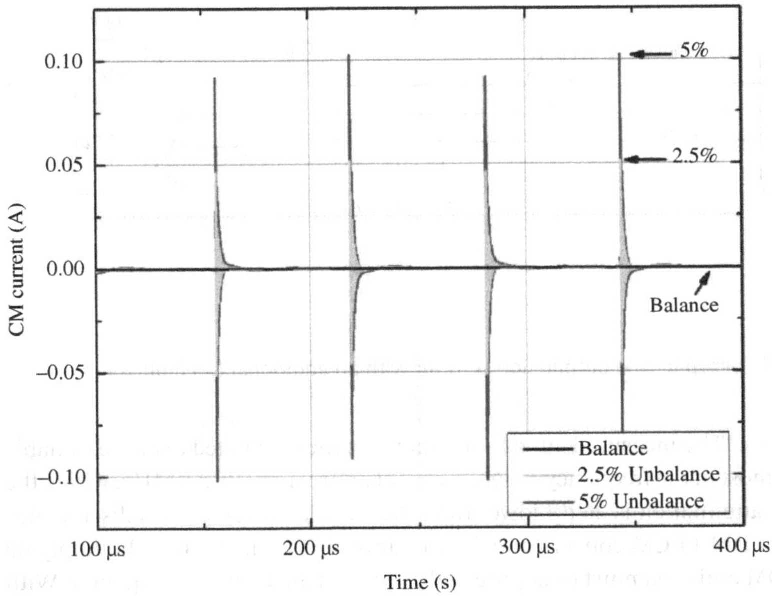

Figure 8.16 Simulated CM current under pure DM excitation with different imbalance levels.

With the online-extracted source impedance, a quantitative evaluation of CM filter attenuation becomes feasible. This approach can also be used to determine whether mode conversion is present and whether it degrades CM filter performance. Based on the previously measured CM noise source and load impedances, the effect of mode conversion on CM filter performance can be evaluated quantitatively. Considering the second-order LC filter consisting of a 1.5 mH CM choke and 1 nF Y capacitors as a representative example, the attenuation provided by the LC filter can be estimated as

$$\mathrm{ATT} = 20\log\left|\frac{Z_L\left(Z_S + Z_{\mathrm{CMC}} + \left(Z_{Y\text{-cap}} + Z_{\mathrm{damping}}\right)\|Z_L\right)}{\left(Z_S + Z_L\right)\left(\left(Z_{Y\text{-cap}} + Z_{\mathrm{damping}}\right)\|Z_L\right)}\right| \tag{8.14}$$

where $Z_{Y\text{-cap}}$ is the CM equivalent impedance of the 1 nF Y capacitor branches and Z_{damping} is the impedance of the damping resistor. The parasitic effects of the components are incorporated using their measured impedance responses. By substituting the measured impedance frequency responses into Equation (8.14), the CM filter attenuation can be calculated.

To eliminate the influence of mode conversion, an appropriately sized DM inductor is introduced, as shown in Figure 8.17. Figure 8.19a compares the calculated and measured CM attenuation of the second-order LC filter without the

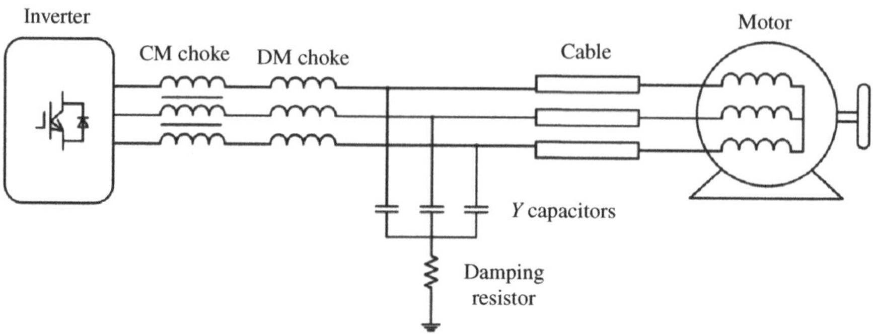

Figure 8.17 Setup for the output-side LC filter with an additional DM inductor.

DM inductor. The measured attenuation matches the calculated result reasonably well over most of the frequency range, except at approximately 2.6 MHz, where the measured attenuation is 26 dB lower than the calculated value. This discrepancy indicates a DM-to-CM conversion of approximately 26 dB at 2.6 MHz, implying that the DM emission must be suppressed by at least 26 dB at this frequency. With an additional design margin of 6 dB, the required DM attenuation becomes 32 dB at 2.6 MHz.

The DM attenuation provided by the additional DM inductor can be estimated by comparing the DM emissions with and without the DM inductor,

$$\mathrm{ATT\_DM} = 20\log\left|\frac{Z_{L\_\mathrm{DM}} + Z_{\sigma} + Z_{S\_\mathrm{DM}} + Z_{\mathrm{cap}}\|Z_{\mathrm{load\_DM}}}{Z_{\sigma} + Z_{S\_\mathrm{DM}} + Z_{\mathrm{cap}}\|Z_{\mathrm{load\_DM}}}\right| \tag{8.15}$$

where Z_{L_DM} is the impedance of the DM inductor, Z_{σ} is the leakage impedance of the CM choke, Z_{S_DM} is the measured DM source impedance, Z_{cap} is the equivalent DM impedance of the Y capacitors, and $Z_{\mathrm{load\_DM}}$ is the DM load impedance. As an illustration, Figure 8.18 presents the measured frequency responses of these impedances for the considered case study.

From Equation (8.15), the required DM inductance is calculated to be approximately 25 μH. A commercially available 30 μH DM inductor is therefore selected. Compared with the CM choke, the DM inductor contributes negligible CM impedance, but it is included in the CM attenuation calculation as
where Z_{DMC} is the CM impedance contribution of the DM inductor.

$$\mathrm{ATT} = 20\log\left|\frac{Z_L\left(Z_S + Z_{\mathrm{CMC}} + Z_{\mathrm{DMC}} + \left(Z_{Y-\mathrm{cap}} + Z_{\mathrm{damping}}\right)\|Z_L\right)}{\left(Z_S + Z_L\right)\left(\left(Z_{Y-\mathrm{cap}} + Z_{\mathrm{damping}}\right)\|Z_L\right)}\right| \tag{8.16}$$

The performance of the LC filter (CM choke with three 1 nF Y capacitors and a 22 Ω damping resistor) without and with the DM inductor is shown in

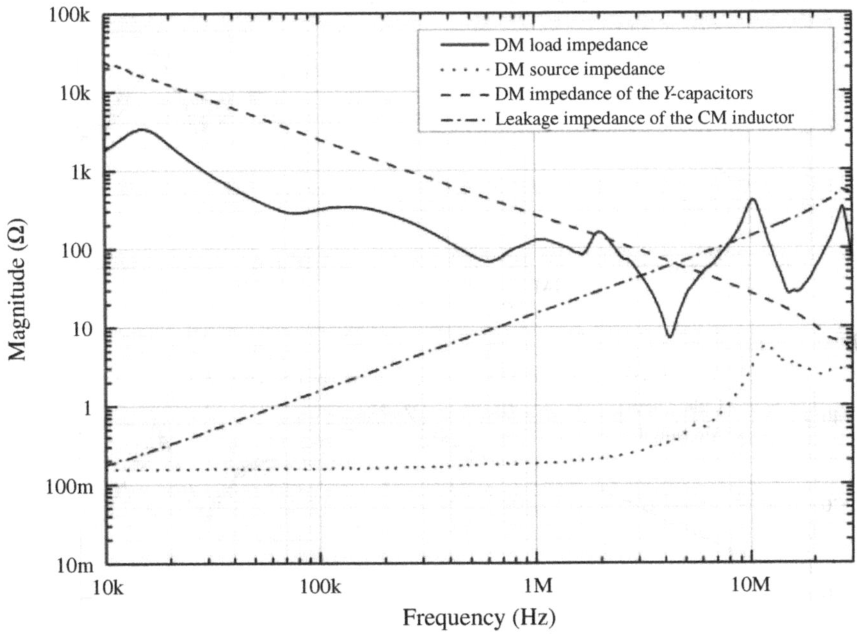

Figure 8.18 DM impedances of source, load, *Y* capacitors, and leakage inductance of the CM inductor.

Figure 8.19a,b, respectively. Although the CM impedance of the DM inductor contributes little to the CM attenuation directly, it plays a crucial role in suppressing DM-to-CM conversion. As a result, the measured CM attenuation in Figure 8.19b is significantly higher than that in Figure 8.19a.

Due to the inherent asymmetry of motor drive systems, DM-to-CM conversion has a significant impact on the performance of second-order LC filters. Consequently, second-order CM filters do not necessarily result in more compact solutions, as the introduction of an additional DM inductor may be required to mitigate mode conversion effects.

This section presented a systematic methodology for CM EMI filter design in inverter-driven motor systems based on contactless online impedance measurement. By extracting the CM noise source and load impedances under actual operating conditions, this approach overcomes the limitations of conventional trial-and-error filter design and enables quantitative prediction of filter attenuation. A step-by-step design procedure for CM chokes was developed, incorporating analytical estimation and FEM-based refinement to balance attenuation performance, component size, and saturation constraints. Through experimental validation on a practical motor drive system, the effectiveness of the impedance-based

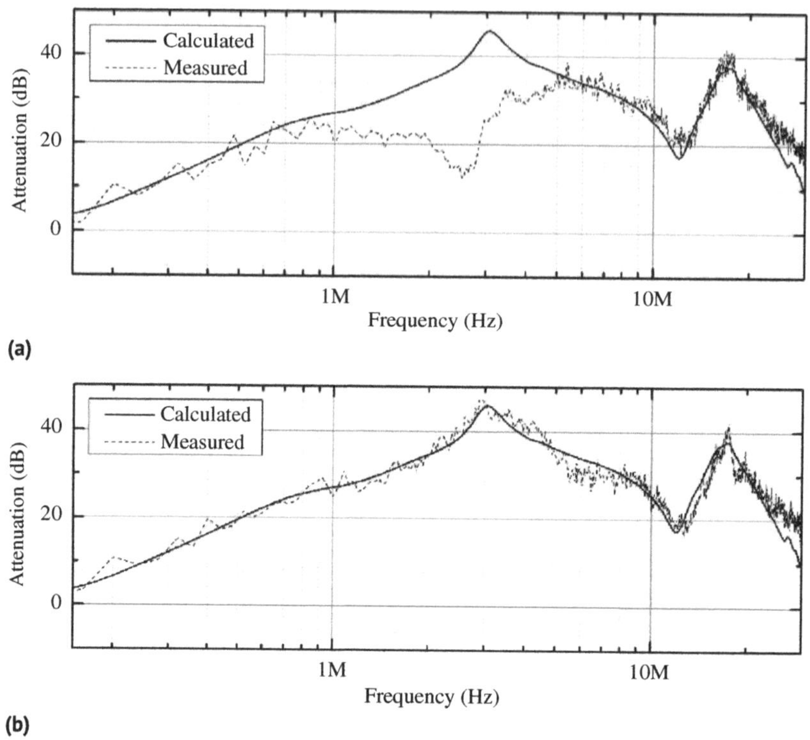

Figure 8.19 Calculated and measured attenuation for the LC filter. (a) Without DM inductor. (b) With additional DM inductor.

design methodology was demonstrated, showing that optimized CM filters can be achieved with reduced design iteration and minimized overdesign. The section further revealed the critical impact of DM-to-CM conversion on second-order LC filters, highlighting that higher filter order does not necessarily guarantee a more compact solution. Overall, this section establishes impedance-informed EMI filter design as a physically grounded and practical tool for improving EMI compliance in motor drive applications.

8.2 Detection of Stator Faults in Motor Drive Systems

With the advancement of fast-switching power electronics and PWM control techniques, inverters have become widely adopted for variable-speed and variable-torque control of induction motors (IMs) due to their high conversion efficiency and excellent control performance [29]. Like other electrical assets, IMs are inevitably subject to faults arising from thermal, electrical, environmental, and mechanical stresses. Among the common failure modes, bearing, stator, and rotor faults account for approximately 41, 36, and 9% of total IM faults, respectively [30]. Stator faults are most commonly caused by insulation degradation in the stator windings. Such degradation can lead to inter-turn short-circuit (ITSC) faults, which in turn result in increased current flowing through the shorted turns and excessive localized heating in adjacent turns. If left undetected, an ITSC fault can progressively evolve into more severe failures, such as coil-to-coil, phase-to-phase, or phase-to-ground short circuits, ultimately leading to complete motor failure [31]. These failures are financially detrimental due to both asset replacement costs and revenue losses associated with unplanned downtime. Consequently, reliable and early detection of stator ITSC faults is essential to prevent catastrophic failures.

A wide range of ITSC fault detection methods have been reported in the literature and can generally be classified into two categories: model-based methods [32–34] and signal-based methods [35–42]. Model-based approaches rely on accurate models of faulty IMs to identify ITSC faults, which can be computationally intensive and may not always be practical for real-time or industrial applications. In contrast, signal-based methods detect ITSC faults by monitoring and analyzing one or more measurable signals, such as motor current or voltage [35, 36], stray flux [37], air-gap torque [38], vibration [39], temperature [40], instantaneous power [41], and sequence-component impedance matrices [42].

Although many signal-based methods have demonstrated effective ITSC detection under specific operating conditions, several limitations remain. For example, monitored signals such as motor current, stray flux, air-gap torque, vibration, and temperature may also be influenced by faults other than stator ITSC faults, including bearing or rotor faults [38, 43–46], which can compromise detection reliability; thus, the effectiveness of the ITSC fault detection can be an issue. Furthermore, voltage sensors used to monitor motor voltage [36], instantaneous power [41], or sequence-component impedance matrices [42] require direct electrical connections to the energized IM. Such physical connections introduce electrical safety risks for on-site personnel and expose sensors to dielectric and thermal stresses, necessitating regular maintenance or replacement and potentially causing unnecessary motor downtime. More recently, artificial intelligence (AI) techniques [47–51] have been integrated with signal-based methods to enhance ITSC fault detection performance. While these approaches can improve detection accuracy, they typically require sufficient training datasets collected under diverse operating conditions to ensure robustness and reliability, which can be challenging to obtain in practice.

In this section, a novel signal-based method for detecting stator ITSC faults in inverter-fed IMs is presented. This approach detects ITSC faults through online monitoring of the CM impedance of the IM, $Z_{CM,IM}$, at a specific frequency of interest (FOI). The TD-TPS shown in Figure 3.19 is employed to monitor $Z_{CM,IM}$ by tracking a known externally injected RF sinusoidal signal within the IM. As a result, the monitored $Z_{CM,IM}$ is immune to supply voltage quality issues, such as voltage unbalance. Importantly, the measurement setup (i.e., TD-TPS) requires no direct electrical connection to the IM. The clamp-on inductive probes used can be easily installed or removed from the power cables between the inverter and the motor, enabling straightforward on-site implementation without shutting down the IM. In addition, signal amplification and surge protection modules are incorporated to enhance system robustness in harsh industrial environments characterized by high electrical noise and power surges. By carefully selecting the FOI, the method eliminates the need for broadband impedance monitoring, thereby reducing detection latency. Fast response is critical for timely decision-making before an ITSC fault escalates into a catastrophic failure.

Through online monitoring of $Z_{CM,IM}$ at the selected FOI, this method achieves reliable ITSC fault detection with high confidence. Moreover, the method is inherently robust, as it is insensitive to variations in motor load and speed, as well as to bearing and rotor faults.

8.2.1 ITSC Fault Analysis

Before analyzing the influence of stator ITSC faults on $Z_{CM,IM}$, the definition of $Z_{CM,IM}$ is briefly introduced. Figure 8.20 shows a CM equivalent circuit model of a three-phase IM [52]. Regardless of the actual stator winding configuration, the equivalent model can be represented in a star-connected form, with circuit parameters derived from available measurements. As a result, the model is universal and remains applicable to different stator winding configurations, such as star, delta, series, or parallel connections.

As illustrated in Figure 8.20, C_{sw1} and C_{sw2} denote the equivalent winding-to-frame parasitic capacitances, while R_{co1} and R_{co2} represent the equivalent copper resistances, including skin- and proximity-effect losses. L_{ls} is the stator winding leakage inductance, and R_e is the eddy-current losses in the stator core. The elements R_t, L_t, and C_t characterize the inter-turn parasitic effects of the stator windings. In addition, L_c represents the combined inductance of the IM's internal feed conductors and the leakage inductance of the first few turns of the stator windings. The CM impedance of the IM, $Z_{CM,IM}$, is defined as $V_{CM,IM}/I_{CM,IM}$, where $V_{CM,IM}$ and $I_{CM,IM}$ are the CM voltage and current at the IM terminals, respectively.

To analyze the influence of ITSC faults on $Z_{CM,IM}$, Figure 8.21 illustrates an ITSC fault occurring in one phase of an IM. Such a fault introduces a very low contact impedance between the shorted turns (Z_{ITSC}), which in turn affects $Z_{CM,IM}$.

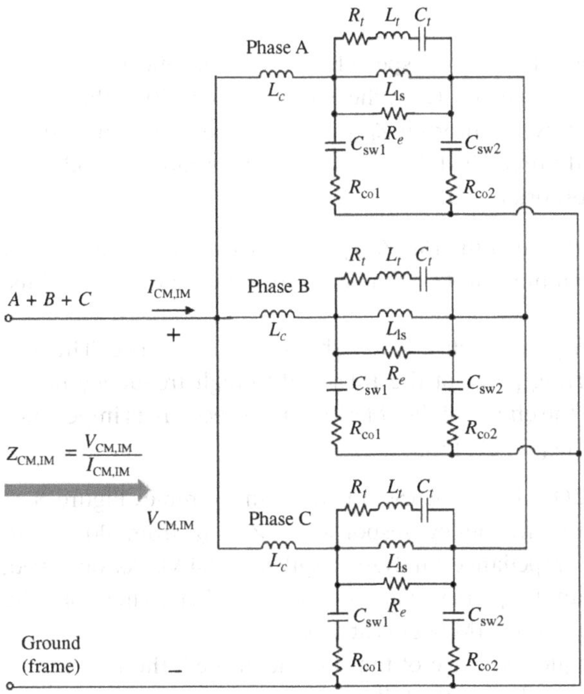

Figure 8.20 A CM equivalent circuit model of a typical three-phase IM.

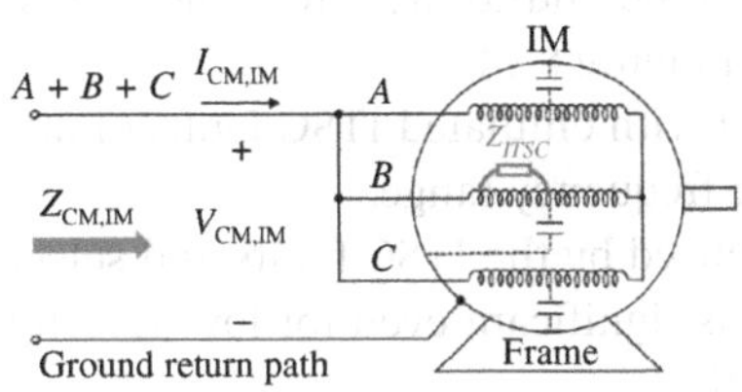

Figure 8.21 Illustration of a stator ITSC fault occurring in one phase of an IM.

It should be noted that, in addition to ITSC faults, other stator faults, such as coil-to-coil, phase-to-phase, and phase-to-ground short-circuit faults, can also influence $Z_{CM,IM}$. Since these more severe stator faults often originate from progressive ITSC faults [31], this section focuses on the detection of stator ITSC faults. Accordingly, the analysis concentrates on the relative variation of $Z_{CM,IM}$ induced by ITSC faults. To avoid long measurement times associated with monitoring $Z_{CM,IM}$ over a wide frequency range, tracking $Z_{CM,IM}$ at a single frequency is preferred to achieve a fast response when an ITSC fault occurs. To ensure reliable fault detection, the procedure for selecting an appropriate FOI for monitoring $Z_{CM,IM}$ is described in Section 8.2.2.

8.2.2 FOI Selection

It has been reported that the IM's load and speed have negligible effects on $Z_{CM,IM}$ in the mid-to-high frequency range [52]. Once the FOI is selected within this range, ITSC faults can be detected by monitoring $Z_{CM,IM}$ with good confidence under varying load and speed conditions. The following procedure is adopted to identify this range and select an appropriate FOI:

1. Measure the DM impedance of the IM, $Z_{DM,IM}$, offline over a wide frequency range (e.g., up to several megahertz) using an impedance analyzer, as illustrated in Figure 8.22.
2. Determine the frequency at which $Z_{DM,IM}$ exhibits self-resonance. This frequency marks the starting point of the IM's mid-to-high frequency range. Beyond this frequency, the effects of the IM's load and speed on its impedance characteristics become negligible [52].

Using a 1/2 hp IM (TECO AEAV1902R500FUS) as an example, Figure 8.23 shows the measured offline frequency response of $Z_{DM,IM}$ from 40 Hz to 5 MHz obtained using an impedance analyzer (Agilent 4294A). As observed, $Z_{DM,IM}$ exhibits a self-resonant frequency at approximately 42 kHz. Therefore, the mid-to-high frequency range of this IM begins at 42 kHz.

Once the mid-to-high frequency range of the IM is identified, the FOI can be selected within this range according to the following steps:

1. Measure $Z_{CM,IM}$ of the healthy IM offline in the mid-to-high frequency range using an impedance analyzer, as shown in Figure 8.24.
2. Measure $Z_{CM,IM}$ of IMs of the same model with emulated ITSC faults of different severity levels offline over the same frequency range.
3. Compute the relative change in $Z_{CM,IM}$ caused by the ITSC faults and select a frequency at which the relative change is significant even for low-severity faults. This frequency is defined as the FOI.

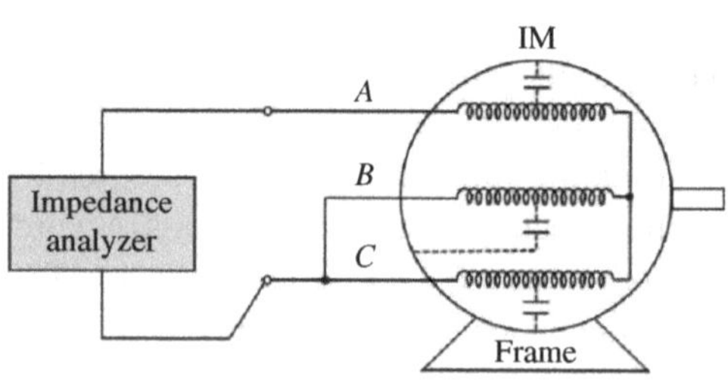

Figure 8.22 Offline measurement of $Z_{DM,IM}$ using an impedance analyzer.

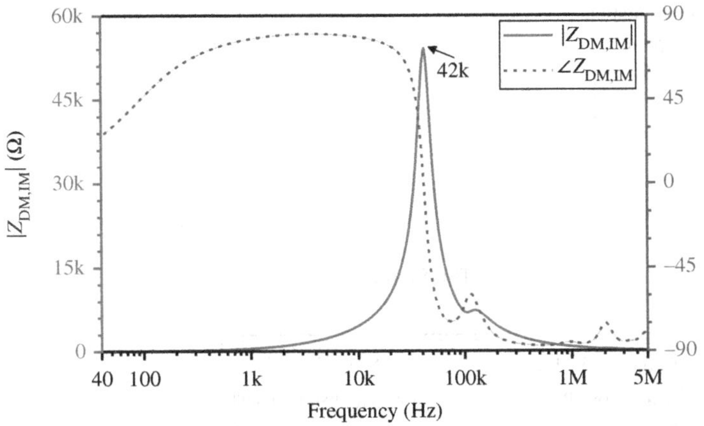

Figure 8.23 Offline measured frequency response of $Z_{DM,IM}$ of a healthy TECO IM from 40 Hz to 5 MHz.

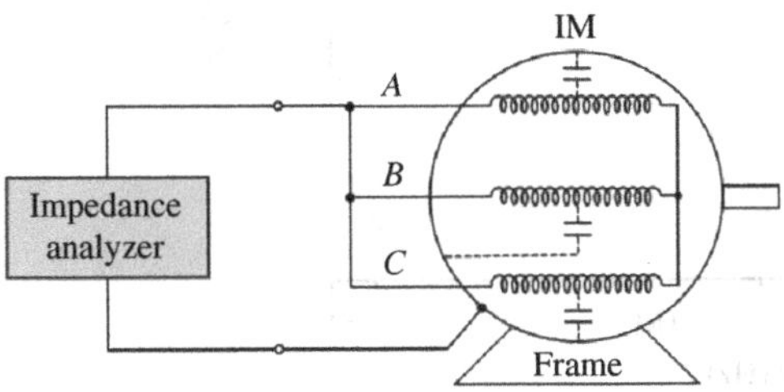

Figure 8.24 Offline measurement of $Z_{CM,IM}$ using an impedance analyzer.

The relative change in $Z_{CM,IM}$ at each frequency is defined as

$$\text{Relative Change} = \frac{|Z_{CM,IM}(\text{faulty}) - Z_{CM,IM}(\text{healthy})|}{|Z_{CM,IM}(\text{healthy})|} \times 100\% \tag{8.17}$$

The criterion for determining whether the relative change is significant is set to 5%, which will be justified in Section 8.2.4.

To emulate ITSC faults with different severity levels, the connections of the stator windings of the IM are modified using various tapping points, as shown in Figure 8.25. These tapping points are provided on the adjacent stator windings in each phase of the IM, enabling controlled emulation of ITSC faults with different numbers of shorted turns.

Figure 8.26 shows the measured offline frequency responses of $Z_{CM,IM}$ of the TECO IM under the healthy condition and with emulated ITSC faults over the

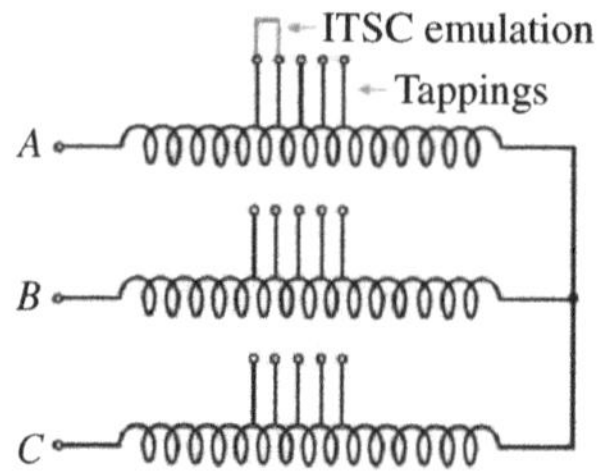

Figure 8.25 Schematic of the emulated ITSC faults.

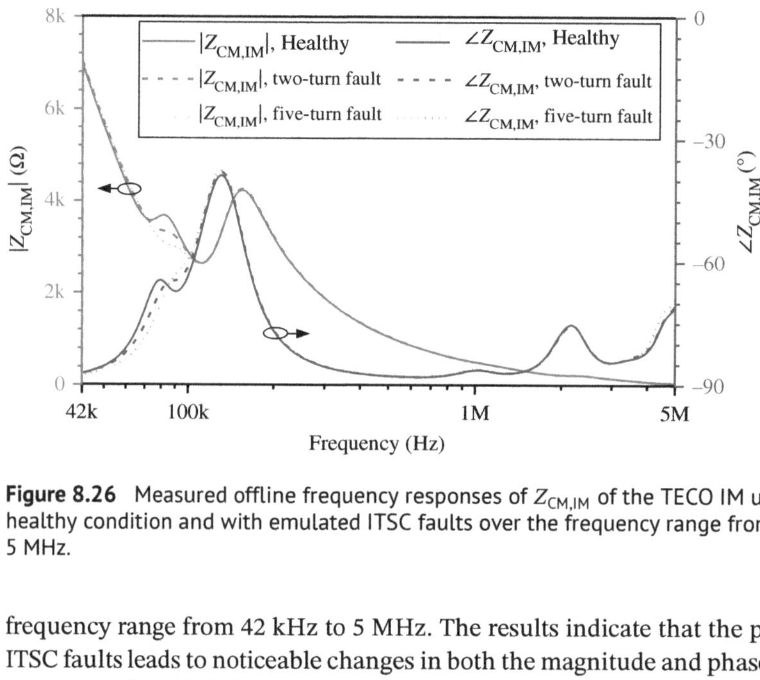

Figure 8.26 Measured offline frequency responses of $Z_{CM,IM}$ of the TECO IM under the healthy condition and with emulated ITSC faults over the frequency range from 42 kHz to 5 MHz.

frequency range from 42 kHz to 5 MHz. The results indicate that the presence of ITSC faults leads to noticeable changes in both the magnitude and phase of $Z_{CM,IM}$ over the mid-to-high frequency range identified earlier.

Taking the frequency response of $Z_{CM,IM}$ of the healthy IM as a reference, Figure 8.27 illustrates the relative change in $Z_{CM,IM}$ caused by two emulated ITSC faults with different severity levels, namely, the two-turn and five-turn faults. As observed in Figure 8.27, the relative change caused by the two-turn fault is already noticeable (>5%) in the frequency range between 66 and 96 kHz, reaching a maximum value of 9.6% at 80 kHz. For the five-turn fault, a more pronounced increase in the relative change is observed across a wider frequency range, particularly between 66 and 96 kHz, where the maximum relative change reaches 17.2%.

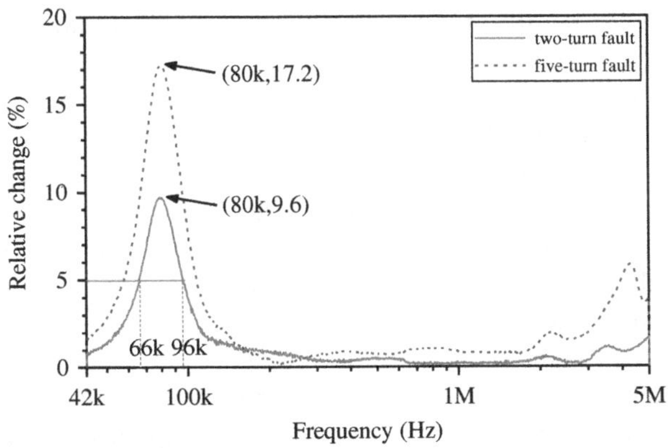

Figure 8.27 Relative change in $Z_{CM,IM}$ of the TECO IM caused by emulated ITSC faults with two severity levels (two-turn and five-turn faults).

The results clearly demonstrate that a more severe ITSC fault leads to a larger relative change in $Z_{CM,IM}$. Therefore, for the given IM, a frequency range between 66 and 96 kHz can be selected as the FOI for monitoring $Z_{CM,IM}$, enabling reliable detection of ITSC faults even at low severity levels (e.g., a two-turn fault).

8.2.3 Online Monitoring of Motor CM Impedance

Sections 8.2.1 and 8.2.2 have demonstrated that ITSC faults can be effectively detected by monitoring the relative change in $Z_{CM,IM}$. In addition, a systematic procedure for selecting an appropriate FOI for monitoring $Z_{CM,IM}$ has been established. Building upon these results, this section focuses on the measurement setup and methodology for online monitoring of $Z_{CM,IM}$ at the selected FOI under practical operating conditions.

As discussed in Chapter 2, three main approaches are commonly used for online impedance extraction: the *V–I* approach, the capacitive coupling approach, and the inductive coupling approach. Among these, the inductive coupling approach is particularly attractive for industrial motor drive applications because it does not require any physical electrical connection to the energized SUT, enabling nonintrusive and safe implementation. This feature is especially important in inverter-fed MDSs, which typically generate significant EMI and are frequently subjected to power surges during operation [31, 53]. To ensure robust and reliable online monitoring of $Z_{CM,IM}$ in such harsh environments, a TD-TPS incorporating signal amplification and surge protection modules is adopted in this study.

Figure 8.28 illustrates the structure of a typical inverter-fed MDS and the associated CM current path. A representative inverter-fed MDS consists of a variable-frequency drive (VFD) and an IM connected by a three-phase power cable. In practice, the AC input of the VFD is usually equipped with a built-in EMI filter to comply with regulatory emission requirements [54]. The high-frequency switching of the power semiconductor devices in the VFD generates CM noise at the inverter output, which gives rise to a CM current loop formed by the three-phase power cable bundle, the parasitic capacitances between the IM stator windings and the motor frame, and the ground return path.

Figure 8.29a shows the TD-TPS employed for online monitoring of $Z_{CM,IM}$. The system consists of a computer-controlled SGAS, a power amplifier (PA), two attenuators (AT1 and AT2), two surge protectors (SP1 and SP2), a clamp-on IIP, and a clamp-on RIP. For online monitoring, the IIP and RIP are clamped onto the three-phase power cable, with their positions denoted as p_1 and p_2, respectively. Figure 8.29b presents the CM equivalent circuit of the inverter-fed MDS corresponding to the TD-TPS measurement configuration. In this model, $V_{CM,NS}$ represents the equivalent CM noise source resulting from the switching action of the VFD, $Z_{CM,VFD}$ denotes the equivalent CM impedance of the VFD, and $Z_{CM,W}$ represents the equivalent CM loop impedance of the three-phase power cable and the ground return path.

Consequently, the resultant impedance Z_X seen between p_1 and p_2 can be expressed as

$$Z_X = Z_{CM,VFD} + Z_{CM,W} + Z_{CM,IM} \tag{8.18}$$

Since the physical length of the power cable is much shorter than the wavelength corresponding to the FOI, the system can be accurately modeled using a lumped-parameter circuit representation [55]. After performing the three-term calibration procedure prior to online measurement, Z_X can be directly extracted using the TD-TPS. The presence of the PA allows the injected test signal to be sufficiently

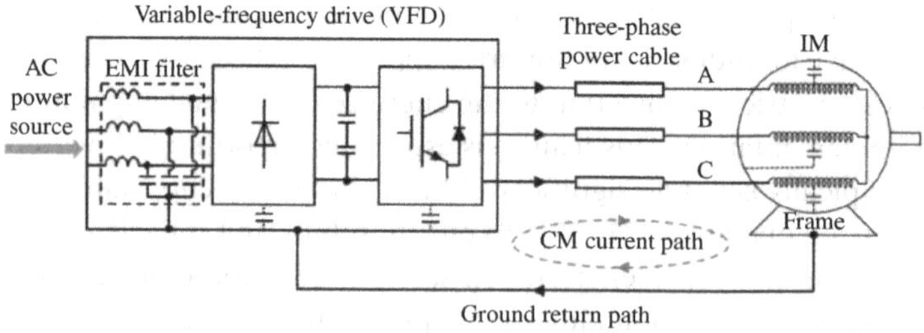

Figure 8.28 Typical inverter-fed MDS and the associated CM current path.

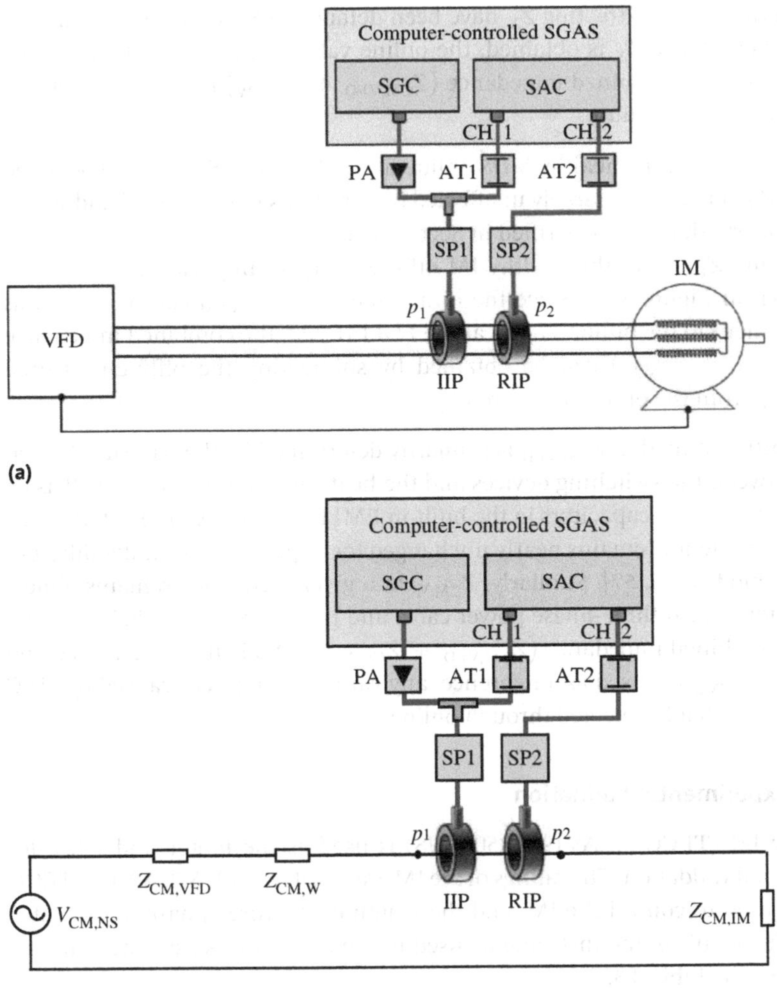

Figure 8.29 (a) TD-TPS for online monitoring of $Z_{CM,IM}$. (b) CM equivalent circuit of the inverter-fed MDS corresponding to the TD-TPS measurement.

amplified such that the induced signal level in the CM path of the MDS is significantly higher than the background noise level at the FOI, thereby ensuring adequate SNR, typically exceeding 40 dB. As the three-term calibration technique and the principle of extracting Z_X have been detailed in Section 3.4, they are not repeated here. Once Z_X is obtained, the online value of $Z_{CM,IM}$ can be extracted provided that the combined impedance $(Z_{CM,VFD} + Z_{CM,W})$ is known, according to the following procedure:

1. Measure Z_X of the healthy MDS online using the TD-TPS. It should be noted that Z_X at the FOI is largely unaffected by variations in motor load and speed, a property that will be verified in Section 8.2.4.
2. Measure $Z_{CM,IM}$ of the healthy IM offline using an impedance analyzer, as shown in Figure 8.24. Since the offline-measured $Z_{CM,IM}$ can be treated as equivalent to the online $Z_{CM,IM}$ at the FOI [51, 55], the combined impedance $(Z_{CM,VFD} + Z_{CM,W})$ can be obtained by subtracting the offline-measured $Z_{CM,IM}$ from the online-measured Z_X.

It is worth noting that $Z_{CM,VFD}$ is primarily determined by the parasitic capacitances between the switching devices and the heat sink in the VFD, as well as by the phase-to-ground capacitors in the built-in EMI filter [56]. As a result, $Z_{CM,VFD}$ at a given frequency remains nearly unchanged for a specific VFD under different operating modes [56, 57]. Similarly, $Z_{CM,W}$ at a given frequency remains almost constant for a fixed three-phase power cable and ground return path. Therefore, once the combined impedance $(Z_{CM,VFD} + Z_{CM,W})$ at the FOI is determined and the healthy $Z_{CM,IM}$ is used as a reference, any variation in $Z_{CM,IM}$ caused by ITSC faults can be reliably detected through online monitoring.

8.2.4 Experimental Validation

The same IM (TECO AEAV1902R500FUS) is used as the motor under test for experimental validation. The ratings of the IM are listed in Table 8.2. A VFD (TECO L510s) is used to control the IM, and the length of the three-phase power cable is 1 m. The details of the instruments used in the actual measurement setup are summarized in Table 8.3.

Table 8.2 Ratings of the three-phase IM under test.

Rating	Value	Rating	Value
Power	1/2-hp	Frequency	50 Hz
Voltage	220–240 V(Δ)	Pole	2
Current	1.85–1.70 A(Δ)	Speed	2765 RPM

Table 8.3 Details of the instruments (inst.) used in the actual measurement.

Inst.	Descriptions	Inst.	Descriptions
SGAS	NI PXI Platform	PA	Mini-Circuits ZHL-32A-S+
SGC	NI PXI-5412	AT1	Pasternack PE7025-20
SAC	NI PXI-5922	AT2	AIM-Cambridge 27-9300-6
IIP	Solar 9144-1N	SP1	Bussmann BSPD5BNCDI
RIP	Solar 9134-1	SP2	Bussmann BSPD5BNCDI

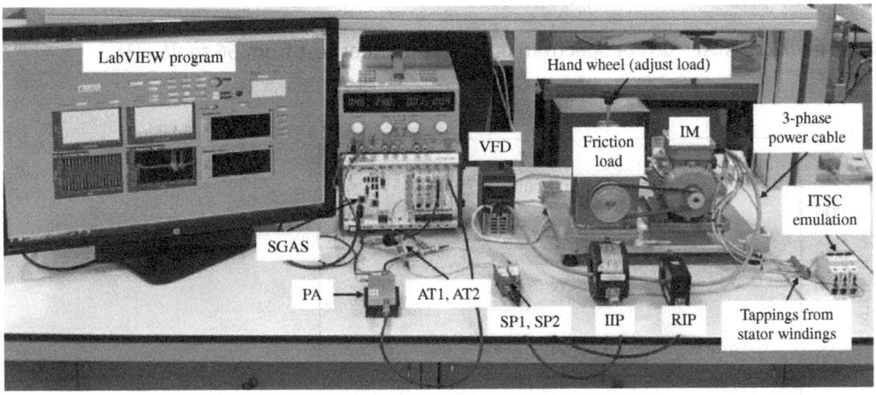

Figure 8.30 Actual TD-TPS with the inverter-fed MDS under test. *Source*: [61]/with permission of IEEE.

Figure 8.30 shows the actual measurement setup of the MDS under test. An adjustable friction load is driven by the IM via a belt. The ITSC faults are emulated by connecting additional low-impedance contacts (<1 Ω) between selected tapping points, as illustrated in Figure 8.25. Using the proposed TD-TPS together with the LabVIEW-based monitoring program, the response time for monitoring and detecting ITSC faults is within 1 second.

As discussed in Section 8.2.2, to ensure reliable detection of ITSC faults, the FOI for monitoring $Z_{CM,IM}$ of the IM lies between 66 and 96 kHz. From a practical perspective, the FOI should also be selected such that the background noise level in the MDS is relatively low in order to achieve a high SNR. Figure 8.31 shows the noise current spectrum of the MDS under test from 66 to 96 kHz, from which frequency regions with relatively low noise levels can be identified. As observed from the enlarged view in Figure 8.31, the noise level at 82 kHz is approximately 2.4 μA. Therefore, the FOI is selected as 82 kHz for online monitoring of $Z_{CM,IM}$ for ITSC fault detection.

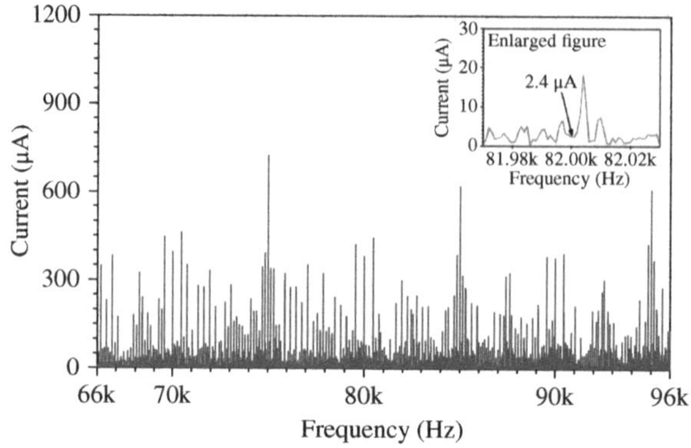

Figure 8.31 Noise current spectrum of the MDS under test from 66 to 96 kHz.

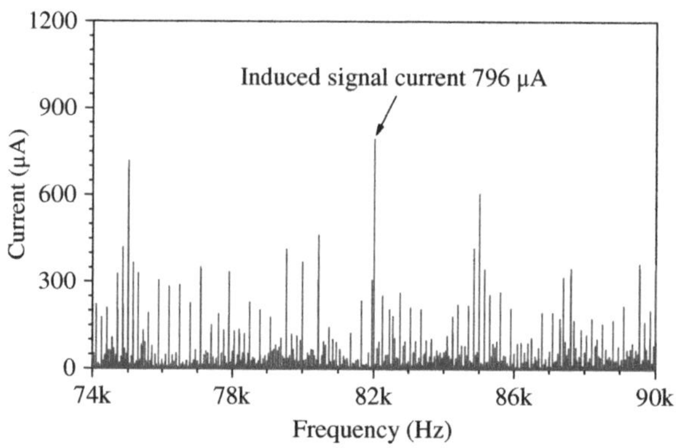

Figure 8.32 Current spectrum around 82 kHz in the MDS under test when injecting an RF sinusoidal excitation signal (600 mV at 82 kHz) into the MDS by using the TD-TPS.

As required in this section, the SNR in the MDS is desired to be larger than 40 dB to ensure accurate extraction of $Z_{CM,IM}$. Figure 8.32 shows the current spectrum around 82 kHz in the MDS under test when an RF sinusoidal excitation signal (600 mV at 82 kHz) is injected using the TD-TPS. The induced signal component at 82 kHz is clearly visible with an amplitude of 796 μA. With a background noise level of 2.4 μA at 82 kHz, the induced signal achieves an SNR of approximately 50 dB. Moreover, the induced current of 796 μA is sufficiently small and does not affect the normal operation of the MDS.

By injecting the RF sinusoidal excitation signal (600 mV at 82 kHz) into the MDS under test, Figures 8.33 and 8.34 show the measured resultant impedance Z_X of the healthy MDS under varying load conditions (at nominal speed) and varying

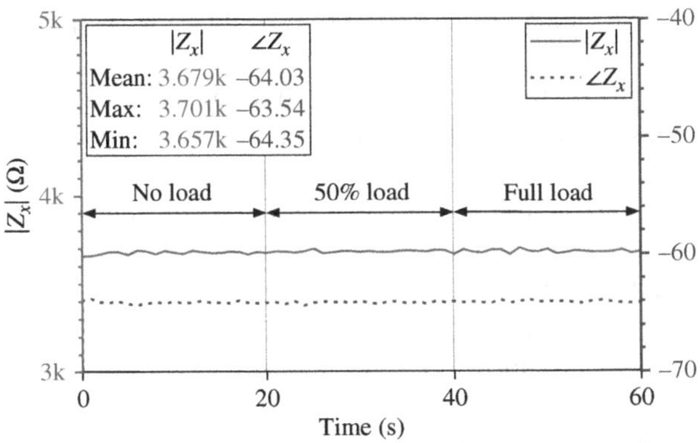

Figure 8.33 Measured Z_X of the healthy MDS under varying load conditions at nominal speed.

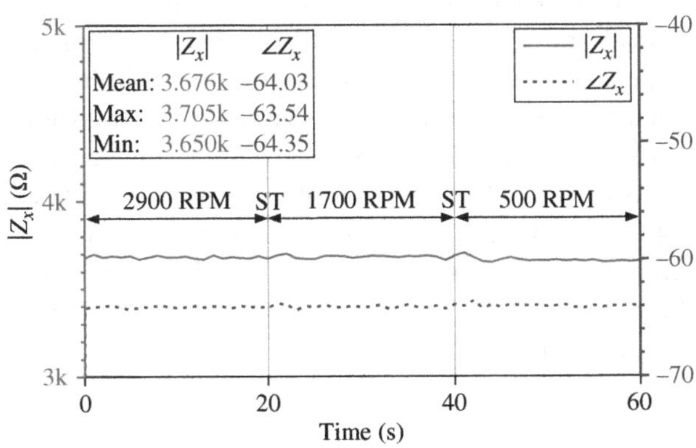

Figure 8.34 Measured Z_X of the healthy MDS under varying speed conditions.

speed conditions, respectively. It should be noted that all impedances Z_X, $Z_{CM,IM}$, and $(Z_{CM,VFD} + Z_{CM,W})$ referred to in this section correspond to the impedance measured at the FOI (i.e., 82 kHz). As shown in Figures 8.33 and 8.34, Z_X remains essentially unchanged under both load and speed variations, even during speed transients (STs).

Using the measurement configuration shown in Figure 8.24, the measured $Z_{CM,IM}$ of the healthy IM is 3.682 k$\Omega \angle -64.98°$. It should be noted that $(Z_{CM,VFD} + Z_{CM,W})$ is very small at 82 kHz due to the short length of the three-phase power cable (1 m) and the built-in EMI filter of the VFD. Since $(Z_{CM,VFD} + Z_{CM,W}) \ll Z_{CM,IM}$, the resultant impedance Z_X can be approximated as $Z_{CM,IM}$ for the MDS under test. Therefore, the relative change in Z_X closely represents the relative change in $Z_{CM,IM}$.

Figure 8.35 shows the relative change in $Z_{CM,IM}$ of the IM under varying load conditions (nominal speed) when the IM is in the healthy condition and in the two-turn ITSC fault condition, respectively. It is observed that the relative change is unaffected by load variations. In addition, when the IM is in the two-turn ITSC fault condition, the relative change remains almost constant at approximately 9.4%. By contrast, only a very small relative change (<1%) is observed when the IM operates in the healthy condition. As shown in Figure 8.33, such a minor variation is caused by the slight normal fluctuation in the monitored impedance.

Figure 8.36 shows the relative change in $Z_{CM,IM}$ of the IM under varying speed conditions when the IM is in the healthy condition and in the two-turn ITSC fault condition, respectively. Again, the relative change is insensitive to speed variations. Under the two-turn ITSC fault condition, the relative change remains nearly constant at about 9.4%, while the minor relative change caused by normal impedance fluctuation is still maintained within 1%.

From Figures 8.35 and 8.36, it is concluded that the relative change in $Z_{CM,IM}$ at the FOI is unaffected by load and speed variations. Since the minor relative change caused by normal impedance fluctuation is consistently kept within 1%, a threshold of 1% can be defined for ITSC fault detection. Based on this threshold, a relative change can be regarded as significant when it exceeds 5%. Furthermore, even the IM with a low-severity two-turn ITSC fault exhibits a noticeable relative change of approximately 9.4%, indicating that the introduced method has good sensitivity for detecting incipient ITSC faults. Table 8.4 summarizes the relative change in $Z_{CM,IM}$ of the IM for ITSC faults with different severity levels, showing a clear increasing trend with fault severity.

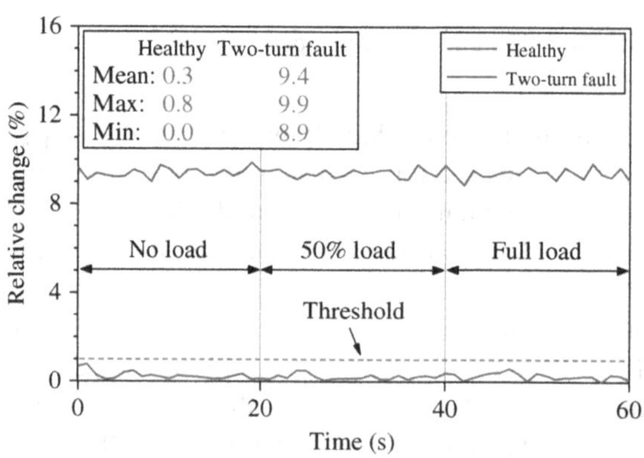

Figure 8.35 Relative change in $Z_{CM,IM}$ of the IM (healthy and two-turn ITSC fault) monitored online using the TD-TPS under varying load conditions at nominal speed.

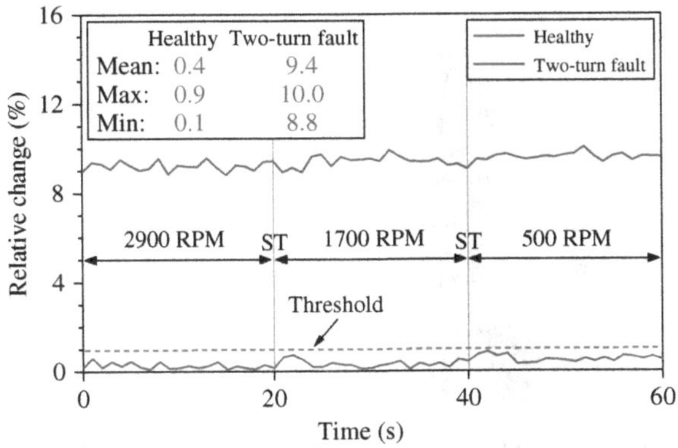

Figure 8.36 Relative change in $Z_{CM,IM}$ of the IM (healthy and two-turn ITSC fault) monitored online using the TD-TPS under varying speed conditions.

Table 8.4 Relative change in $Z_{CM,IM}$ of the IM for ITSC faults with different severity levels.

Fault Severity Level	Relative Change in $Z_{CM,IM}$		
	Mean Value (%)	**Maximum Value (%)**	**Minimum Value (%)**
Two-turn fault	9.4	10.0	8.8
Three-turn fault	10.5	11.2	10.1
Four-turn fault	12.1	12.9	11.5
Five-turn fault	17.4	18.1	16.6

Besides stator faults, rotor and bearing faults are also among the most prevalent fault types in IMs. Bearing faults are commonly associated with surface roughness degradation [31], while rotor cage faults (RCFs) represent a typical rotor fault mode [30, 58]. As a proof of concept, abnormal bearing friction (AFB) and RCFs are emulated, as shown in Figure 8.37, to examine their potential influence on ITSC fault detection.

Figure 8.38 presents the relative change in $Z_{CM,IM}$ of the IM under AFB and RCF conditions. It is observed that the relative changes in $Z_{CM,IM}$ for both fault types remain well below 1%, which is negligible and does not interfere with ITSC fault detection.

Based on these results, the presented method demonstrates strong reliability and robustness for stator ITSC fault detection under varying load and speed conditions, as well as in the presence of rotor and bearing faults.

Figure 8.37 Emulated AFB and RCF of the IM. *Source*: [61]/with permission of IEEE.

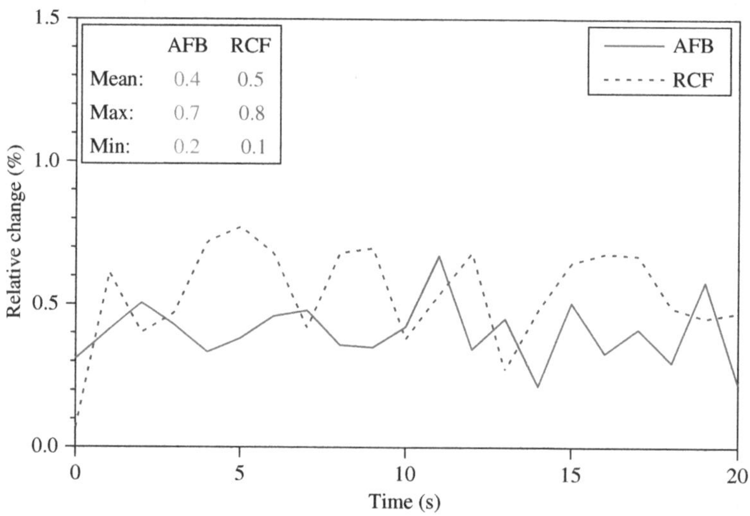

Figure 8.38 Relative change in $Z_{CM,IM}$ of the IM under AFB and RCF conditions.

This section presents an online CM impedance monitoring method for detecting stator ITSC faults in an inverter-fed MDS. By analyzing the relative change in $Z_{CM,IM}$, a fault indicator is established that is insensitive to load and speed variations. Besides, a systematic procedure is introduced to select an appropriate FOI in the mid-to-high frequency range, ensuring reliable fault sensitivity even at low fault severities. A TD-TPS is then developed to monitor $Z_{CM,IM}$ in real time without physical electrical contact. Experimental results demonstrate that the TD-TPS-based method can reliably detect early-stage ITSC faults while remaining immune to rotor and bearing faults, validating its effectiveness and robustness for motor condition monitoring.

8.3 Current Collector Health Monitoring of LRT Trains

The current collector is a key component of most LRT trains, responsible for drawing electric power from the catenary or power rail during operation. Due to wear, tear, and harsh operating environments, the current collector is prone to malfunction, which can lead to unexpected power interruptions and directly degrade service reliability. This section presents contactless online impedance measurement methods for monitoring the health of current collectors in LRT trains powered by three-phase AC power rails. The monitoring system developed can be readily integrated into trains, enabling it to function as a train-borne health monitoring system (TBHMS) during revenue service. The TBHMS monitors the electrical contact quality (ECQ) between the current collector shoe (CCS) and the power rail in a noncontact and electrically safe manner. By doing so, it eliminates the need for manual inspections during engineering hours, which are labor-intensive and time-consuming. In such trains, the current collector consists of a current collector assembly (CCA) that accommodates the CCSs. Two performance indices, the contact quality index (CQI) and the contact loss index (CLI), are defined to quantitatively assess the health conditions of the CCS and the CCA, respectively.

8.3.1 Current Collector of LRT Trains Powered by Three-phase AC Power Rail

Figure 8.39 shows the front view of an LRT train guided by a central guideway and powered by a 600 V, 50 Hz, three-phase AC power rail. The power rail and the signal/ground rail are supported by insulated support brackets. The LRT train taps the three-phase power from the power rail through two current collectors, one mounted at the front of the train and the other at the rear, as illustrated in Figure 8.40. The use of two current collectors provides redundancy; that is, when one current collector temporarily loses contact with the power rail, the other can continue supplying electrical power to the train.

A closer examination of the current collector in Figure 8.39 shows that each CCA holds three CCSs, one for each phase of the three-phase AC power rail. The power rail is a fixed structure supported by insulated support brackets, which are designed to minimize alignment issues. However, due to harsh tropical environments as well as long-term wear and tear, the holding spring within the CCA may gradually lose its mechanical stiffness. As a result, the CCSs may sag over time. If the sagging becomes severe, a CCS may come into contact with an adjacent phase of the power rail, leading to an unexpected phase-to-phase electrical short circuit. Such a fault can cause tripping of the traction power supply and disruption of LRT services. In addition, the intense arc flashes generated during the electrical short

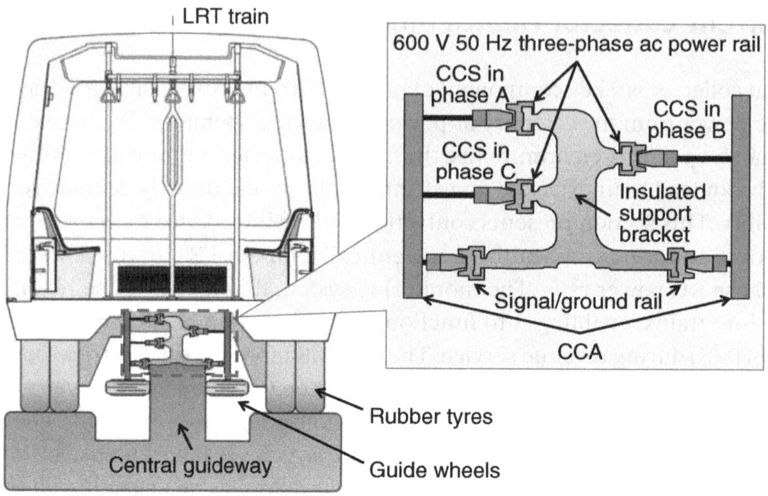

Figure 8.39 Front view of the LRT train.

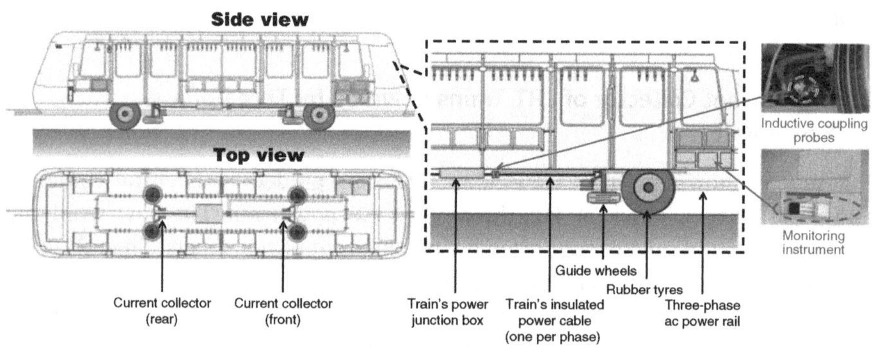

Figure 8.40 Mounting locations of the monitoring instrument and inductive probes of TBHMS on an LRT train.

can cause irreversible damage to the CCS. To prevent the aforementioned issues, the health conditions of both the CCSs and the CCA should be monitored.

8.3.2 TBHMS Based on Inductive Coupling Method

The health monitoring system for the CCS and CCA is fitted on the LRT train and functions as a TBHMS. The system can be retrofitted onto an existing train or integrated into a new train during the assembly process. With the TBHMS installed, the train can additionally serve as an inspection vehicle during revenue service, enabling continuous health monitoring of the CCS and CCA without interrupting normal operation and replacing unproductive inspections conducted during engineering hours. As shown in Figure 8.40, the TBHMS consists of a dedicated monitoring instrument and a set of inductive coupling probes interconnected by shielded coaxial cables. The monitoring instrument is housed in a metallic enclosure and mounted within the under-compartment beneath a train seat. It is powered by the existing onboard power supply of the LRT train and is supported by an uninterruptible power supply (UPS) to ensure continuous operation and automatic data storage in the event of power outages.

To monitor the ECQ between the CCS and the power rail for a specific phase (phase A, B, or C), a pair of inductive coupling probes, one IIP and one RIP, are clamped onto the insulated power cable connected to the CCS of that phase. The probes are neither in close proximity to nor in direct electrical contact with the current collector. If simultaneous monitoring of ECQs for all three phases is required, three such probe pairs can be deployed, with one pair assigned to each phase. Since the ECQ monitoring principle is identical for all phases, only the ECQ monitoring of phase B is presented and discussed in the following for clarity.

Figure 8.41 provides an overview of the TBHMS for ECQ monitoring of phase B. A low-power RF signal generator produces an excitation signal, which is amplified by a signal amplifier and then injected into a closed RF signal path via the IIP. This closed path is formed by the phase B power cable, the power rail, the front CCS, and the rear CCS. As illustrated in Figure 8.40, each LRT train is equipped with two current collectors, one located at the front of the train and the other at the rear. A RIP, clamped onto the insulated power cable, is used to capture the RF response signal propagating in the closed path. A data acquisition (DAQ) device collects both the RF excitation signal at the output of attenuator ATT1 and the RF response signal at the output of attenuator ATT2. The attenuators ATT1 and ATT2 ensure that the measured RF signal levels remain within the permissible input range of the DAQ device. Two surge protectors, SP1 and SP2, are incorporated to protect the monitoring instrument from transient surges that may occur during train operation.

The functions of the individual components in the TBHMS are summarized in Table 8.5. In addition to the RF excitation and response signals, the tachometer signal of the train is also recorded by the DAQ device to determine the exact position of the train along the LRT line. The collected signals are subsequently processed to evaluate the ECQ. The definitions of ECQ and the two performance indices (CQI and CLI) are introduced in Section 8.3.3. Based on data obtained from comprehensive field trials, thresholds for CQI and CLI are established to support decision-making for condition-based maintenance of the CCS and CCA.

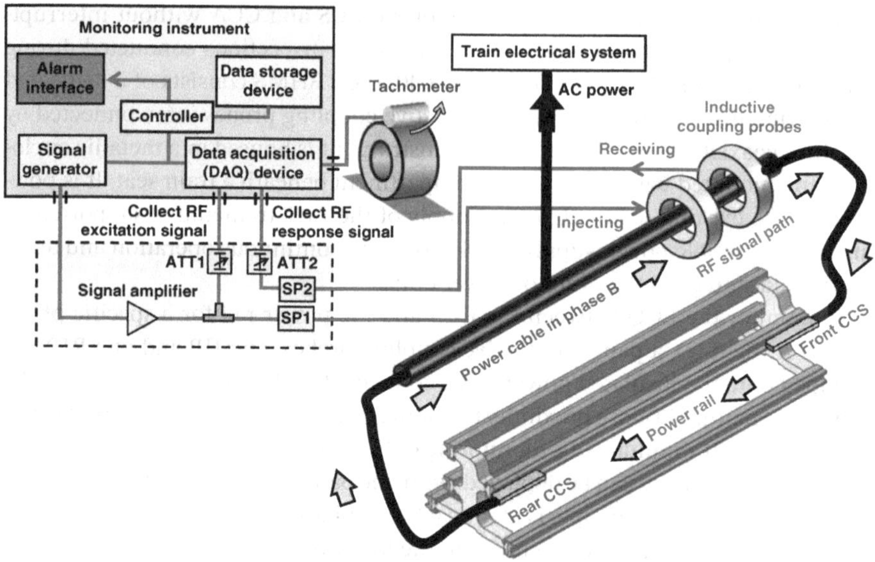

Figure 8.41 Overview of the TBHMS for ECQ monitoring of phase B.

Table 8.5 Functions of individual components in the TBHMS.

Component		Function
Inductive coupling probes		Transducers to inject the RF excitation signal and to receive the RF response signal
Monitoring instrument	Signal generator	To excite a low-power RF signal
	Signal amplifier	To amplify the RF excitation signal
	DAQ device	To receive RF excitation and response signals, and to collect the tachometer signal
	ATT1 and ATT2	To keep measured signal levels within the permissible input range of the DAQ device
	SP1 and SP2	To protect the monitoring instrument from transient surges
	Controller	To process collected signals, to manage the transfer of stored data, and to generate an alarm based on a predefined decision-making process
	Data storage device	To store the time stamp and the data

In addition to the radio-frequency (RF) excitation and response signals, the tachometer signal of the train is also collected by the DAQ device for the exact position of the train on the LRT line. The collected signals are processed to evaluate the ECQ. The definitions of ECQ and the two performance indexes, CQI and CLI, will be described in Section 8.3.3. Based on the data collected from comprehensive field trials, thresholds for CQI and CLI are established in the decision-making process for condition-based maintenance of CCS and CCA.

8.3.3 Health Monitoring Based on CQI and CLI

The ECQ is defined as the quality of the electrical contact between the CCS and the power rail of a specific phase while the train is moving along the LRT line. As shown in Figure 8.41, the RF impedance of the closed path formed by the power cable, power rail, front CCS, and rear CCS is measured in real time when the train is in motion. The impedance contributions from the power cable and the power rail remain unchanged, as they are fixed structures. Therefore, the only time-varying component of the RF impedance is the dynamic contact impedance between the CCS and the power rail, since the CCS moves together with the train.

This RF contact impedance is highly sensitive to the contact quality between the CCS and the power rail. When the CCS and the power rail are in good contact, the RF contact impedance is small; however, it becomes significantly larger when a gap exists between them [59]. To establish the relationship between the RF contact impedance and the measured RF excitation and response signals acquired by the DAQ device of the TBHMS, Figure 8.42 presents the equivalent network representation of Figure 8.41. The equivalent network is constructed based on the two-port and three-port network concepts described in [60, 61].

In Figure 8.42, V_{SG} and Z_{SG} denote the equivalent source voltage and internal impedance of the signal generator, respectively. $\boldsymbol{N}_{ATT1}$ represents the two-port network of ATT1, which is characterized by *ABCD* parameters. SP1, SP2, ATT2, and the inductive probes together with the clamped cable are modeled as a three-port network $\boldsymbol{N}$, which is described using impedance (Z-) parameters. V_{CH1} and V_{CH2} are the RF excitation signal voltage at the output of ATT1 and the RF response signal voltage at the output of ATT2, respectively. V_{CH_i} $(i = 1,\ 2)$ is acquired by the DAQ device, where Z_{CH_i} denotes the termination impedance of channel i of the DAQ device. Z_{cable} and Z_{rail} represent the power cable impedance and the power rail impedance, respectively. $Z_{contacts}$ denotes the RF contact impedances between the front CCS and the power rail as well as the rear CCS and the power rail. The resultant impedance of the closed path, denoted as Z_{path}, is given by

$$Z_{path} = Z_{cable} + Z_{rail} + Z_{contacts} \tag{8.19}$$

Based on the network analysis theory [61], the relationship between Z_{path} and the RF excitation and response signal voltages (V_{CH1} and V_{CH2}) can be established as follows:

$$\frac{V_{CH2}}{V_{CH1}} = \frac{Z_{path} + k_1}{k_2 \cdot Z_{path} + k_3} \tag{8.20}$$

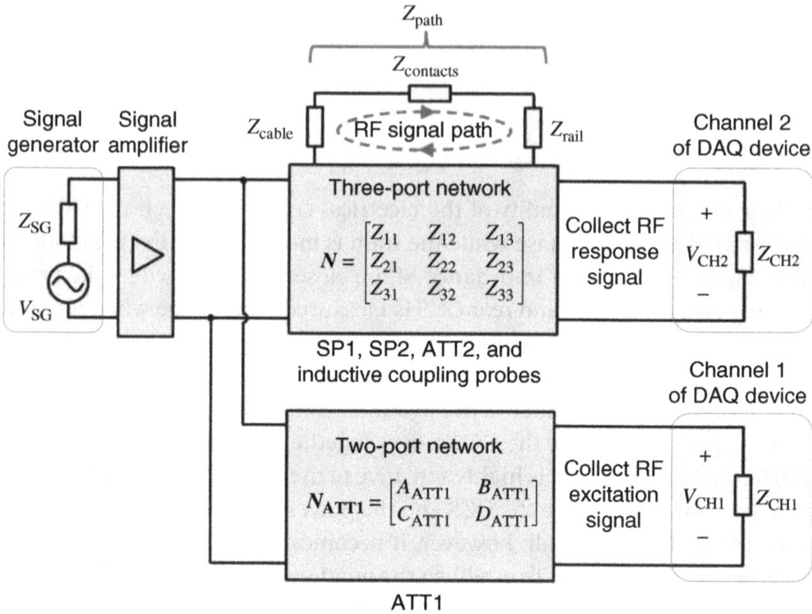

Figure 8.42 Equivalent network representation of Figure 8.41.

where

$$k_1 = \frac{Z_{31} \cdot Z_{23}}{Z_{21}} - Z_{33} \tag{8.21}$$

$$k_2 = \frac{\left(\frac{Z_{11}}{Z_{21}}\right) \cdot \left(1 - \frac{Z_{22}}{Z_{CH2}}\right) + \frac{Z_{12}}{Z_{CH2}}}{A_{ATT1} + \frac{B_{ATT1}}{Z_{CH1}}} \tag{8.22}$$

$$k_3 = \frac{\begin{array}{l}[Z_{31}/Z_{21} \cdot (1 - Z_{22}/Z_{CH2}) + Z_{32}/Z_{CH2}] \\ \cdot [Z_{13} - (Z_{11} \cdot Z_{23})/Z_{21}] - [Z_{33} - (Z_{31} \cdot Z_{23})/Z_{21}] \\ \cdot [Z_{11}/Z_{21} \cdot (1 - Z_{22}/Z_{CH2}) + Z_{12}/Z_{CH2}]\end{array}}{A_{ATT1} + B_{ATT1}/Z_{CH1}} \tag{8.23}$$

Since the detailed derivations of Equations (8.20)–(8.23) have been presented in [61], they are not repeated here. Inspection of Equations (8.21)–(8.23) shows that k_1, k_2, and k_3 depend only on the Z-parameters (Z_{ij}, $i, j = 1, 2, 3$) of the three-port network N, the $ABCD$ parameters (A_{ATT1} and B_{ATT1}) of the two-port network $\boldsymbol{N}_{\mathbf{ATT1}}$, and the termination impedances Z_{CH_i} ($i = 1, 2$) of the DAQ device.

These parameters remain unchanged for a given TBHMS. Therefore, according to Equation (8.20), any variation in V_{CH2}/V_{CH1} directly reflects a change in Z_{path}. As discussed earlier, Z_{cable} and Z_{rail} correspond to fixed installations and are thus time-invariant. Consequently, $Z_{contacts}$ is the only time-varying impedance, as it dynamically changes with the train's movement. Hence, a variation in $Z_{contacts}$ in Equation (8.19) leads to a corresponding variation in V_{CH2}/V_{CH1}.

When the CCS maintains good electrical contact with the power rail during train operation, V_{CH2}/V_{CH1} remains relatively stable, indicating a low and stable $Z_{contacts}$. Conversely, when the contact quality degrades, $Z_{contacts}$ increases and fluctuates significantly, resulting in a highly unstable and fluctuating V_{CH2}/V_{CH1}. Therefore, by tracking the variation of V_{CH2}/V_{CH1} as the train moves, the ECQ between the CCS and the power rail can be effectively evaluated. It is worth noting that high-speed ramps (HSR) and dummy rails (DR) are located at specific sections of the power rail. When the train passes over HSR and DR, large vertical movements of the CCA occur, which can lead to momentary loss of electrical contact between the CCS and the power rail for a visible duration. This phenomenon is clearly observed as sudden changes in V_{CH2}/V_{CH1}, as shown in Figure 8.43.

For power rail sections without HSR and DR, smaller variations in V_{CH2}/V_{CH1} are detected, which are mainly caused by the dynamic electrical contact impedance between the CCS and the power rail. To evaluate the health conditions of the CCS and the CCA, two performance indices, CQI and CLI, are introduced and discussed in detail in the following.

Figure 8.43 is used as an illustration to show a typical measured V_{CH2}/V_{CH1} signal between stations, which enables differentiation between faults originating from a degraded CCS and those caused by a degraded CCA.

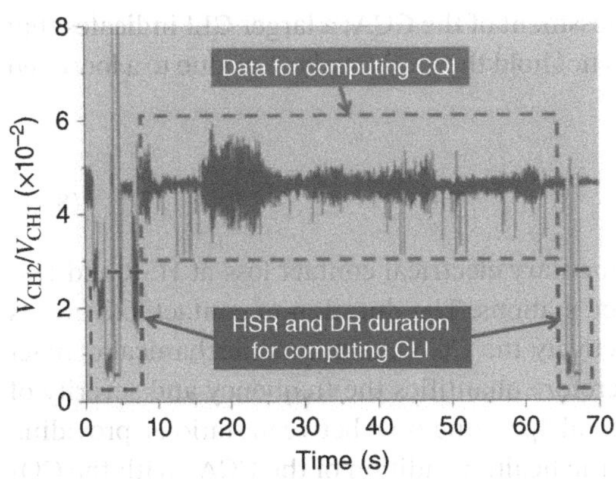

Figure 8.43 Typical measured V_{CH2}/V_{CH1} between the stations.

8.3.3.1 CQI

1. Only the data of V_{CH2}/V_{CH1} when the CCS is in electrical contact with the power rail are selected for processing, as indicated in Figure 8.43.
2. The statistical standard deviation of the selected data is computed according to Equation (8.24) and defined as the CQI.
3. The CQI provides a health assessment of the CCS; that is, a larger CQI indicates poorer health condition of the CCS,

$$\mathrm{CQI} = \sqrt{\frac{1}{n}\sum_{i=1}^{n}\left(\frac{V_{\mathrm{CH2}}}{V_{\mathrm{CH1}}} - \mu\right)^2} \tag{8.24}$$

where n and μ denote the data size and the mean value of the data between stations, respectively.

The standard deviation is chosen as the CQI metric because it quantifies the dispersion of the measured V_{CH2}/V_{CH1} when the CCS is in electrical contact with the power rail. A larger standard deviation implies a higher degree of fluctuation in the ECQ. In practice, this behavior can be attributed to factors such as uneven wear or damage of the CCS, which lead to increased variation in contact impedance. Therefore, the CQI strongly correlates with the health condition of the CCS.

8.3.3.2 CLI

1. The time duration of momentary electrical contact loss at HSR and DR, as indicated in Figure 8.43, together with the total time duration between stations, is used for processing.
2. The ratio of these time durations is computed according to Equation (8.25) and defined as the CLI.
3. The CLI provides a health assessment of the CCA; a larger CLI indicates that the CCA is degraded and does not hold the CCS firmly (e.g., due to a loosened mechanical spring),

$$\mathrm{CLI} = \frac{T_{\mathrm{loss}}}{T} \tag{8.25}$$

where T_{loss} is the duration of momentary electrical contact loss at HSR and DR, and T is the total duration between stations. The duration of contact loss events at HSR and DR reflects how effectively the CCA maintains mechanical contact pressure on the CCS. The CLI therefore quantifies the frequency and severity of contact loss events relative to the total operating time between stations, providing a meaningful metric for assessing the health condition of the CCA. With the CQI and CLI defined, the health conditions of the CCS and the CCA can be assessed quantitatively and independently.

8.3.4 Experimental Validation

Two controlled field trials were conducted for experimental validation. The first trial aimed to verify the capability of the TBHMS to quantitatively monitor the health condition of the CCS. In this trial, a train equipped with a new CCA was selected. The trial was conducted over several nights, during which measurement data were collected for new CCS, used CCS, and worn CCS, all mounted on the same CCA. Here, the term "worn CCS" refers to a CCS that is due for replacement. For consistency, only the CCS of phase B was replaced with new, used, and worn CCSs in each test run. The second trial aimed to verify the capability of the TBHMS to detect faulty CCA. In this trial, a train was selected in which all CCSs were new. The trial was conducted over several nights, during which measurement data were collected for a worn CCA and a reconditioned CCA. For consistency, the CCA under test was installed at the front of the train, while the CCA at the rear of the train was new. Both field trials were conducted on the same test route along the BPLRT line, which is a fully automated LRT system operating on elevated tracks in the neighborhoods of Bukit Panjang and Choa Chu Kang in Singapore. The BPLRT line comprises a total of 13 LRT stations, namely BP1–BP13. As a feeder service to the MRT network, it connects Choa Chu Kang MRT Station (NS4) on the North-South MRT Line and Bukit Panjang MRT Station (DT1) on the Downtown MRT Line, as shown in Figure 8.44. The trains deployed on the BPLRT line are Bombardier Innovia APM 100 vehicles, which draw electric power from a 600 V, 50 Hz, three-phase AC power rail, as shown in Figure 8.45.

Figure 8.46 plots the CQIs for new, used, and worn CCSs between stations along the complete test route in the first trial, which was conducted to verify the monitoring system's capability in detecting the health condition of the CCS. The results clearly differentiate the health conditions of the CCSs. The new CCS consistently exhibits the lowest CQI, falling below 2.0×10^{-3}, while the used CCS displays a moderate CQI ranging between 2.0×10^{-3} and 6.0×10^{-3}. In contrast, the worn CCS shows the largest CQI, exceeding 8.0×10^{-3}. These results demonstrate

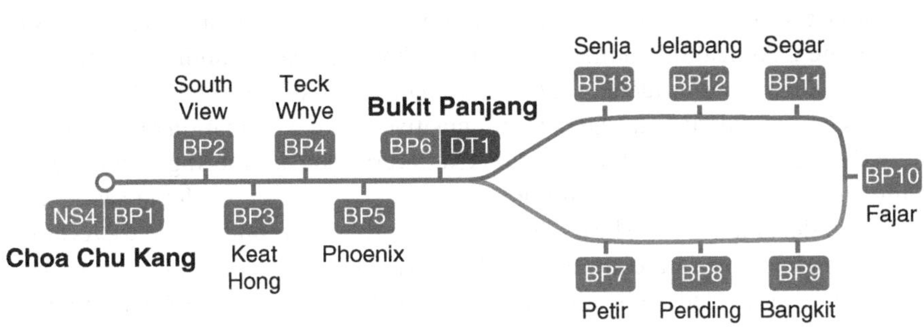

Figure 8.44 Singapore's BPLRT line map.

Figure 8.45 Rubber-tired train deployed in the BPLRT line.

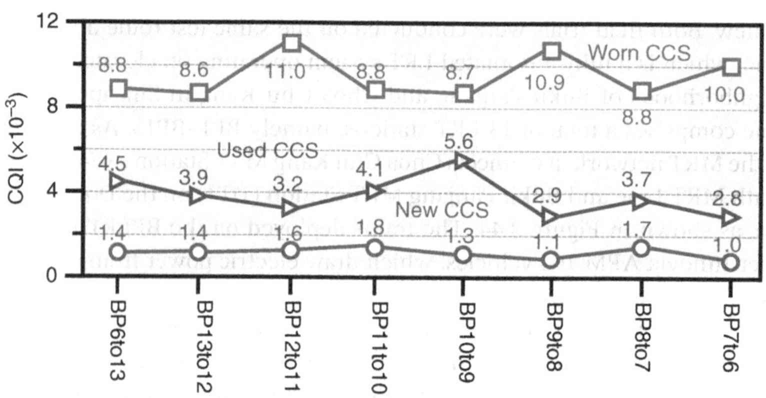

Figure 8.46 CQIs for new, used, and worn CCSs in the first trial.

that the CQI is an effective index for tracking the health condition of the CCS. Figure 8.47 shows the CLIs for new, used, and worn CCSs between stations along the same test route. As observed, only minor variations in CLI are present regardless of the CCS health condition, indicating that the health condition of the CCS has a much smaller influence on CLI. Based on the results shown in Figures 8.46 and 8.47, it is evident that CQI is more sensitive to the degradation of the CCS.

In the second trial, which was conducted to verify the monitoring system's capability in detecting the health condition of the CCA, the CLI was adopted as a quantitative index. Figure 8.48 shows the CLIs for worn and reconditioned CCAs along the test route. The data collected for the worn CCA show that a CLI as high as 4.0×10^{-1} is recorded, indicating that the CCA fails to hold the CCS securely. In contrast, the reconditioned CCA exhibits a significantly lower CLI, with a maximum value of 0.6×10^{-1}, demonstrating that the reconditioned CCA performs

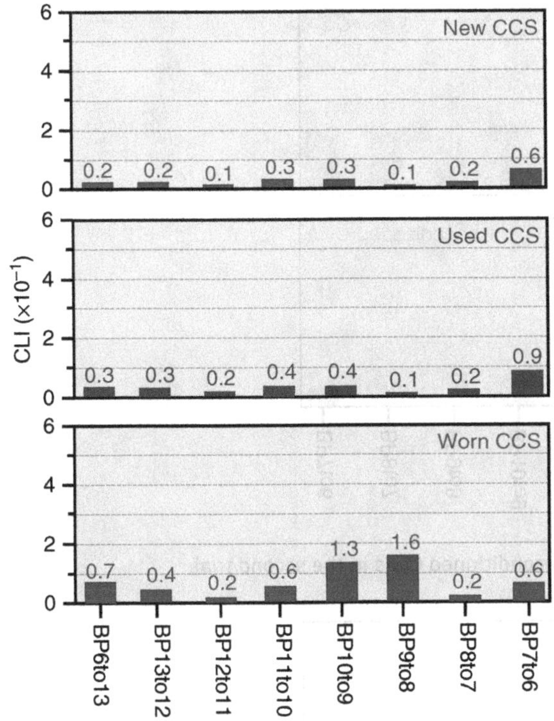

Figure 8.47 CLIs for new, used, and worn CCSs in the first trial.

comparably to a new CCA in firmly holding the CCS. Figure 8.48 thus confirms that CLI is an effective index for differentiating the health condition of the CCA. Figure 8.49 plots the CQIs for worn and reconditioned CCAs along the test route. The results show that the CQIs for worn and reconditioned CCAs are similar, with only minor variations.

Through periodic manual inspections and accumulated experience in parts replacement, the train operator has established an internal maintenance and decision-making process for the replacement of the CCS and CCA. This process is based on physical metric measurements, including the wear rate of the CCS and the sagged vertical distance between the power rail and the CCA, respectively. Worn CCSs and worn CCAs are scheduled for replacement according to the measured wear and sag. Accordingly, the trial data collected for worn CCSs and worn CCAs are used to establish the CQI and CLI thresholds for condition-based maintenance, thereby eliminating the need for manual inspections. With these thresholds, Figure 8.50 illustrates a decision-tree example for maintenance or replacement of the aforementioned CCS and CCA.

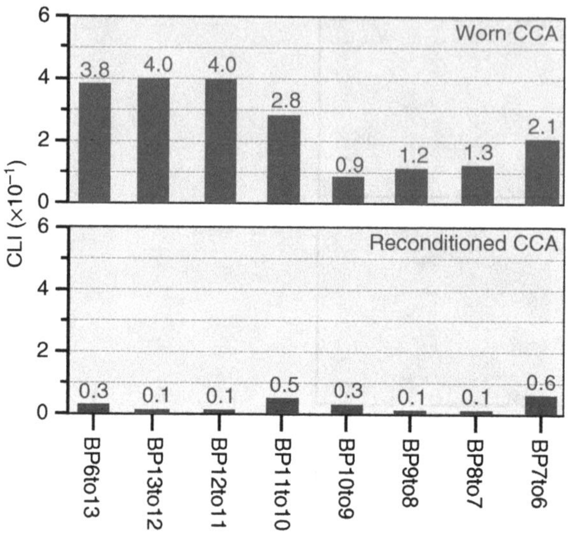

Figure 8.48 CLIs for worn and reconditioned CCAs in the second trial.

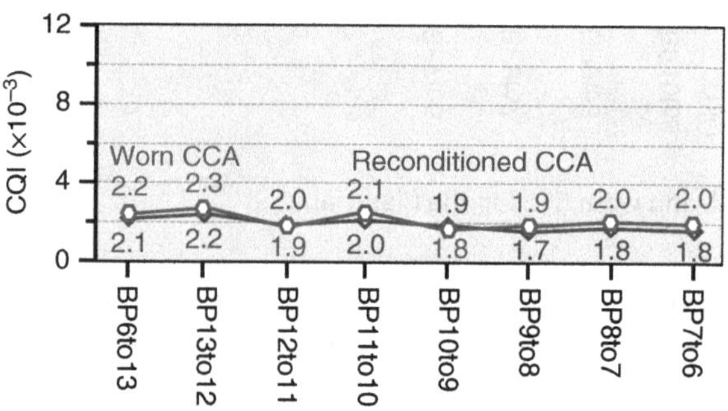

Figure 8.49 CQIs for worn and reconditioned CCAs in the second trial.

This section presents a contactless TBHMS for assessing the health conditions of current collectors in LRT trains powered by a three-phase AC power rail. By leveraging inductive coupling-based online impedance monitoring, the TBHMS enables continuous monitoring of the ECQ between the CCSs and the power rail during normal revenue operation, eliminating the need for manual inspection during engineering hours. Two quantitative performance indices, CQI and CLI, are formulated to separately evaluate the health conditions of the CCS and CCA, respectively. Field trials conducted on an operational LRT line demonstrate that CQI is sensitive to CCS wear, while CLI effectively captures degradation in the mechanical holding capability of the CCA. The experimental results confirm

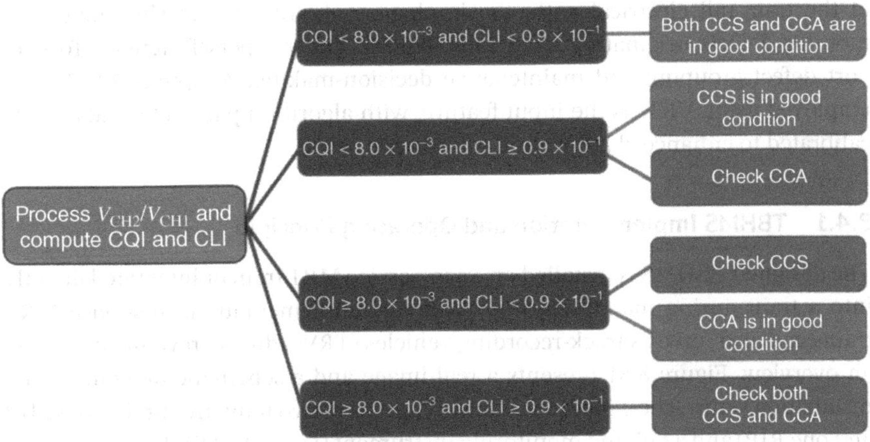

Figure 8.50 A decision tree example employing CQI and CLI for maintenance or replacement of the aforementioned CCS and CCA.

that the TBHMS can reliably distinguish between healthy, degraded, and faulty CCSs and CCAs, supporting condition-based maintenance and informed replacement decisions. Overall, this work establishes a practical and robust framework for nonintrusive and real-time current collector health monitoring in modern LRT systems.

8.4 Defect Detection of Railway Track System for MRT

Railway track systems are vital components of urban mobility and intelligent transportation networks. Defects in railway track systems, such as rails, ballast, and sleepers/fasteners, not only threaten the safety of train operations but also lead to increased maintenance costs and service disruptions [62]. Specifically, due to uneven wheel–rail friction, material fatigue, and temperature fluctuations, various rail defects can occur, including wear, cracks, and corrugation [63]. These defects weaken the load-bearing capacity of rails, making them more susceptible to breakage. Moreover, reduced rail flatness accelerates the aging of MRT components and shortens their service life. Critical defects in ballast include settlement, loss, and fouling, which may result from both operational factors (e.g., vibration) and environmental influences (e.g., precipitation) [64]. As for sleepers and fasteners, typical defects include sleeper degradation, fastener loosening, and elastic failure [65, 66]. Collectively, defects in rails, ballast, sleepers, and fasteners compromise MRT operational safety and reduce passenger comfort [63–66]. Based on the inductive coupling approach, by assessing the voltage health ratio (VHR) of the train-rail electrical path, the implemented TBHMS can simultaneously monitor defects originating from rails, ballast, and sleepers/fasteners. To support defect grouping and maintenance decision-making, Adaptive-DBSCAN is employed using VHR as the input feature, with algorithm parameters adaptively calibrated to enhance classification robustness.

8.4.1 TBHMS Implementation and Operating Principle

The designed TBHMS is installed on an in-service MRT train or integrated directly into a train during manufacturing. With this implementation, in-service MRT trains can also serve as track-recording vehicles (TRVs) during revenue hours. As an overview, Figure 8.51 presents a real image and a schematic diagram of the installation setup. The TBHMS primarily consists of two inductive probes (one IIP and one RIP) and a tailored monitoring instrument (i.e., TD-TPS). The monitoring instrument in the TBHMS contains a signal generator for producing a controllable excitation signal, a DAQ device for collecting the excitation and response signals, a main controller, and a predefined data analysis module for storing the collected data, estimating defect locations using the tachometer signal, and issuing alarms. Furthermore, to ensure a sufficient SNR, a signal amplifier is employed. Accordingly, two signal attenuators (ATT1 and ATT2) are installed at the two channels of the DAQ device (CH1 and CH2), ensuring that the signal levels of both excitation and response remain within the permissible input range. In addition, to prevent transient surges during MRT operation from damaging the monitoring instrument, two surge protectors, labeled SP1 and SP2, are connected at the probe terminals.

In the field trials conducted in this study, the monitoring instrument, amplifier, attenuators, and surge protectors are placed under the train seat within the

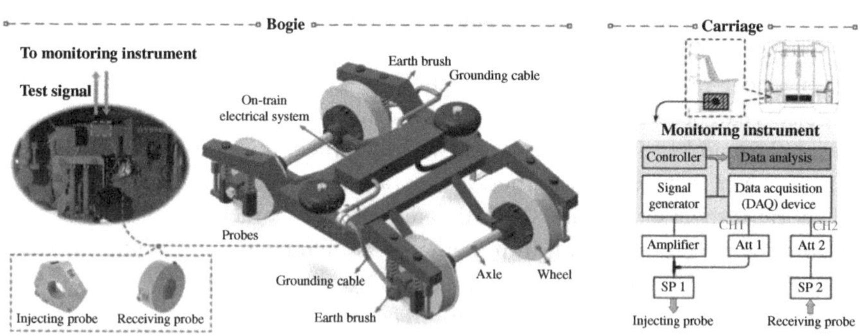

Figure 8.51 Overview of the designed TBHMS for defect detection in an MRT railway track system, including the installation and monitoring topology.

carriage, as illustrated in Figure 8.51. The inductive probes are clamped onto the grounding cable of the on-train electrical system located in the bogie and are connected to the surge protectors via shielded coaxial cables. Consequently, the monitoring train-rail electrical path can be simplified as shown in Figure 8.52, where W_i ($i \in [1,\ 4]$) denotes the wheels in the bogie. Specifically, both paths W1–W2 and W3–W4 are mainly composed of wheel axles, while paths W1–W4 and W2–W3 represent the running rails between the corresponding wheels. Furthermore, the composition of path W1–W3 is relatively complex, containing two earth brushes, the on-train electrical system, and interconnection cables.

All earth brushes are installed in sliding contact with the wheels. Therefore, defects in the rails, ballast, and sleepers/fasteners can degrade the contact quality and subsequently influence the response of the injected test signal by altering the train-rail electrical path. As shown in Figure 8.51, the excitation and response signals are respectively collected by the two channels of the DAQ device (i.e., CH1 and CH2) as the train moves along the MRT railway track system. Consequently, the entire signal path of the designed TBHMS can be represented by an equivalent electrical network based on port-network theory [60], as illustrated in Figure 8.53.

Specifically, the test signal is generated by the signal generator within the monitoring instrument and amplified before injection. The amplified excitation signal is monitored by CH1 of the DAQ device, where ATT1 can be modeled as a two-port network $\boldsymbol{N}_{\mathbf{CH1}}$ characterized by impedance parameters $\left(Z_{11}^{1}, Z_{12}^{1}, Z_{21}^{1}, Z_{22}^{1}\right)$. As shown in Figure 8.53, Port 1 corresponds to the output terminal of the amplifier, with terminal voltage V_A and current I_A. Port 2 corresponds to the input of CH1, where V_{CH1}, I_{CH1}, and Z_{CH1} represent the voltage, current, and internal equivalent impedance of CH1, respectively. The excitation signal then passes through SP1 and is injected into the train-rail electrical path via the injecting probe (i.e., IIP). The response signal is subsequently picked up by the receiving probe (i.e., RIP) and transmitted to CH2 through SP2 and ATT2. This signal collection process can be modeled as a three-port network $\boldsymbol{N}_{\mathbf{CH2}}$, represented by impedance parameters

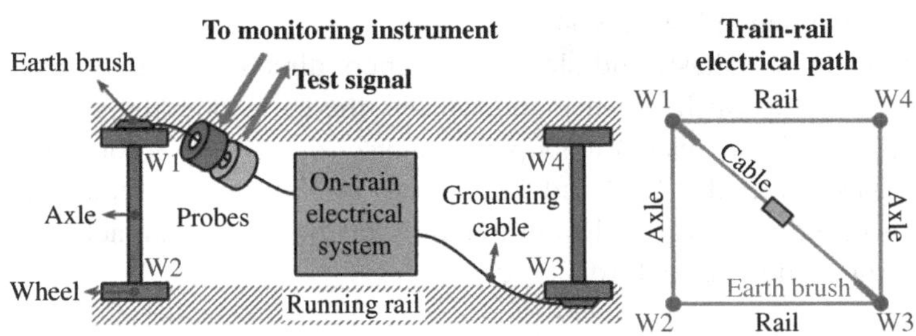

Figure 8.52 Schematic diagram of the train-rail electrical path.

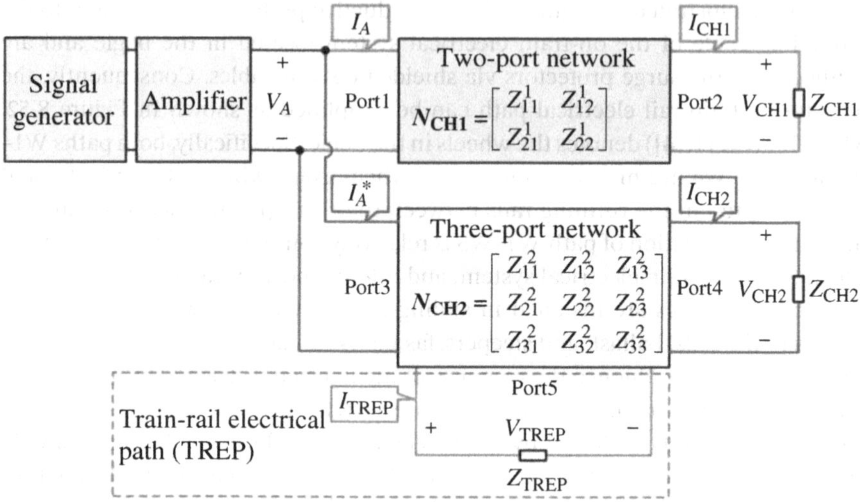

Figure 8.53 Equivalent electrical network representation of the TBHMS.

Z^2_{jk} $(j, k \in [1,\ 3])$. Here, V_A^* and I_A^* denote the voltage and current at Port 3, respectively, while V_{CH2} and I_{CH2} correspond to the voltage and current at CH2 (Port 4), associated with the internal equivalent impedance Z_{CH2}. In addition, V_{TREP} and I_{TREP} at Port 5 denote the voltage and current of the train-rail electrical path, and Z_{TREP} represents the resultant loop impedance of this path.

Based on circuit theory, the train-rail electrical path shown in Figure 8.52 can be simplified to the equivalent circuit in Figure 8.54, in which Z_{TREP} is given by

$$\begin{aligned} Z_{TREP} &= Z_{Contact} + Z_{Cable} + Z_{OTES} \\ &\quad + (Z'_{Wheel} + Z'_{Axle} + Z'_{Rail}) || (Z''_{Wheel} + Z''_{Axle} + Z''_{Rail}) \end{aligned} \tag{8.26}$$

where || denotes the parallel combination, and the definitions of each impedance component are illustrated in Figure 8.54.

Since defects in rails, ballast, and sleepers/fasteners directly affect the contact quality between the earth brushes and wheels, the corresponding contact impedance $Z_{contact}$ and the resultant loop impedance Z_{TREP} vary accordingly. Because the signals collected at CH1 and CH2 of the DAQ device are voltage signals, the VHR is introduced as an indicator to monitor variations in contact quality. The VHR, expressed in decibels (dB), is defined as

$$\text{VHR} = 20 \log\left(\frac{V_{CH2}}{V_{CH1}}\right) \tag{8.27}$$

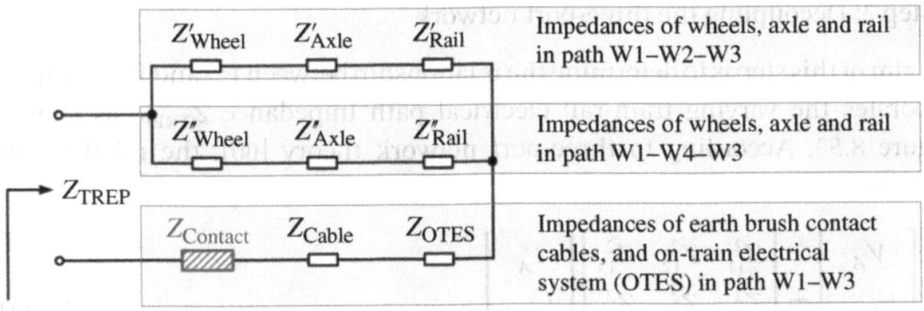

Figure 8.54 Equivalent circuit of the train-rail electrical path, including impedance definition of each component.

To correlate the monitored VHR with variations in the train-rail electrical path impedance Z_{TREP} or the earth-brush contact impedance Z_{contact} for defect detection and classification, the corresponding derivation is presented as follows:

- Step 1: Decoupling the two-port network

The objective of this step is to determine the relationship between V_A and V_{CH1}. Based on two-port network theory and the definition of the impedance matrix [60], the relationship can be expressed as

$$\begin{bmatrix} V_A \\ V_{\text{CH1}} \end{bmatrix} = \begin{bmatrix} Z_{11}^1 & Z_{12}^1 \\ Z_{21}^1 & Z_{22}^1 \end{bmatrix} \begin{bmatrix} I_A \\ I_{\text{CH1}} \end{bmatrix} \tag{8.28}$$

The expanded form of Equation (8.28) is given by

$$V_A = Z_{11}^1 I_A + Z_{12}^1 I_{\text{CH1}} \tag{8.29}$$

$$V_{\text{CH1}} = Z_{21}^1 I_A + Z_{22}^1 I_{\text{CH1}} \tag{8.30}$$

By substituting Equation (8.30) into Equation (8.29) and applying Ohm's Law (i.e., $I_{\text{CH1}} = V_{\text{CH1}}/Z_{\text{CH1}}$), the two-port network $\boldsymbol{N}_{\textbf{CH1}}$ can be decoupled, yielding the following relationship between V_A and V_{CH1}:

$$\frac{V_A}{V_{\text{CH1}}} = \alpha \tag{8.31}$$

where

$$\alpha = \frac{Z_{11}^1}{Z_{21}^1} \cdot \left(1 - \frac{Z_{22}^1}{Z_{\text{CH1}}}\right) + \frac{Z_{12}^1}{Z_{\text{CH1}}} \tag{8.32}$$

Since Z_{11}^1, Z_{12}^1, Z_{21}^1, Z_{22}^1, and Z_{CH1} remain constant at a specific frequency for a given TBHMS configuration, the coefficient α is also invariant.

- Step 2: Decoupling the three-port network

The aim of this step is to determine the relationship between V_A and V_{CH2}, which incorporates the varying train-rail electrical path impedance Z_{TREP}, as shown in Figure 8.53. According to three-port network theory [60], the relationship is given by

$$\begin{bmatrix} V_A \\ V_{\mathrm{CH2}} \\ V_{\mathrm{TREP}} \end{bmatrix} = \begin{bmatrix} Z_{11}^2 & Z_{12}^2 & Z_{13}^2 \\ Z_{21}^2 & Z_{22}^2 & Z_{23}^2 \\ Z_{31}^2 & Z_{32}^2 & Z_{33}^2 \end{bmatrix} \begin{bmatrix} I_A^* \\ I_{\mathrm{CH2}} \\ I_{\mathrm{TREP}} \end{bmatrix} \tag{8.33}$$

The expanded form of Equation (8.33) is

$$V_A = Z_{11}^2 I_A^* + Z_{12}^2 I_{\mathrm{CH2}} + Z_{13}^2 I_{\mathrm{TREP}} \tag{8.34}$$

$$V_{\mathrm{CH2}} = Z_{21}^2 I_A^* + Z_{22}^2 I_{\mathrm{CH2}} + Z_{23}^2 I_{\mathrm{TREP}} \tag{8.35}$$

$$V_{\mathrm{TREP}} = Z_{31}^2 I_A^* + Z_{32}^2 I_{\mathrm{CH2}} + Z_{33}^2 I_{\mathrm{TREP}} \tag{8.36}$$

Applying Ohm's Law, $I_{\mathrm{CH2}} = V_{\mathrm{CH2}}/Z_{\mathrm{CH2}}$ and $I_{\mathrm{TREP}} = V_{\mathrm{TREP}}/Z_{\mathrm{TREP}}$. Substituting these expressions into Equations (8.34)–(8.36) and dividing both sides by V_{CH2}, the equations can be reformulated as

$$\frac{V_A}{V_{\mathrm{CH2}}} = \frac{Z_{11}^2 I_A^*}{V_{\mathrm{CH2}}} + \frac{Z_{12}^2}{Z_{\mathrm{CH2}}} + \frac{Z_{13}^2 V_{\mathrm{TREP}}}{V_{\mathrm{CH2}} Z_{\mathrm{TREP}}} \tag{8.37}$$

$$1 = \frac{Z_{21}^2 I_A^*}{V_{\mathrm{CH2}}} + \frac{Z_{22}^2}{Z_{\mathrm{CH2}}} + \frac{Z_{23}^2 V_{\mathrm{TREP}}}{V_{\mathrm{CH2}} Z_{\mathrm{TREP}}} \tag{8.38}$$

$$\frac{V_{\mathrm{TREP}}}{V_{\mathrm{CH2}}} = \frac{Z_{31}^2 I_A^*}{V_{\mathrm{CH2}}} + \frac{Z_{32}^2}{Z_{\mathrm{CH2}}} + \frac{Z_{33}^2 V_{\mathrm{TREP}}}{V_{\mathrm{CH2}} Z_{\mathrm{TREP}}} \tag{8.39}$$

From Equation (8.38), I_A^*/V_{CH2} can be expressed as

$$\frac{I_A^*}{V_{\mathrm{CH2}}} = \frac{1}{Z_{21}^2} \cdot \left(1 - \frac{Z_{22}^2}{Z_{\mathrm{CH2}}} - \frac{Z_{23}^2 V_{\mathrm{TREP}}}{V_{\mathrm{CH2}} Z_{\mathrm{TREP}}}\right) \tag{8.40}$$

Substituting Equation (8.40) into Equation (8.39), we obtain

$$\frac{V_{\mathrm{TREP}}}{V_{\mathrm{CH2}}} = \frac{\frac{Z_{31}^2}{Z_{21}^2} \cdot \left(1 - \frac{Z_{22}^2}{Z_{\mathrm{CH2}}}\right) + \frac{Z_{32}^2}{Z_{\mathrm{CH2}}}}{1 + \frac{Z_{31}^2 Z_{23}^2}{Z_{21}^2 Z_{\mathrm{TREP}}} - \frac{Z_{33}^2}{Z_{\mathrm{TREP}}}} \tag{8.41}$$

Then, by combining Equations (8.40) and (8.41) and substituting them into Equation (8.37), the final expression of V_A/V_{CH2} can be derived as

$$\frac{V_A}{V_{\text{CH2}}} = \beta_1 - \frac{\beta_2}{Z_{\text{TREP}} + \beta_3} \tag{8.42}$$

$$\beta_1 = \frac{Z_{11}^2}{Z_{21}^2} \cdot \left(1 - \frac{Z_{22}^2}{Z_{\text{CH2}}}\right) + \frac{Z_{12}^2}{Z_{\text{CH2}}} \tag{8.43}$$

$$\beta_2 = \left(\frac{Z_{11}^2 Z_{23}^2}{Z_{21}^2} - Z_{13}^2\right) \cdot \left[\frac{Z_{31}^2}{Z_{21}^2} \cdot \left(1 - \frac{Z_{22}^2}{Z_{\text{CH2}}}\right) + \frac{Z_{23}^2}{Z_{\text{CH2}}}\right] \tag{8.44}$$

$$\beta_3 = \frac{Z_{31}^2 Z_{23}^2}{Z_{21}^2} - Z_{33}^2 \tag{8.45}$$

Since all Z_{jk}^2 ($j, k \in [1, 3]$) and Z_{CH2} remain constant at a given frequency for a specific TBHMS, the parameters β_1, β_2, and β_3 are also invariant. Consequently, the ratio V_A/V_{CH2} is directly correlated with the train-rail electrical path impedance Z_{TREP}.

- Step 3: Calculation of VHR

According to the definition of VHR in Equation (8.27), it can be calculated using V_A/V_{CH1} and V_A/V_{CH2}, as given in Equations (8.31) and (8.42), respectively. Therefore, VHR is expressed as

$$\begin{aligned} \text{VHR} &= 20 \log\left(\frac{V_{\text{CH2}}}{V_{\text{CH1}}}\right) = 20 \log\left(\frac{V_A}{V_{\text{CH1}}} \cdot \frac{V_{\text{CH2}}}{V_A}\right) \\ &= 20 \log \alpha - 20 \log\left(\beta_1 - \frac{\beta_2}{Z_{\text{TREP}} + \beta_3}\right) \end{aligned} \tag{8.46}$$

where α is defined in Equation (8.32), and β_1, β_2, and β_3 are defined in Equations (8.43)–(8.45). This expression explicitly shows that VHR is a function of the train-rail electrical path impedance Z_{TREP}.

When a train passes over defects in the railway track system, the contact quality between the earth brushes and the wheels is affected, leading to variations in Z_{contact} and consequently in Z_{TREP}. According to Equation (8.46), such variations directly result in measurable changes in VHR, demonstrating the theoretical effectiveness of the TBHMS.

As illustrated in Figure 8.51, the designed TBHMS interacts with the train-rail electrical path via two inductive probes. Owing to their snap-on design, these probes can be clamped onto the grounding cables without disassembling the bogie, thereby simplifying on-site deployment. Moreover, the contactless measurement approach enhances electrical safety by eliminating direct electrical connections. Since the test signal is both injected and received through inductive coupling, the TBHMS is capable of monitoring VHR for defect detection while remaining insensitive to wheel–rail rolling noise. In addition, environmental factors such as cleanliness, temperature, precipitation, visibility, and obstructions have negligible influence on the contact impedance Z_{contact} and the resulting Z_{TREP}. This indicates the strong environmental robustness and practical applicability of the TBHMS developed from the TD-TPS.

Although defects in railway track systems can be detected through abnormal VHR values, classification based solely on direct observation remains challenging. To address this limitation, Section 8.4.2 introduces an Adaptive-DBSCAN algorithm for defect classification of rails, ballast, and sleepers/fasteners.

8.4.2 Defects Classification Using Adaptive-DBSCAN

To facilitate defect classification using the monitored VHR data, an Adaptive-DBSCAN algorithm is developed. This algorithm is derived from the conventional DBSCAN method and introduces adaptive functionality by predefining analytical expressions for its two key parameters, namely the neighborhood radius ϵ and the minimum number of points minPts.

DBSCAN is a density-based clustering algorithm that identifies clusters by searching for regions of high data density while treating sparsely distributed points as noise [67]. Compared with partition-based clustering methods, DBSCAN does not require a predefined number of clusters and is capable of identifying clusters with arbitrary shapes, making it particularly suitable for railway track defect classification where defect patterns are irregular and data labeling is difficult. To illustrate the fundamental principles of DBSCAN, the input dataset, denoted as $\mathfrak{D}$ (i.e., the collected VHR data), is introduced.

The key definitions underpinning DBSCAN are summarized as follows:

- Definition 1: ϵ-Neighborhood

  For a data point $P_l \in \mathfrak{D}$, its ϵ-neighborhood $N_\epsilon(P_l)$ is defined as the set of data points whose distances from P_l are not greater than the predefined neighborhood radius ϵ, as illustrated in Figure 8.55a. Formally, it is expressed as

$$N_\epsilon(P_l) = \{P_m \in \mathcal{D} | \text{dist}(P_l, P_m) \leq \epsilon\} \tag{8.47}$$

  where dist(·) denotes the distance metric, and P_m represents another data point in the dataset $\mathfrak{D}$.

- Definition 2: Point classes

  In DBSCAN, each data point P (i.e., a VHR sample) is classified into one of three categories: core point, border point, or noise point, as shown in Figure 8.55b. Specifically, if the number of points in the ϵ-neighborhood of

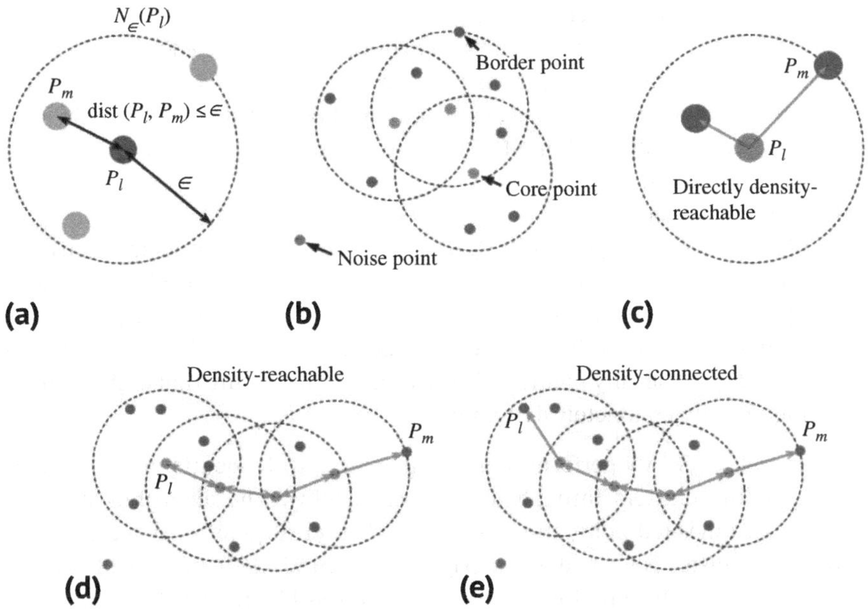

Figure 8.55 Key definitions in DBSCAN. (a) ϵ-Neighborhood. (b) Point classes. (c) Directly density-reachable. (d) Density-reachable. (e) Density-connected.

P satisfies $|N_\epsilon(P)| \geq \text{minPts}$, P is classified as a core point, where minPts is a user-defined dense region. If P is not a core point but lies within the ϵ-neighborhood of a core point, it is classified as a border point. Otherwise, P is regarded as a noise point. This classification mechanism enables effective separation of noise from meaningful defect-related patterns.

- Definition 3: Directly density-reachable

  A data point P_m is said to be directly density-reachable from a core point P_l with respect to ϵ if

$$\begin{cases} |N_\epsilon(P_l)| \geq \text{minPts} \\ P_m \in N_\epsilon(P_l) \end{cases} \tag{8.48}$$

  This definition is utilized in cluster determination, and Figure 8.55c illustrates a diagrammatic representation.

- Definition 4: Density-reachable

  A data point P_m is considered density-reachable from another point P_l if there exists a sequence of points $\{P_1, P_2, \dots, P_N\}$ in $\mathcal{D}$ such that $P_1 = P_l$ and $P_N = P_m$, and each point P_n (for $n \in [2, N]$) is directly density-reachable from P_{n-1}. This concept is illustrated in Figure 8.55d.

- Definition 5: Density-connected

  Two data points P_l and P_m are said to be density-connected if there exists at least one data point that is density-reachable from both P_l and P_m. As illustrated in Figure 8.55e, density connectivity allows clusters to extend beyond direct reachability, enabling DBSCAN to identify clusters with complex and arbitrary shapes.
- Definition 6: Cluster

  A cluster $\boldsymbol{C}_x(x \in [1, N])$ in DBSCAN is defined as a nonempty subset of $\mathfrak{D}$ that satisfies both maximality and connectivity. Specifically, maximality requires that if a data point $P_m \in \mathfrak{D}$ is density-reachable from any point $P_l \in \boldsymbol{C}_x$, then P_m must also belong to $\boldsymbol{C}_x$. Connectivity requires that any two points within $\boldsymbol{C}_x$ are density-connected. These properties ensure that each cluster is uniquely and completely determined.

In general, DBSCAN performs clustering by iteratively identifying core points and expanding clusters through density reachability. The algorithm starts by selecting an unvisited point and evaluating whether it qualifies as a core point. If so, a new cluster is formed and expanded by recursively including all density-reachable points. Once no further expansion is possible, the cluster is finalized, and the algorithm proceeds to process the remaining unvisited points. Data points that are not assigned to any cluster are labeled as noise.

Since DBSCAN is an unsupervised method, it does not require predefined labels, making it suitable for railway track system applications where data acquisition and labeling are difficult. Additionally, DBSCAN is particularly effective at identifying clusters of arbitrary shapes and distributions, and thus facilitates the defect classification of rails, ballast, and sleepers/fasteners, exhibiting significant value for intelligent transportation. Considering that ϵ and $\min\text{Pts}$ are two key parameters in realizing DBSCAN, optimizing their selection is of great importance.

In the algorithm, ϵ represents the neighborhood radius that determines which points are considered close enough to be grouped together; $\min\text{Pts}$ denotes the threshold to define a core point for cluster formation. Typically, there are two ways to determine their values: the empirical approach and automated retrieval. In the empirical paradigm, ϵ and $\min\text{Pts}$ are identified based on the rule of thumb, introducing inherent uncertainties that can affect classification accuracy. As an alternative, automated retrieval iterates over various values of ϵ and $\min\text{Pts}$ to execute DBSCAN and evaluate performance using relevant cluster validity indices, such as the Silhouette Coefficient or the Davies–Bouldin index. However, the complexity and time-consuming nature of this retrieval process compromise its practicality.

Given these limitations associated with ϵ and minPts, this subsection introduces an adaptive parameter determination method, expressed as

$$\epsilon = \frac{\sqrt{\frac{1}{N_{\mathcal{D}}} \cdot \sum_{y=1}^{N_{\mathcal{D}}} \left(\text{VHR}_y - \text{VHR}_{\text{avg}}\right)^2}}{\text{VHR}_{\text{avg}}} \tag{8.49}$$

$$\text{minPts} = \left\lfloor \frac{N_{\mathcal{D}}}{d_m} \cdot l_b \right\rfloor \tag{8.50}$$

where $N_{\mathcal{D}}$ denotes the length of dataset $\mathcal{D}$ with VHR as inputs. VHR_y ($y \in [1, N_{\mathcal{D}}]$) represents an individual data point in $\mathcal{D}$, and VHR_{avg} is their corresponding average value. $\lfloor \cdot \rfloor$ denotes a floor function. d_m denotes the distance traveled by the train, and l_b represents the distance between the center points of the front and rear wheels of the bogie.

Specifically, the expression of ϵ listed in Equation (8.49) represents the coefficient of variation, which is widely used in the financial industry and experimental science to quantify the discreteness within datasets. By adopting this approach, the designed ϵ can adapt to data volatility with good robustness, effectively preventing over- or under-fitting during manual parameter adjustment. As for minPts, it is formulated based on the physical properties of the TBHMS. $N_{\mathcal{D}}/d_m$ represents the number of acquired data points per meter, while l_b determines the minimum impact distance of a defect location. Using Equation (8.50), minPts becomes adaptable to various working conditions, including different sections and sampling strategies.

Consequently, the Adaptive-DBSCAN algorithm enables defect detection and classification of railway track systems in an easy-to-implement fashion. For illustration, the pseudocode is presented in Table 8.6.

8.4.3 Experimental Validation

Through field trials conducted in Singapore, typical voltage signals collected at CH2 (V_{CH2}) are illustrated in Figure 8.56a. Specifically, Tests 1–3 were carried out on different days under varying weather conditions, including a midnight period with intense rainfall, a cloudy midnight, and a sunny daytime condition. The horizontal axis represents the distance traveled by the train under test, which is determined using the built-in tachometer of the train. The monitored route covers an approximate distance of 1.35 km.

The vertical axis in Figure 8.56a denotes the detected voltage V_{CH2} at CH2 of the TBHMS, which fluctuates between several hundred microvolts and tens of millivolts. Since the detected voltage V_{CH1} at CH1 of the TBHMS remains essentially constant at 1 V, any variation in V_{CH2} directly indicates changes in the train-rail

Table 8.6 Adaptive-DBSCAN algorithm.

Step	Algorithm Adaptive-DBSCAN
1	**Dataset construction:** Collect VHR information to form $\mathscr{D}$
2	**Input:** Dataset $\mathscr{D}$ with $\boldsymbol{N}_{\mathscr{D}}$ points
3	**Initialization:** Calculate ϵ based on (8.49) Calculate minPts based on (8.50). Set all points in $\mathscr{D}$ as unvisited Set ClusterID $\mathscr{C}_x$ to $\mathscr{C}_1$
4	**For** each unvisited point $\boldsymbol{P}$ in $\mathscr{D}$
	Mark $\boldsymbol{P}$ as visited
	Find the neighborhood N_ϵ of $\boldsymbol{P}$
	If $\lvert N_\epsilon \rvert <$ minPts
	Mark $\boldsymbol{P}$ as the noncore point
	Else
	Identify $\boldsymbol{P}$ as the core point
	Assign current ClusterID $\mathscr{C}_x$ to $\boldsymbol{P}$
	ExpandCluster ($\boldsymbol{P}$, N_ϵ, $\mathscr{C}_x$, ϵ, minPts) via density-connected
	Mark all points in this cluster as visited
	Update ClusterID via $\mathscr{C}_x :=$ nextId($\mathscr{C}_x$)
	End
	End
5	**For** each noncore point $\boldsymbol{Q}$ in $\mathscr{D}$
	If $Q \in \forall\mathscr{C}_x$
	Identify $\boldsymbol{Q}$ as border point
	Else
	Identify $\boldsymbol{Q}$ as noise point
	End
	End
6	**Output:** ClusterID or Cluster labels $\mathscr{C}_x$

electrical path impedance Z_{TREP} and, consequently, the VHR. Based on Equation (8.27), the VHRs for the three tests are calculated from the collected voltage signals and are presented in Figure 8.56b–d, together with their corresponding defect distributions. Since the field trials were conducted on different days, slight differences in the starting positions of the train and in the health conditions of the track are observed. Considering the worst-case scenario, namely, the night with heavy rainfall, a detailed defect detection and classification analysis of Test 1 is selected for further evaluation.

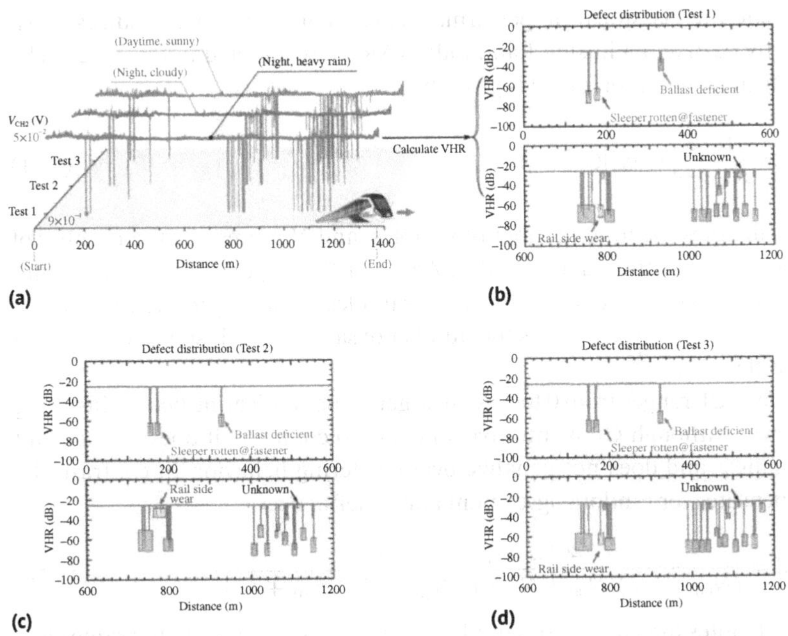

Figure 8.56 (a) Collected signal V_{CH2} on different days under varying weather conditions. (b) Defect distributions and calculated VHR versus distance using the worst case considering the night with heavy rain. (c) Defect distributions and calculated VHR versus distance on a cloudy night. (d) Defect distributions and calculated VHR versus distance on a sunny day.

Figure 8.56b shows the defect distribution together with the calculated VHR for Test 1. The defect information and ground-truth distribution are provided by the SMRT Corporation and mainly include four categories: (a) sleeper rot or fastener issues, (b) ballast deficiency, (c) rail side wear, and (d) unknown defects. Representative images of these common defect types can be found in [63–66]. The unknown defects (e.g., track deformation in this study) correspond to fault types that are not explicitly categorized, which is consistent with practical diagnosis and maintenance scenarios. Considering operational noise, the collected signals can be grouped into five characteristic types. The VHR data shown in Figure 8.56b are subsequently imported into the Adaptive-DBSCAN algorithm for verification, with an input signal length of 10,300 samples. Given that the route length is approximately 1.35 km (d_m) and the distance between the front and rear wheels is $l_b = 2.5$ m for the train under test, the algorithm parameters ϵ and minPts are determined to be 0.14 and 190, respectively.

To evaluate the classification performance, two commonly used indices, clustering purity (CP) and adjusted Rand index (ARI), are adopted in this study. The mathematical formulation of CP is given by

$$\mathrm{CP} = \frac{1}{\boldsymbol{N_\mathcal{D}}} \sum_{i=1}^{N} \max_j \left| C_i \cap T_j \right| \tag{8.51}$$

where $\boldsymbol{N_D}$ denotes the total number of samples, and N represents the number of clusters generated by the Adaptive-DBSCAN algorithm. C_i denotes the set of samples in cluster i, while T_j denotes the set of samples belonging to the true class j. Consequently, $\left|C_i \cap T_j\right|$ represents the number of samples in cluster C_i that belong to the true class T_j.

The value of CP ranges from 0 to 1, with larger values indicating better clustering performance. Although CP is intuitive and easy to compute, it does not account for randomness and does not penalize over-clustering behavior. In contrast, the ARI compensates for random agreement and is defined as

$$\mathrm{ARI} = \frac{2(N_{00}N_{11} - N_{01}N_{10})}{(N_{00} + N_{01})(N_{00} + N_{10}) + (N_{10} + N_{11})(N_{01} + N_{11})} \tag{8.52}$$

where N_{00} denotes the number of sample pairs that do not belong to the same cluster and do not share the same category in both the clustering results and the ground truth labels. N_{11} represents the number of sample pairs that belong to the same cluster and the same category. N_{01} denotes the number of sample pairs belonging to the same category but assigned to different clusters, while N_{10} represents the opposite case. The ARI value ranges from −1 to 1, with values closer to 1 indicating better clustering performance.

All experiments were conducted on a laptop equipped with an Intel® Core™ i7-7700HQ processor and 16 GB of RAM, using MATLAB 2021b. The processing time for each run was approximately 20 seconds. For the given VHR dataset and the Adaptive-DBSCAN algorithm, the average CP and ARI were found to be 98.57% and 0.81, respectively. To verify the effectiveness of parameter selection for ϵ and minPts in DBSCAN, a series of comparative experiments was conducted. Specifically, ϵ was varied among 0.1, 0.2, 0.3, 0.4, and 0.5, while minPts was selected from 50, 100, 150, 200, and 250. The corresponding results are summarized in Tables 8.7 and 8.8.

In addition, an exhaustive evaluation was performed by sweeping ϵ over 100 unevenly spaced values in the range [0.01, 1] and minPts over 100 values in the range [5, 205], resulting in a total of 20,000 parameter combinations. This exhaustive process required more than four days of computation.

Table 8.7 Comparison of CP under different values of ϵ and minPts.

CP		ϵ				
		0.1 (%)	**0.2 (%)**	**0.3 (%)**	**0.4 (%)**	**0.5 (%)**
minPts	50	98.57	98.53	98.47	98.46	98.43
	100	98.58	98.58	98.56	98.52	98.50
	150	98.55	98.59	98.57	98.56	98.54
	200	98.53	98.59	98.58	98.57	98.55
	250	98.51	98.59	98.59	98.58	98.56

Table 8.8 Comparison of ARI under different values of ϵ and minPts.

ARI		ϵ				
		0.1	**0.2**	**0.3**	**0.4**	**0.5**
minPts	50	0.78	0.74	0.72	0.70	0.69
	100	0.80	0.78	0.77	0.74	0.73
	150	0.81	0.79	0.77	0.77	0.75
	200	0.80	0.80	0.78	0.77	0.76
	250	0.80	0.80	0.79	0.78	0.77

The results indicate that the best achievable classification performance under exhaustive search is 98.59% for CP and 0.81 for ARI. The proposed Adaptive-DBSCAN achieves nearly identical performance while significantly reducing computational complexity and time consumption. These findings confirm that the adaptive parameter selection strategy effectively eliminates the need for exhaustive parameter traversal.

To further validate the superiority of the Adaptive-DBSCAN algorithm, several classic unsupervised clustering methods, including K-means, DPC, GMM, and agglomerative clustering, were employed for comparison. The classification results are summarized in Table 8.9, using VHR data from the three tests shown in Figure 8.56b–d. To mitigate random variation, all clustering results were averaged over 10 independent runs. The comparative analysis demonstrates that Adaptive-DBSCAN consistently achieves the highest CP and ARI among all evaluated methods, highlighting its superior performance in railway track system defect classification. Moreover, the method exhibits stable CP and ARI values across different weather conditions (Tests 1–3), further confirming the robustness and reliability of the proposed approach.

Table 8.9 Classification performance comparison of different clustering methods using CP and ARI.

Method	CP			ARI		
	Test 1 (%)	Test 2 (%)	Test 3 (%)	Test 1	Test 2	Test 3
K-means	87.68	81.70	87.34	0.48	0.63	0.68
DPC	41.79	35.12	35.05	0.01	0.01	0.01
GMM	72.16	83.86	76.20	0.09	0.11	0.08
Agglomerative	79.08	69.05	74.45	0.06	0.01	0.03
Adaptive-DBSCAN	98.57	98.78	99.23	0.81	0.79	0.88

This section presents a noncontact TBHMS for detecting and classifying defects in MRT railway track systems, including rails, ballast, and sleepers/fasteners. By injecting and sensing RF signals through inductive coupling, the TBHMS monitors variations in the train-rail electrical path impedance, which are quantified using the VHR. A detailed circuit and network-based analysis establishes the theoretical relationship between VHR and track defects, demonstrating the sensitivity of the presented method to changes in contact quality caused by structural degradations. To enable robust defect classification under varying operating and environmental conditions, an Adaptive-DBSCAN algorithm is introduced, featuring data-driven and physically informed parameter selection for $\epsilon$and minPts. Field trials conducted on an operational MRT line in Singapore validate the effectiveness, robustness, and environmental adaptability of this approach. Overall, this section demonstrates a practical, scalable, and intelligent solution for continuous railway track condition monitoring and defect diagnosis.

8.5 Chapter Summary

This chapter demonstrates the versatility and practical significance of electrically contactless online impedance measurement in transport electrification. Through the systematic design of EMI filters for motor drive systems, the presented method enables quantitative prediction of attenuation performance and reveals critical phenomena, such as DM-to-CM conversion, that are difficult to capture using conventional trial-and-error approaches. In motor health monitoring applications, online CM impedance tracking provides a robust and nonintrusive solution for early detection of stator ITSC faults, exhibiting strong immunity to variations in load, speed, rotor condition, and bearing health. Beyond motor drives, the chapter extends online impedance-based monitoring concepts to rail transit systems.

Individual TBHMS is developed to assess LRT current collector integrity and MRT railway track conditions during normal revenue service. Collectively, these case studies confirm that electrically contactless online impedance measurement, or its extended forms, is not only a powerful diagnostic and design tool but also a scalable sensing paradigm capable of enhancing reliability, safety, and maintenance efficiency in next-generation transportation systems.

References

1 J. Xue and F. Wang, "Modeling and design of common-mode inductor for conductive EMI noise suppression in DC-fed motor drive system," in *Proc. IEEE Energy Conver. Congr. Expo. (ECCE)*, USA, 2012, pp. 645–651.

2 C. R. Sullivan and A. Muetze, "Simulation model of common-mode chokes for high-power applications," *IEEE Trans. Ind. Appl.*, vol. 46, no. 2, pp. 884–891, Feb. 2010.

3 D. Han, C. Morris, W. Lee et al., "Determination of CM choke parameters for SiC MOSFET motor drive based on simple measurements and frequency domain modeling," in *Proc. IEEE Applied Power Electron. Conf. Expo. (APEC)*, USA, 2016, pp. 2861–2867.

4 D. Liu and X. Jiang, "High frequency model of common mode inductor for EMI analysis based on measurements," in *Proc. 3rd Int. Symp. Electromag. Compat.*, China, 2002, pp. 462–465.

5 I. Boldea and S. A. Nasar, *Electric drives*. Boca Raton, FL, USA: CRC Press, 2016.

6 F. Luo, S. Wang, F. Wang et al., "Analysis of CM volt-second influence on CM inductor saturation and design for input EMI filters in three-phase DC-fed motor drive systems," *IEEE Trans. Power Electron.*, vol. 25, no. 7, pp. 1905–1914, 2010.

7 L. Xing, M. Shen, M. Li et al., "Common mode choke optimization for three-phase motor drive systems," in *Proc. 2012 IEEE Energy Conver. Congr. Expo. (ECCE)*, IEEE, 2012, pp. 1399–1405.

8 S. Negri, G. Spadacini, F. Grassi et al., "Inductively coupled in-circuit measurement of two-port admittance parameters," *IEEE Trans. Ind. Electron.*, vol. 71, no. 10, pp. 13351–13360, Feb. 2024.

9 A. Muetze and C. R. Sullivan, "Simplified design of common-mode chokes for reduction of motor ground currents in inverter drives," *IEEE Trans. Ind. Appl.*, vol. 47, no. 6, pp. 2570–2577, 2011.

10 S. Ogasawara and H. Akagi, "Modeling and damping of high-frequency leakage currents in PWM inverter-fed AC motor drive systems," *IEEE Trans. Ind. Appl.*, vol. 32, no. 5, pp. 1105–1114, 1996.

11 A. Roc'h and F. Leferink, "Nanocrystalline core material for high-performance common mode inductors," *IEEE Trans. Electromagn. Compat.*, vol. 54, no. 4, pp. 785–791, 2012.

12 Y. Yoshizawa, S. Oguma, and K. Yamauchi, "New Fe-based soft magnetic alloys composed of ultrafine grain structure," *J. Appl. Phys.*, vol. 64, no. 10, pp. 6044–6046, 1988.

13 Vacuumschmelze. (2011). *Nanocrystalline material in common mode chokes.* Available from: https://www.vacuumschmelze.com.

14 Vacuumschmelze. (2016). *EMC products based on nanocrystalline vitroperm.* Available from: https://www.mouser.com/pdfdocs/VACChokesandCoresDatasheet.pdf.

15 M. L. Heldwein, L. Dalessandro, and J. W. Kolar, "The three-phase common-mode inductor: modeling and design issues," *IEEE Trans. Ind. Electron.*, vol. 58, no. 8, pp. 3264–3274, 2011.

16 D. Han, C. T. Morris, W. Lee et al., "Three-phase common mode inductor design and size minimization," in *Proc. 2016 IEEE* Trans. Electr. *Conf. Expo (ITEC)*, IEEE, 2016, pp. 1–8.

17 I. Stevanovic, S. Skibin, M. Masti et al., "Behavioral modeling of chokes for EMI simulations in power electronics," *IEEE Trans. Power Electron.*, vol. 28, no. 2, pp. 695–705, 2013.

18 M. K. Jutty, V. Swaminathan, and M. K. Kazimierczuk, "Frequency characteristics of ferrite core inductors," in *Proc. Electr./Electron. Insulation Conf.*, 1993, pp. 369–372.

19 A. Massarini and M. K. Kazimierczuk, "Self-capacitance of inductors," *IEEE Trans. Power Electron.*, vol. 12, no. 4, pp. 671–676, 1997.

20 H. Chen, Z. Qian, S. Yang et al., "Finite-element modeling of saturation effect excited by differential-mode current in a common-mode choke," *IEEE Trans. Power Electron.*, vol. 24, no. 3, pp. 873–877, 2009.

21 J. Xue, F. Wang, and B. Guo, "EMI noise mode transformation due to propagation path unbalance in three-phase motor drive system and its implication to EMI filter design," in *Proc. 2014 IEEE Applied Power Electron. Conf. Expo.*, 2014, pp. 806–811.

22 D. Bellan, "Mode transformation of EMI noise due to unbalanced filter capacitors in three-phase motor drive systems," in *Proc. 43rd Annual Conf. IEEE Ind. Electron. Soc.*, 2017, pp. 7024–7027.

23 D. Drozhzhin, G. Griepentrog, A. Sauer et al., "Investigation on differential to common mode coupling in the output cable of AC drive for more electric aircraft," in *Proc. 2017 19th European Conf. Power Electron. Appl.*, 2017, pp. P.1–P.9.

24 S. Wang and F. C. Lee, "Investigation of the transformation between differential-mode and common-mode noises in an EMI filter due to unbalance," *IEEE Trans. Electromagn. Compat.*, vol. 52, no. 3, pp. 578–587, 2010.

25 J. Meng and W. Ma, "A new technique for modeling and analysis of mixed-mode conducted EMI noise," *IEEE Trans. Power Electron.*, vol. 19, no. 6, pp. 1679–1687, 2004.

26 S. Qu and D. Chen, "Mixed-mode EMI noise and its implications to filter design in offline switching power supplies," *IEEE Trans. Power Electron.*, vol. 17, no. 4, pp. 502–507, 2002.

27 J. Xue and F. Wang, "Mixed-mode EMI noise in three-phase DC-fed PWM motor drive system," in *Proc. 2013 IEEE Energy Conver. Congr. Expo.*, 2013, pp. 4312–4317.

28 F. Fan, K. Y. See, J. K. Banda et al., "Simulation of three-phase motor drive system with bearing discharge process," in *Proc. 2017 Prog. Electromagn. Research Symp.*, 2017, pp. 2091–2097.

29 H. Karkkainen, L. Aarniovuori, M. Niemela et al., "Converter-fed induction motor efficiency: practical applicability of IEC methods," *IEEE Ind. Electron. Mag.*, vol. 11, no. 2, pp. 45–57, Jun. 2017.

30 P. F. Albrecht, J. C. Appiarius, R. M. McCoy et al., "Assessment of the reliability of motors in utility applications - Updated," *IEEE Trans. Energy Convers.*, vol. EC-1, no. 1, pp. 39–46, Mar. 1986.

31 P. Zhang, Y. Du, T. G. Habetler et al., "A survey of condition monitoring and protection methods for medium-voltage induction motors," *IEEE Trans. Ind. Appl.*, vol. 47, no. 1, pp. 34–46, Jan./Feb. 2011.

32 C. Rojas, M. Melero, M. Cabanas et al., "Finite element model for the study of inter-turn short circuits in induction motors," in *Proc. IEEE Int. Symp. Diagnostics Electr. Mach., Power Electron. Drives*, Cracow, Poland, 2007, pp. 415–419.

33 D. Diaz, M. Amaya, and A. Paz, "Inter-turn short-circuit analysis in an induction machine by finite elements method and field tests," in *Proc. IEEE Int. Conf. Electr. Mach (ICEM)*, Marseille, France, 2012, pp. 1757–1763.

34 K. Prasob, N. P. Kumar, and T. Isha, "Inter-turn short circuit fault analysis of PWM inverter fed three-phase induction motor using finite element method," in *Proc. Int. Conf. Circuit, Power Comput. Technol. (ICCPCT)*, Kollam, India, 2017, pp. 1–6.

35 W. T. Thomson, "On-line MCSA to diagnose shorted turns in low voltage stator windings of 3-phase induction motors prior to failure," in *Proc. Int. Electric. Machines and Drives Conf. (IEMDC)*, Cambridge, MA, USA, 2001, pp. 891–898.

36 J. Sottile, F. C. Trutt, and J. L. Kohler, "Condition monitoring of stator windings in induction motors. II. Experimental investigation of voltage mismatch detectors," *IEEE Trans. Ind. Appl.*, vol. 38, no. 5, pp. 1454–1459, Sep./Oct. 2002.

37 G. N. Surya, Z. J. Khan, M. S. Ballal et al., "A simplified frequency-domain detection of stator turn fault in squirrel-cage induction motors using an observer coil technique," *IEEE Trans. Ind. Electron.*, vol. 64, no. 2, pp. 1495–1506, Feb. 2017.

38 J. S. Hsu, "Monitoring of defects in induction motors through air-gap torque observation," *IEEE Trans. Ind. Appl.*, vol. 31, no. 5, pp. 1016–1021, Sep./Oct. 1995.

39 F. B. Batista, P. C. Filho, R. Pederiva et al., "An empirical demodulation for electrical fault detection in induction motors," *IEEE Trans. Instrum. Meas.*, vol. 65, no. 3, pp. 559–569, Mar. 2016.

40 A. Mohammed, J. I. Melecio, and S. Djurović, "Stator winding fault thermal signature monitoring and analysis by *in situ* FBG sensors," *IEEE Trans. Ind. Electron.*, vol. 66, no. 10, pp. 8082–8092, Oct. 2019.

41 S. F. Legowski, A. H. M. S. Ula, and A. M. Trzynadlowsh, "Instantaneous power as a medium for the signature analysis of induction motors," *IEEE Trans. Ind. Appl.*, vol. 32, no. 4, pp. 904–909, Jul./Aug. 1996.

42 S. B. Lee, R. M. Tallam, and T. G. Habetler, "A robust, on-line turn-fault detection technique for induction machines based on monitoring the sequence component impedance matrix," *IEEE Trans. Power Electron.*, vol. 18, no. 3, pp. 865–872, May 2003.

43 G. Betta, C. Liguori, A. Paolillo et al., "A DSP-based FFT-analyzer for the fault diagnosis of rotating machine based on vibration analysis," *IEEE Trans. Instrum. Meas.*, vol. 51, no. 6, pp. 1316–1322, Dec. 2002.

44 R. Romary, R. Pusca, J. P. Lecointe et al., "Electrical machines fault diagnosis by stray flux analysis," in *Proc. IEEE Workshop Electr. Mach. Des. Control Diagn. (WEMDCD)*, Paris, France, 2013, pp. 245–256.

45 T. Yang, H. Pen, Z. Wang et al., "Feature knowledge based fault detection of induction motors through the analysis of stator current data," *IEEE Trans. Instrum. Meas.*, vol. 65, no. 3, pp. 549–557, Mar. 2016.

46 A. Glowacz and Z. Glowacz, "Diagnosis of the three-phase induction motor using thermal imaging," *Infrared Phys. Technol.*, vol. 81, pp. 7–16, 2017.

47 R. M. Tallam, T. G. Habetler, and R. G. Harley, "Self-commissioning training algorithms for neural networks with applications to electric machine fault diagnostics," *IEEE Trans. Power Electron.*, vol. 17, no. 6, pp. 1089–1095, Nov. 2002.

48 J. Seshadrinath, B. Singh, and B. K. Panigrahi, "Vibration analysis based interturn fault diagnosis in induction machines," *IEEE Trans. Ind. Inform.*, vol. 10, no. 1, pp. 340–350, Feb. 2014.

49 G. H. Bazan, P. R. Scalassara, W. Endo et al., "Stator short-circuit diagnosis in induction motors using mutual information and intelligent systems," *IEEE Trans. Ind. Electron.*, vol. 66, no. 4, pp. 3237–3246, Apr. 2019.

50 A. Stief, J. R. Ottewill, J. Baranowski et al., "A PCA and two-stage Bayesian sensor fusion approach for diagnosing electrical and mechanical faults in induction motors," *IEEE Trans. Ind. Electron.*, vol. 66, no. 12, pp. 9510–9520, Dec. 2019.

51 M. Singh and A. G. Shaik, "Incipient fault detection in stator windings of an induction motor using Stockwell Transform and SVM," *IEEE Trans. Instrum. Meas.*, vol. 69, no. 12, pp. 9496–9504, Dec. 2020.

52 M. S. Toulabi, L. Wang, L. Bieber et al., "A universal high-frequency induction machine model and characterization method for arbitrary stator winding connections," *IEEE Trans. Energy Convers.*, vol. 34, no. 3, pp. 1164–1177, Sep. 2019.

53 X. Song, Z. Wang, S. Li et al., "Sensorless speed estimation of an inverter-fed induction motor using the supply-side current," *IEEE Trans. Energy Convers.*, vol. 34, no. 3, pp. 1432–1441, Sep. 2019.

54 P. Baudesson, J. Lafontaine, and A. Nourrisson, "EMC filtering device in a variable speed drive," U.S. Patent 7 738 268 B2, Jun. 15, 2010.

55 K. Y. See and J. Deng, "Measurement of noise source impedance of SMPS using a two probes approach," *IEEE Trans. Power Electron.*, vol. 19, no. 3, pp. 862–868, May 2004.

56 E. Mazzola, F. Grassi, and A. Amaducci, "Novel measurement procedure for switched-mode power supply modal impedances," *IEEE Trans. Electromagn. Compat.*, vol. 62, no. 4, pp. 1349–1357, Aug. 2020.

57 D. Zhao, J. A. Ferreira, A. Roc'h et al., "Common-mode DC-bus filter design for variable-speed drive system via transfer ratio measurements," *IEEE Trans. Power Electron.*, vol. 24, no. 2, pp. 518–524, Feb. 2009.

58 X. Liang, M. Z. Ali, and H. Zhang, "Induction motors fault diagnosis using finite element method: a review," *IEEE Trans. Ind. Appl.*, vol. 56, no. 2, pp. 1205–1217, Mar./Apr. 2020.

59 A. J. Bahr, V. R. Frank, J. P. Petro et al., *Analysis of radar detection of agitated metals (RADAM)*. Menlo Park, CA, USA: SRI International, 1978.

60 D. M. Pozar, *Microwave engineering*. Hoboken, NJ, USA: Wiley, 2012.

61 Z. Zhao, F. Fan, W. Wang et al., "Detection of stator interturn short-circuit faults in inverter-fed induction motors by online common-mode impedance monitoring," *IEEE Trans. Instrum. Meas.*, vol. 70, p. 3513110, 2021.

62 A. P. Gonzalo, R. Horridge, H. Steele et al., "Review of data analytics for condition monitoring of railway track geometry," *IEEE Trans. Intell. Transp. Syst.*, vol. 23, no. 12, pp. 22737–22754, Dec. 2022.

63 L. Zhuang, H. Qi, and Z. Zhang, "The automatic rail surface multi-flaw identification based on a deep learning powered framework," *IEEE Trans. Intell. Transp. Syst.*, vol. 23, no. 8, pp. 12133–12143, Aug. 2022.

64 J. Sadeghi, M. M. Najar, J. Zakeri et al., "Development of railway ballast geometry index using automated measurement system," *Measurement*, vol. 138, pp. 132–142, May 2019.

65 B. Omondi, D. G. Aggelis, H. Sol et al., "Improved crack monitoring in structural concrete by combined acoustic emission and digital image correlation techniques," *Struct. Health Monit.*, vol. 15, no. 3, pp. 359–378, May 2016.

66 J. Liu, Y. Qiu, X. Ni et al., "Fast detection of railway fastener using a new lightweight network Op-YOLOv4-Tiny," *IEEE Trans. Intell. Transp. Syst.*, vol. 25, no. 1, pp. 133–143, Jan. 2024.

67 M. Ester, H. P. Kriegel, J. Sander et al., "A density-based algorithm for discovering clusters in large spatial databases with noise," in *Proc.* 2nd Int. *Conf. Knowl. Discov. Data Min.*, 1996, pp. 226–231.

9

Applications in Power and Energy

Power and energy systems form the backbone of modern electrified society. The continuous push toward higher power density, the widespread deployment of wide-bandgap devices, and increasingly complex multiscale interconnections introduce growing challenges related to EMC, system reliability, and operational safety. Traditional measurement and modeling techniques often rely on offline testing or intrusive probing, which can be impractical or even infeasible in high-power, high-voltage, and continuously operating environments. This chapter demonstrates how electrically contactless online impedance measurement provides a powerful alternative for real-world power and energy applications. By enabling in situ characterization, continuous condition monitoring, and early fault diagnosis without disrupting normal operation, the proposed techniques bridge the gap between theoretical impedance analysis and field-deployable engineering solutions. Representative applications are presented across three major domains: condition monitoring and defect detection in power transformers, fault diagnosis in high-power DC feeder systems with complex cable infrastructures, and EMI analysis and mitigation in PV systems. Through these case studies, the chapter illustrates how contactless impedance-based methods can be effectively applied to address practical challenges in modern power and energy systems, enhancing reliability, safety, and maintenance efficiency.

9.1 Online Condition Monitoring of Transformers

Power transformers are among the most critical and capital-intensive assets in electrical power systems, and their reliable operation is essential for ensuring continuity and stability of the power supply. Degradation or incipient faults in transformers can lead to severe failures, prolonged outages, and high maintenance costs. Consequently, condition monitoring of transformers has long been an important research and industrial focus, aiming to detect early-stage abnormalities and to support condition-based maintenance strategies.

A wide range of transformer condition monitoring techniques have been reported in the literature, including dissolved gas analysis, partial discharge monitoring, thermal monitoring, and frequency response analysis [1]. Among these methods, impedance- and frequency-response-based techniques are particularly attractive, as changes in winding geometry, insulation condition, or core characteristics are directly reflected in the transformer's electrical impedance over frequency [2, 3]. Conventional frequency response analysis methods, however, are predominantly performed offline and require the transformer to be de-energized, which limits their applicability for continuous monitoring under actual operating conditions. Several online extensions have been proposed, often relying on capacitive coupling or signal injection through bushings [4, 5], but these approaches typically involve additional electrical connections to high-voltage components, raising safety concerns and potentially altering the original impedance characteristics of the transformer.

These limitations motivate the development of online condition monitoring techniques that can extract transformer impedance information under energized conditions without direct electrical contact. In this context, inductive coupling-based contactless measurement approaches provide a promising alternative, enabling online impedance monitoring while avoiding intrusive connections and associated safety risks. This section introduces an online transformer condition monitoring methodology based on electrically contactless impedance measurement, demonstrating how transformer impedance variations can be extracted under normal operating conditions and used as effective indicators of structural or parametric changes related to transformer health.

9.1.1 Case Study and Transformer Equivalent Circuit Model

For demonstration purposes, a 2:1 step-down 1.5 kW transformer is selected as the case study. The online condition monitoring setup was selected as the FD-TPS, as illustrated in Figure 9.1a. Figure 9.1b depicts the equivalent circuit model of the measurement configuration together with the transformer, following established transformer modeling approaches [6–9]. The primary winding of the transformer (b–b') is connected to a 230 V/50 Hz AC mains supply (a–a'). Both the injecting and receiving inductive probes are clamped onto the power cable between the AC mains and the primary winding, enabling contactless excitation and response acquisition without interrupting normal operation.

For a nonideal transformer, hysteresis and eddy-current losses occur in the magnetic core. These losses are modeled using a core loss resistance R_C, while the

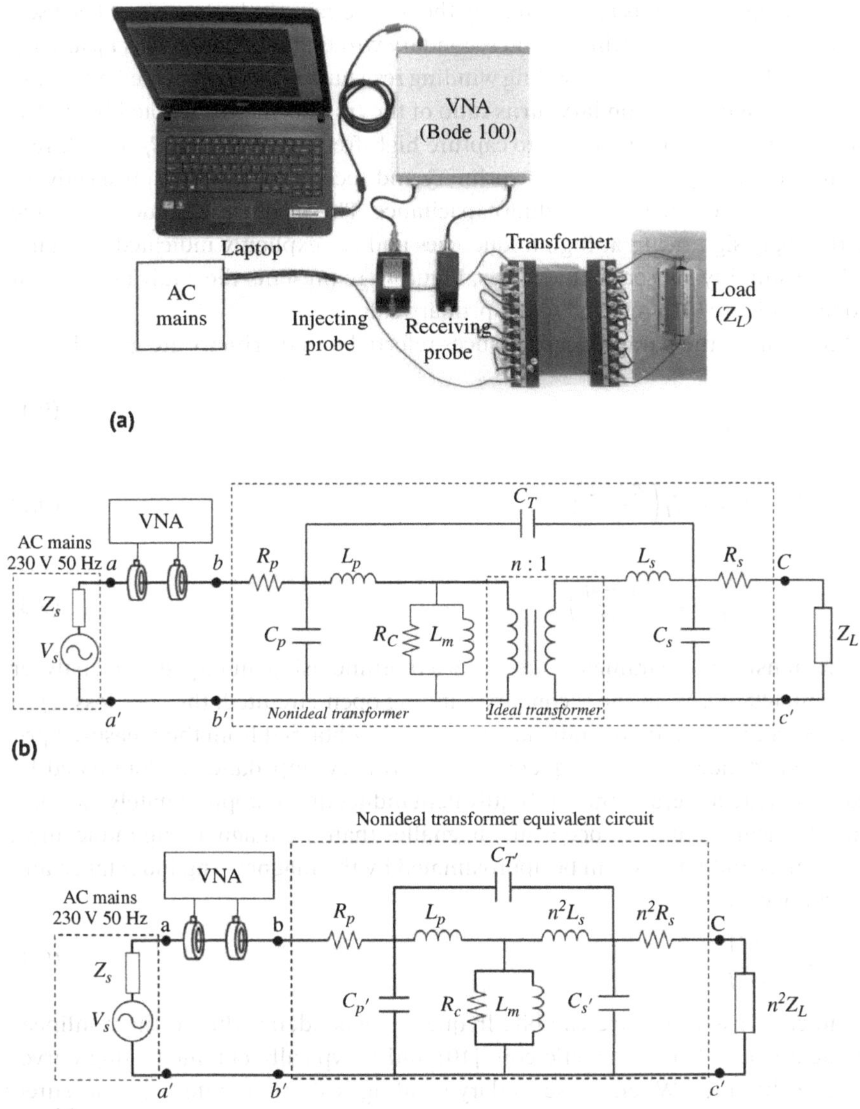

Figure 9.1 Online monitoring setup for online impedance extraction of the transformer primary impedance. (a) Experimental setup. *Source*: [3]/with permission of John Wiley & Sons. (b) Equivalent circuit model of the nonideal transformer and measurement configurations. (c) Equivalent circuit model referred to the primary side.

magnetizing behavior is represented by the magnetizing inductance L_m. The leakage inductances of the primary and secondary windings are denoted as L_P and L_S, respectively, and the corresponding winding resistances are represented by R_P and R_S. The primary-to-secondary turns ratio of the transformer is denoted by n. Parasitic capacitances are included to capture high-frequency effects: C_P and C_S represent the self-capacitances of the primary and secondary windings, respectively, while C_T denotes the inter-winding capacitance. These parasitic elements become increasingly significant at high frequencies and are explicitly indicated in Figure 9.1b. To simplify subsequent analysis, Figure 9.1c presents the equivalent circuit model of Figure 9.1b referred to the primary side.

Accordingly, the parasitic capacitances referred to the primary are given by

$$C_{T'} = \frac{C_T}{n} \tag{9.1}$$

$$C_{P'} = C_P + C_T\left(\frac{n-1}{n}\right) \tag{9.2}$$

$$C_{S'} = \frac{C_S}{n^2} + C_T\left(\frac{1-n}{n^2}\right) \tag{9.3}$$

The transformer parameters are extracted offline using an impedance analyzer (Agilent 4294A). With the secondary winding open-circuited, the core loss resistance R_C and magnetizing inductance L_m can be obtained from the measured primary impedance. At low frequencies, the primary impedance is dominated by inductive reactance, and the total equivalent inductance is approximately $L_P + L_m$. Since the leakage inductance is much smaller than the magnetizing inductance, the primary inductance can be approximated by the magnetizing inductance and determined as

$$L = \frac{|Z|}{2\pi f} \tag{9.4}$$

The core loss resistance exhibits frequency dependence due to the nonlinear characteristics of the magnetic core [10] and is typically obtained using curve-fitting techniques. When the secondary winding is short-circuited, the measured primary impedance is dominated by leakage inductance and winding resistance. Under this condition, the leakage inductance can be calculated using Equation (9.4) at low frequencies, yielding $L_P + n^2 L_S$, while the low-frequency resistance corresponds to $R_P + n^2 R_S$.

Similarly, the secondary inductance can be obtained from the measured secondary impedance with the primary short-circuited. Once the inductance is known, the corresponding parasitic capacitance is calculated from the resonant frequency using

$$\omega = \frac{1}{\sqrt{LC}} \tag{9.5}$$

The extracted equivalent circuit parameters of the 1.5 kW transformer are given in Table 9.1.

To validate the extracted equivalent circuit parameters of the transformer, a 150 Ω resistive load is connected to the secondary winding as the load Z_L, as shown in Figure 9.1a. The primary impedance is then measured using an impedance analyzer and compared with the simulated primary impedance obtained from the established equivalent circuit model under the same loading condition. Figure 9.2 presents the comparison between the measured and simulated primary impedance magnitude, demonstrating close agreement across the frequency range of interest. Parasitic capacitances and winding inductances of a transformer are expected to deviate from their nominal values when winding deformation or displacement occurs. Such structural changes alter the electromagnetic coupling within the transformer and consequently affect its impedance frequency response. To illustrate the sensitivity of the primary impedance frequency response to these parameter variations, simulations are performed with the transformer terminated by a 150 Ω secondary load, while systematically varying the primary winding capacitance and inductance around their nominal healthy values of 219 pF and 1.08 mH, respectively.

The simulated primary impedance frequency responses under different primary capacitance values are shown in Figure 9.3a, while the responses under varying primary inductance values are presented in Figure 9.3b. These results show that any slight change of these parameters due to deformation of winding can be detected through the impedance frequency response analysis.

Table 9.1 Extracted equivalent circuit parameters of the 1.5 kW transformer.

Parameter	Description	Value
C_P	Primary capacitance	219 pF
C_S	Secondary capacitance	304 pF
C_T	Winding to winding capacitance	258 pF
R_C	Core loss	25 kΩ
L_m	Magnetizing inductance	58 mH
$R_P + n^2R_S$	Total winding resistance referred to primary	272 Ω
$L_P + n^2L_S$	Total leakage inductance referred to primary	1.08 mH

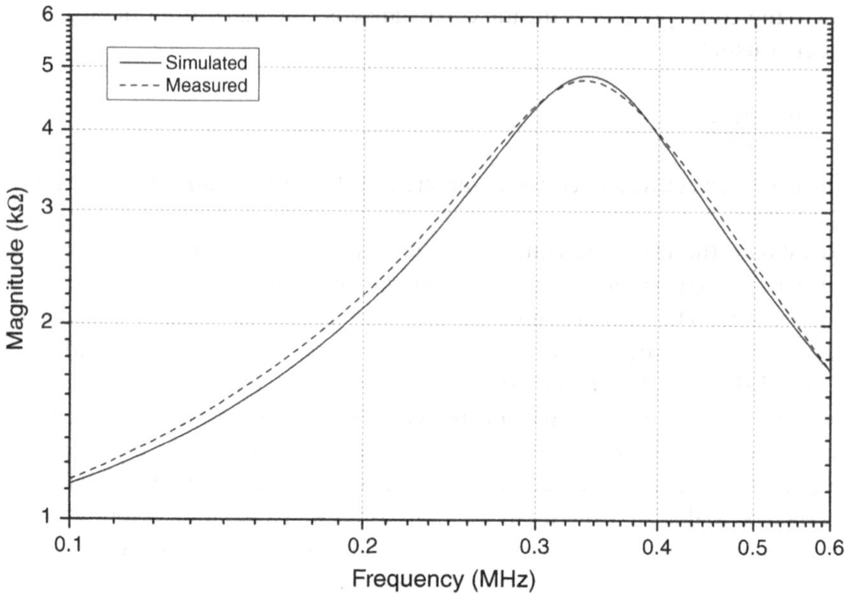

Figure 9.2 Measured and simulated primary impedance magnitude of the transformer with a 150 Ω resistive secondary load.

9.1.2 Online Impedance Measurement of Transformer

To demonstrate the capability of extracting the online impedance of a transformer under varying operating conditions, the primary impedance of the transformer is measured under two loading conditions, with $Z_L = 150\ \Omega$ and $Z_L = 270\ \Omega$. The online impedance, which consists of the transformer primary impedance and the power source impedance Z_S, is monitored and measured using an FD-TPS.

Before connecting the transformer to the AC mains, Z_S is extracted by adding a bypass capacitor across terminals a–a'. To accurately determine the impedance of the bypass capacitor, it is first measured offline using an impedance analyzer. The bypass capacitor is then connected across a–a', and the resultant impedance of the source and bypass capacitor is measured online. In this manner, the impedance contribution of the bypass capacitor can be de-embedded from the measured in-circuit impedance to extract Z_S. Based on the extracted Z_S as a function of frequency, as shown in Figure 9.4, the source impedance exhibits inductive behavior and corresponds to an inductance of approximately 1.8 μH.

Once Z_S is obtained, the bypass capacitor is removed, and the transformer is connected according to the configuration shown in Figure 9.1a. In this configuration, the online impedance to be monitored corresponds to the resultant impedance of the power source, wiring, and transformer primary. Similarly, the

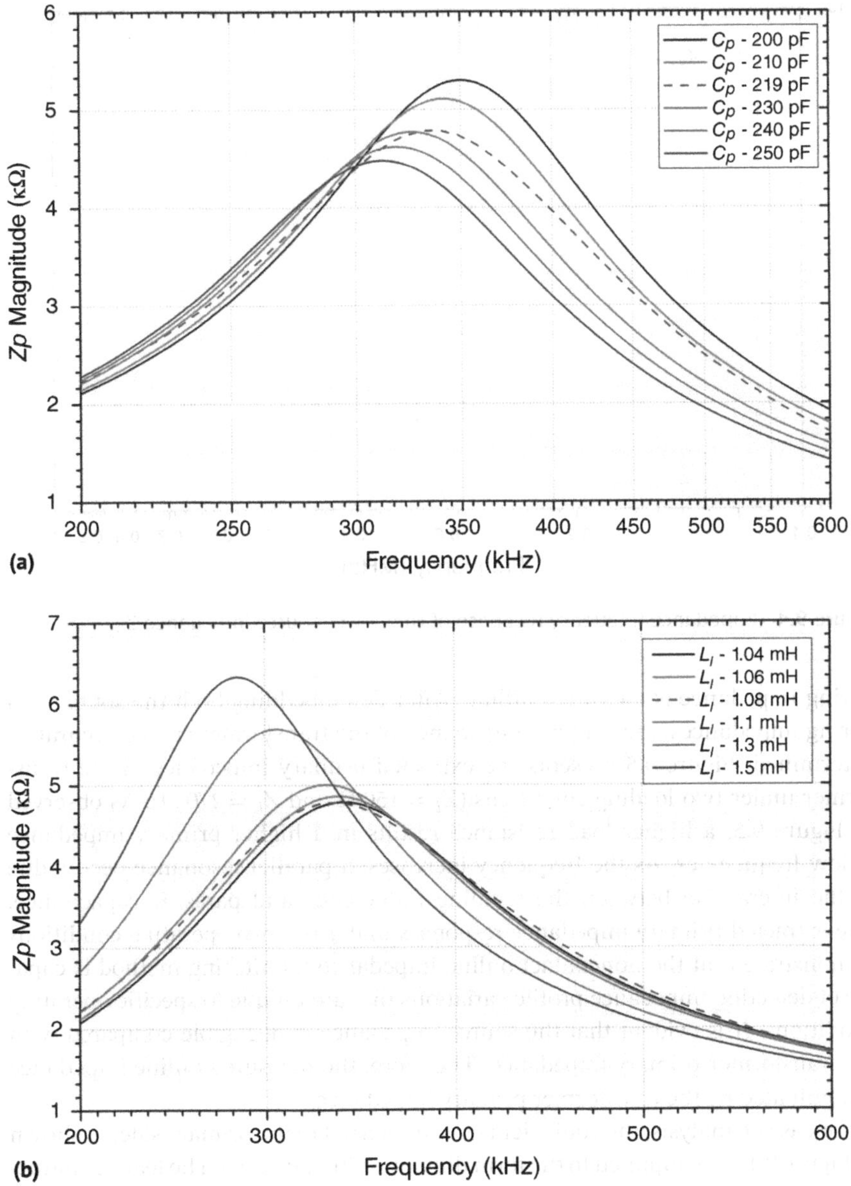

Figure 9.3 Primary impedance frequency responses with parameter variations. (a) Varying capacitance. (b) Varying inductance.

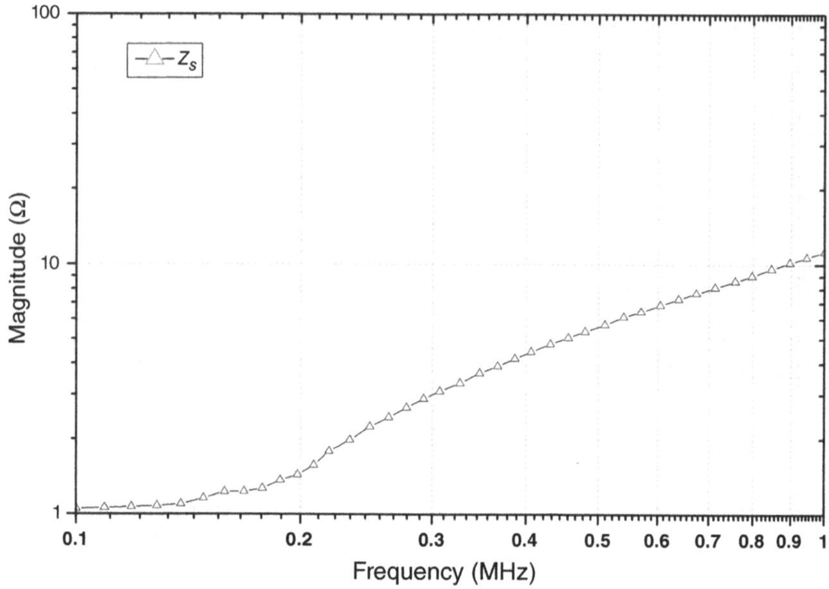

Figure 9.4 Impedance frequency response of the power source impedance Z_S.

wiring impedance is measured offline. After de-embedding both the source and wiring impedances, the primary impedance of the transformer can be accurately determined. Figure 9.5 presents the extracted primary impedance of the transformer under two loading conditions ($Z_L = 150\ \Omega$ and $Z_L = 270\ \Omega$). As observed in Figure 9.5, a higher load resistance results in a higher primary impedance at low frequencies. As the frequency increases, a parallel resonance occurs due to the interaction between the winding inductance and parasitic capacitance. The extracted primary impedance responses under the two operating conditions demonstrate that the noncontact online impedance monitoring method is capable of detecting impedance profile variations that are unique to specific operating conditions. It is evident that the source impedance is negligible compared with the transformer primary impedance. Therefore, the measured online impedance is dominated by the transformer primary impedance.

For ease of analysis, the equivalent circuit referred to the primary side, as shown in Figure 9.1c, is simplified to the form illustrated in Figure 9.6. The leakage inductances of the primary and secondary windings are much smaller than the magnetizing inductance and are therefore neglected. In addition, the inter-winding capacitance C_T is also negligible because the voltage drop across the leakage inductances is insignificant [11]. Consequently, the overall parasitic capacitance can be represented by a shunt capacitance C_e across the transformer primary, as shown in

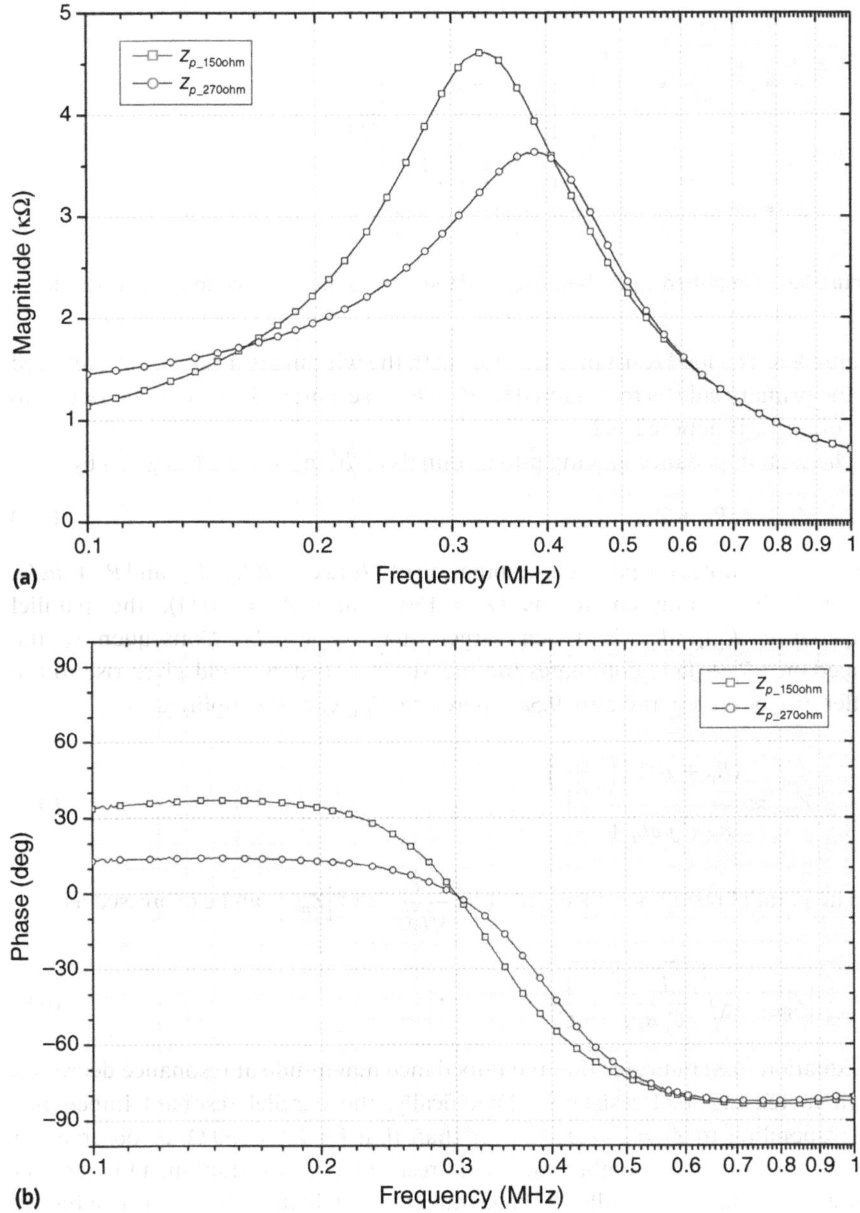

Figure 9.5 Online primary impedance frequency response with $Z_L = 150\ \Omega$ and $Z_L = 270\ \Omega$. (a) Magnitude. (b) Phase.

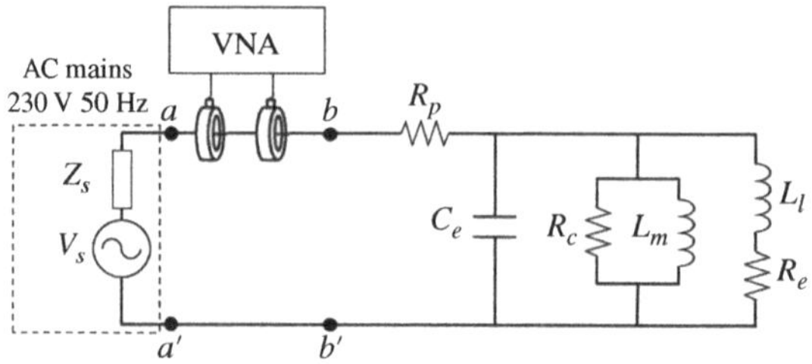

Figure 9.6 Simplified equivalent circuit of the transformer referred to the primary side.

Figure 9.6. The load resistance in series with the winding resistance, both referred to the primary side, is represented by R_e. The resultant leakage inductance due to L_P and n^2L_S is denoted as L_l.

The total impedance looking into terminals b–b' in Figure 9.6 is given by

$$Z_{b-b'} = R_P + Z_{\text{par}} \tag{9.6}$$

where Z_{par} denotes the parallel combination of $1/(j\omega C_e)$, $R_c \| j\omega L_m$, and $R_e + j\omega L_l$.

For both loading conditions ($Z_L = 150\ \Omega$ and $Z_L = 270\ \Omega$), the parallel impedance $R_C \| j\omega L_m$ is much larger than $R_e + j\omega L_l$. Consequently, the impedance $R_e + j\omega L_l$ dominates the measured impedance and gives rise to the difference observed in Figure 9.5a. Therefore, Z_{par} can be simplified as

$$Z_{\text{par}} \approx \frac{(R_e + j\omega L_l)\left(\frac{1}{j\omega C_e}\right)}{R_e + j\omega L_l + \frac{1}{j\omega C_e}} \tag{9.7}$$

The parallel resonance occurs at $\omega_o = \frac{1}{\sqrt{L_l C_e}}$ and $|Z_{\text{par}}|$ can be expressed as

$$|Z_{\text{par}}| = \sqrt{\left(\frac{L_l}{C_e R_e}\right)^2 + \frac{L_l}{C_e}} \tag{9.8}$$

Equation (9.8) indicates that the impedance magnitude at resonance decreases with increasing load resistance. Specifically, the parallel resonant impedance corresponding to $Z_L = 270\ \Omega$ is lower than that for $Z_L = 150\ \Omega$, as observed in Figure 9.5a, due to the higher equivalent resistance R_e. In addition, a higher load resistance results in a smaller leakage inductance, which in turn leads to a higher resonant frequency, as described by $\omega_o = \frac{1}{\sqrt{L_l C_e}}$.

9.1.3 Online Defects Detection

To demonstrate the capability of the introduced method for online monitoring of the transformer impedance frequency response as a defect detection tool, a customized 1.5 kW step-down transformer, shown in Figure 9.7, is designed and fabricated as the equipment under test (EUT). Extra one, two, and three turns are wound on both the primary and secondary windings, and these extra turns can be short-circuited externally to emulate defects within the transformer, as illustrated in Figure 9.8.

A 100 Ω resistor (Z_L) is connected to the transformer secondary as a nominal load, as shown in Figure 9.9. The primary terminals of the transformer (d–d')

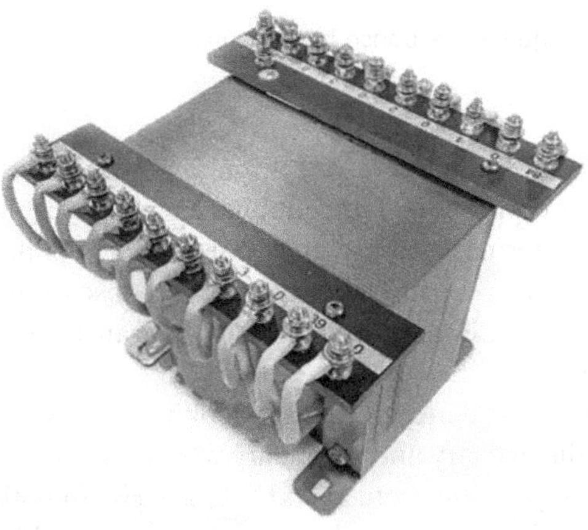

Figure 9.7 Fabricated transformer used as the EUT for online monitoring.
Source: [3]/with permission of John Wiley & Sons.

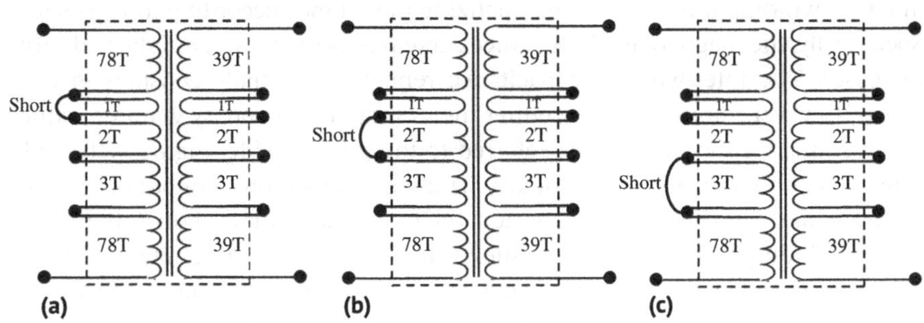

Figure 9.8 Turn-to-turn short-circuit configurations in the primary winding. (a) One-turn shorted. (b) Two-turn shorted. (c) Three-turn shorted.

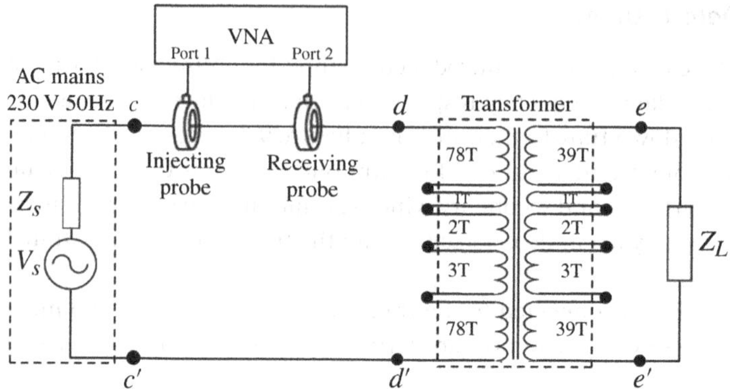

Figure 9.9 Setup for online impedance frequency response measurement for inter-turn short-circuit detection.

are connected to a 230 V/50 Hz AC power supply (c–c') through power cables. The injecting and receiving inductive probes are clamped onto the power cable between the power supply and the transformer primary, enabling extraction of the impedance response looking into terminals d–d'. The frequency response of the primary impedance is thus obtained to monitor any deviation of the transformer impedance from its normal condition.

Figure 9.10a,b show the measured impedance responses looking into d–d' under inter-turn short conditions in the primary and secondary windings, respectively. The measurement frequency range spans from 100 kHz to 2 MHz. In both cases, the impedance deviation caused by inter-turn shorts is clearly detectable. This deviation originates from changes in the transformer winding inductance and parasitic capacitance relative to their normal conditions. Because the primary and secondary windings are wound on the same magnetic core, shorted turns in either winding influence the magnetizing inductance according to Lenz's law. Specifically, the magnetizing inductance decreases as the number of shorted turns increases. The inter-winding capacitance remains essentially constant in both cases. However, shorted turns in either the primary or secondary winding reduce the overall parasitic capacitance referred to the primary side. Figure 9.10a indicates that shorted turns in the primary winding have a more pronounced impact on the primary impedance. As expected, the influence of shorted turns in the secondary winding is less evident, as shown in Figure 9.10b. A resonance around 250 kHz, caused by the interaction between the winding inductance and parasitic capacitance, is clearly observed. Below 250 kHz, the impedance is dominated by the transformer primary inductance as well as the reflected inductance and load impedance from the secondary to the primary side. With an increasing number of

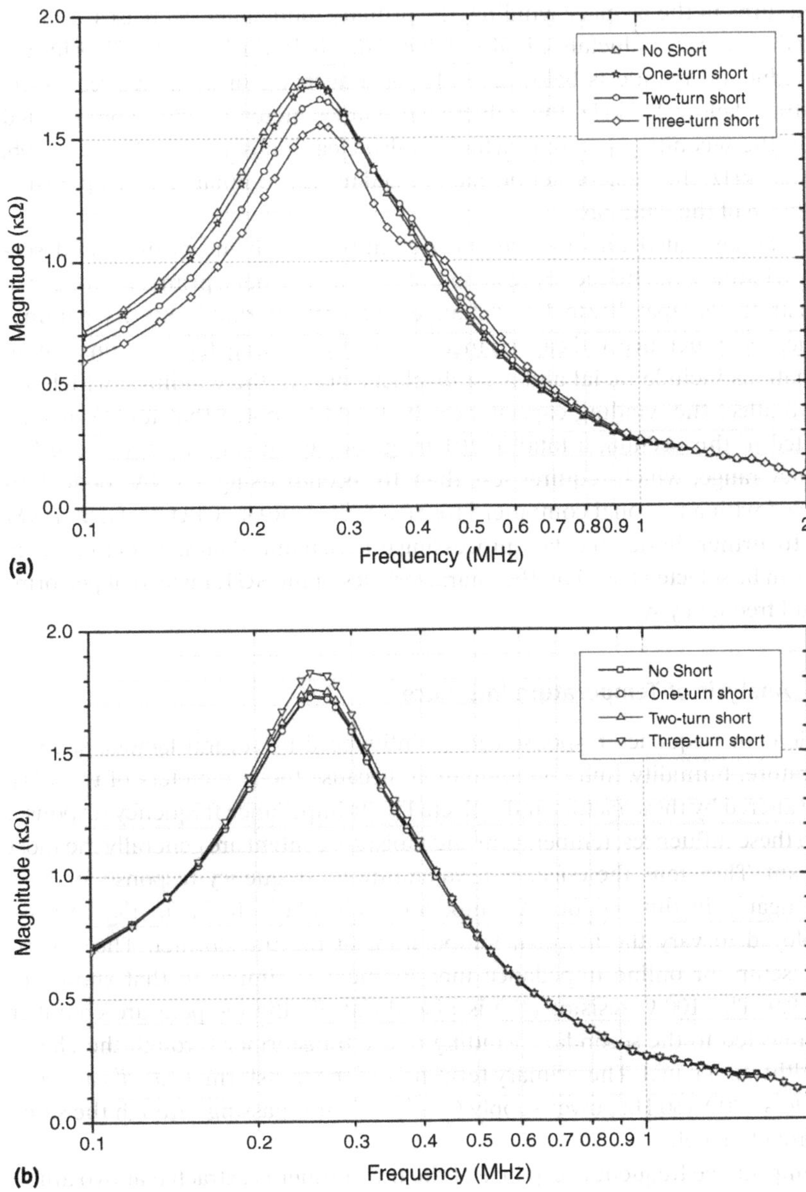

Figure 9.10 Measured online impedance frequency responses of transformer under inter-turn short-circuit conditions. (a) Short circuit in the primary winding. (b) Short circuit in the secondary winding.

shorted turns in the primary winding, the primary inductance decreases, and the reflected secondary inductance is also reduced due to lower turn ratio. This inductance reduction manifests below 250 kHz as a decrease in inductive reactance. In contrast, below 250 kHz, the inductive reactance increases with more shorted turns in the secondary winding, which is also clearly observed in Figure 9.10b. Above 250 kHz, the transformer primary impedance is dominated by the parasitic capacitance of the windings.

With the emulated turn-to-turn short circuits in both the primary and secondary windings, the frequency response of the transformer's primary impedance demonstrates its capability to detect changes in the transformer's internal parameters, such as inductance and capacitance, caused by winding deformations. These deformations include radial and axial displacements of the windings, which are likely to affect the winding capacitance. It should be noted that for the results presented in this section, a total of 101 frequency points are used over the full frequency range, which requires less than 10 seconds using a VNA (Bode 100) interfaced with a personal computer (Toshiba Portege R930, CPU 2.9 GHz, RAM 8 GB). To further shorten the data-processing time, a reduced number of frequency points can be selected based on the characteristics of the SUT, instead of performing a full frequency sweep.

9.1.4 Analysis of Temperature Influence

Deviations in frequency response can be influenced by several factors, such as temperature, humidity, and contamination, because the parameters of the EUT may be altered by these factors and reflected in its impedance frequency response. Among these influences, temperature and moisture content are generally the most significant. Therefore, the effect of temperature on frequency response analysis is investigated in this section. A temperature chamber, shown in Figure 9.11, is employed to vary the ambient temperature of the transformer. The experimental setup for online impedance measurement is similar to that shown in Figure 9.9. The 100 Ω resistor (Z_L) is placed outside the temperature chamber and connected to the secondary winding of the transformer through the chamber feedthrough panel. The primary terminals of the transformer (d–d') are connected to a 230 V/50 Hz power supply (c–c') via wires passing through the same feed through panel.

The impedance frequency response of the transformer is extracted at two ambient temperatures, 30 and 75 °C, as shown in Figure 9.12.

Above the resonant frequency, it is clearly observed that the equivalent capacitance of the transformer decreases with increasing temperature. This behavior is expected, as the permittivity of enamel-based winding insulation decreases at elevated temperatures [12]. In contrast, the influence of temperature on inductance

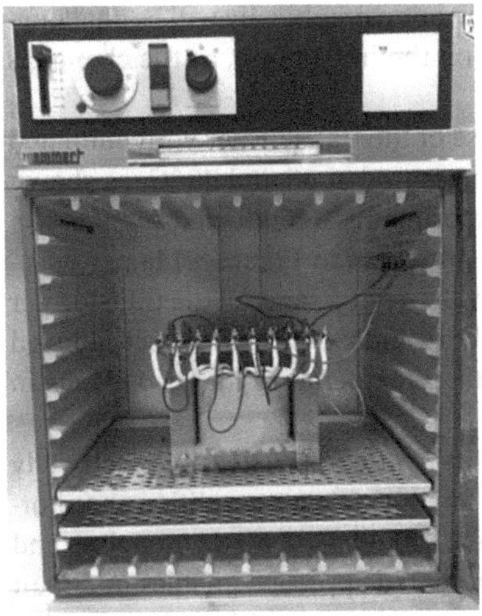

Figure 9.11 Temperature chamber.

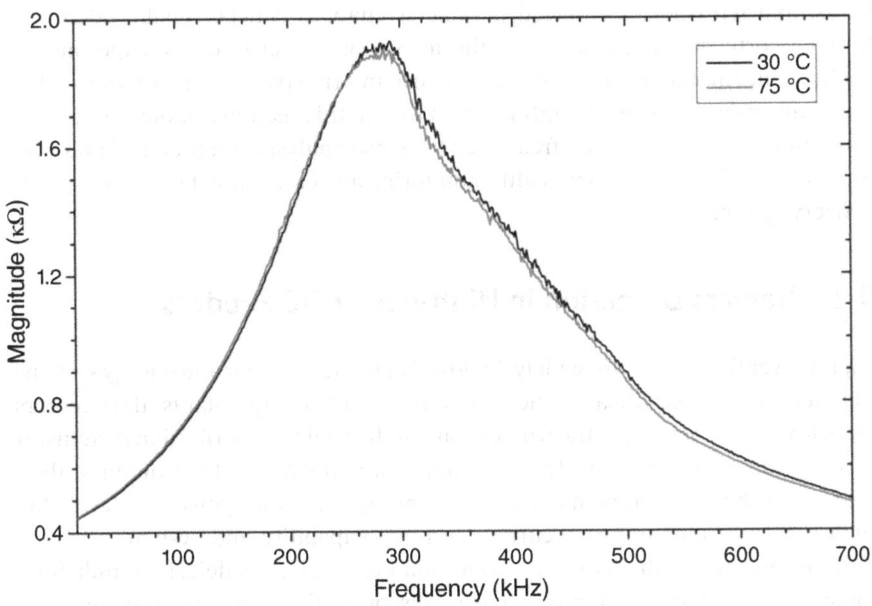

Figure 9.12 Extracted online impedance frequency response of the transformer at 30 and 75 °C.

is relatively insignificant. By comparing the deviation graphically will be challenging to distinguish the temperature influence on the impedance, as the change is hardly noticeable.

Similarly, the impedance frequency response of the transformer with emulated inter-turn shorts in the primary winding is measured at 30 and 40 °C, as shown in Figure 9.13a,b, respectively. The same experiment is conducted for the secondary winding at 30 and 40 °C, with results presented in Figure 9.14a,b, respectively. In both cases, a clear change in the equivalent inductance of the transformer is observed due to the emulated inter-turn short, indicating that defect-induced impedance variations are more pronounced than temperature-induced effects in this case study.

This section has demonstrated the feasibility and effectiveness of electrically contactless online impedance measurement for transformer condition monitoring under representative operating conditions. Through systematic modeling and controlled experimental investigations, the presented approach enables in-circuit extraction of the transformer impedance frequency response without interrupting normal operation or requiring direct electrical connections. The results show that variations in loading conditions, winding deformation, inter-turn short-circuit faults, and temperature-induced parameter changes can be reliably identified through corresponding changes in the impedance spectrum. The experimental studies confirm that the method is sensitive to internal parameter variations while remaining robust to external influences. Overall, this section establishes contactless online impedance-based frequency response analysis as a practical and non-intrusive tool for transformer health monitoring and early fault detection in power delivery systems.

9.2 Defects Detection in High-power DC Feeders

High-power DC feeders are widely deployed in modern power and energy systems, such as railway electrification networks, renewable energy plants, data centers, electric vehicle charging infrastructure, and industrial DC distribution systems. To accommodate large current demands and ensure operational redundancy, these feeders are typically implemented using multiple parallel power cables. While such configurations improve current-carrying capability and system reliability, they also introduce challenges for condition monitoring, as defects in individual cables, such as degraded joints or partial disconnections, may remain concealed during normal operation.

Previous studies on high-power DC feeder monitoring have reported that high transient currents, unbalanced mutual inductances, and installation constraints

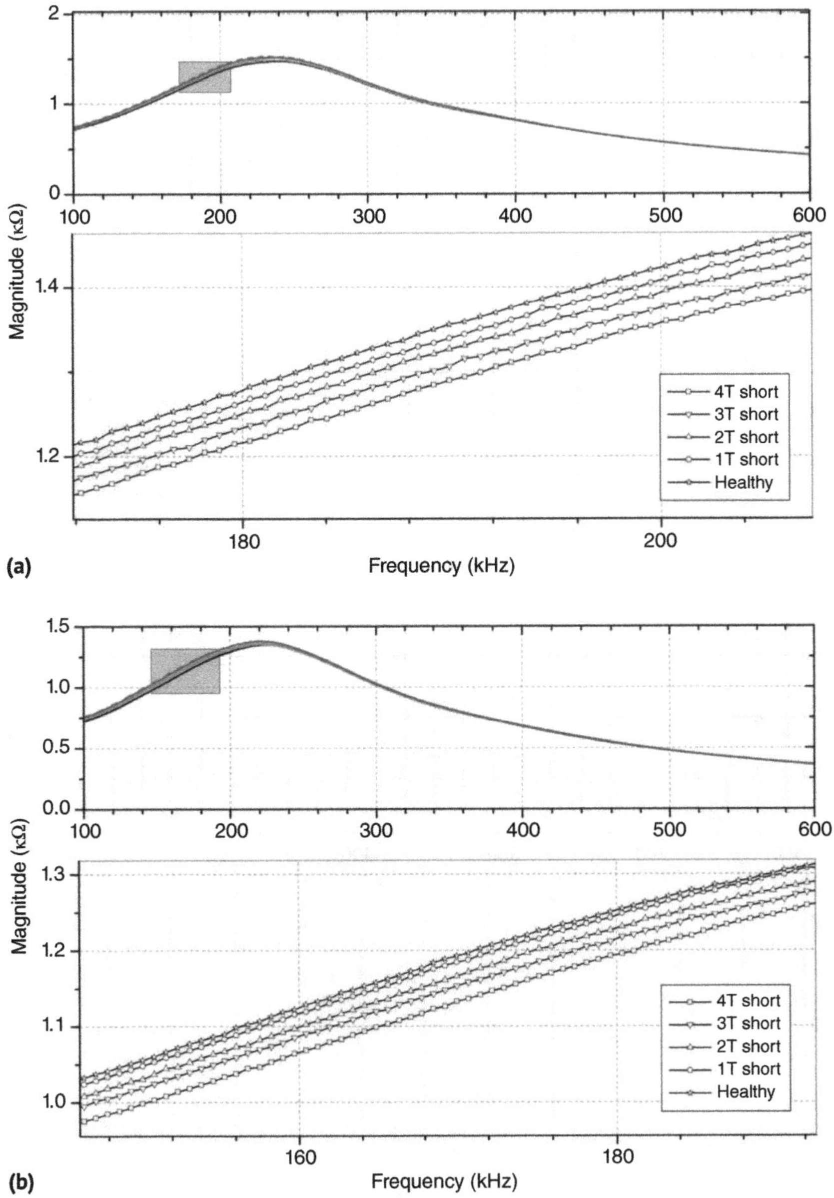

Figure 9.13 Extracted frequency response with an inter-turn short circuit in the primary winding at (a) 30 °C and (b) 40 °C.

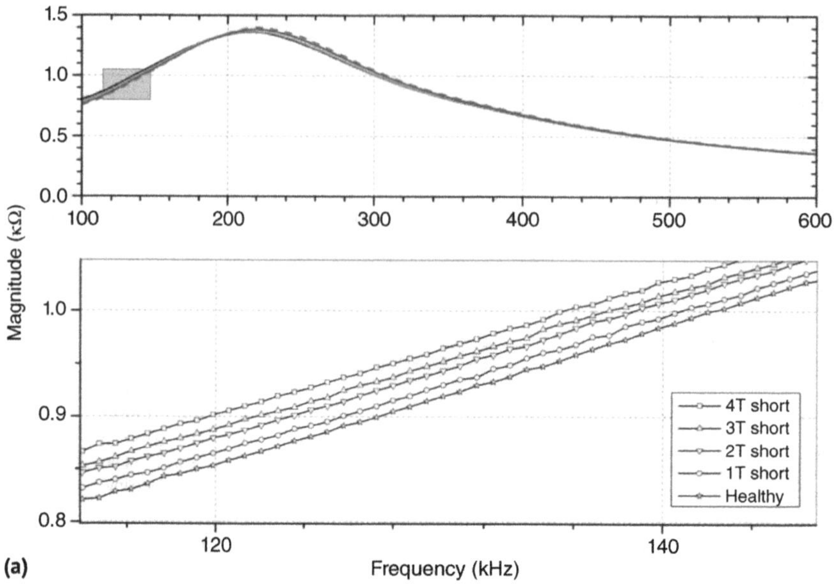

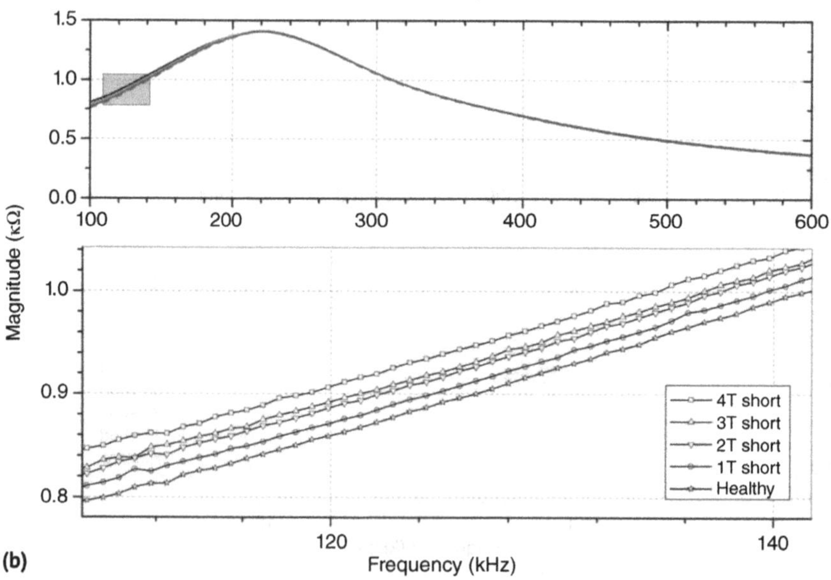

Figure 9.14 Extracted frequency response with an inter-turn short circuit in the secondary winding at (a) 30 °C and (b) 40 °C.

can lead to unequal current sharing among parallel cables, resulting in localized overheating and accelerated aging of specific conductors [13–15]. In principle, such defects could be identified through current measurements or offline inspection. However, in some applications such as practical railway and industrial DC systems, load currents are highly time-varying and contaminated by harmonics generated by traction converters, which significantly limits the reliability of current-based diagnostic approaches under operating conditions. Moreover, several existing online monitoring techniques for high-power DC feeders rely on direct electrical contact with energized conductors, increasing installation complexity and raising safety concerns in field deployment. These challenges motivate the development of nonintrusive and online defect detection methods for high-power DC feeders that are compatible with continuous operation.

Impedance-based monitoring approaches are particularly attractive, as defects in parallel cables manifest as measurable changes in the resultant feeder impedance, reflecting contact degradation or impedance imbalance among conductors. This section introduces a nonintrusive defect detection methodology for high-power DC feeders based on electrically contactless online impedance measurement, enabling early identification of defective cables in multi-parallel feeder systems without interrupting normal system operation.

9.2.1 Emulated DC Feeders and Their Online Monitoring Setup

To illustrate how the inductive coupling approach can be applied to defect detection in multiple parallel cables through online impedance extraction, an emulated DC feeder system is constructed as a case study. Three single-core cables (model: General Cable 01779-35T.01; insulation material: EPDM rubber; conductor material: copper; conductor diameter: 8.251 mm; overall cable diameter: 19–22 mm) together with two copper busbars are used to emulate a high-power DC feeder. The selected configuration represents a typical parallel-cable arrangement commonly employed in high-current power delivery systems. The FD-TPS is adopted as the measurement setup to monitor the online impedance characteristics of the emulated feeder. The feeder configuration consists of three parallel cables, each 6 m in length with a cable-to-cable spacing of 5 cm, and two copper busbars, as illustrated in Figure 9.15. The three cables are used to emulate parallel DC positive feeder paths, while the two copper busbars represent the DC source output point and the load connection point, respectively. Both ends of each cable are mechanically fastened to the busbars using screw connections to ensure reliable electrical contact.

As indicated by the dashed arrows in Figure 9.15, the injected RF current I_1 introduced by the injecting probe initially flows along Cable 1. The current then divides into two RF currents, I_2 and I_3, which propagate along Cables 2 and 3,

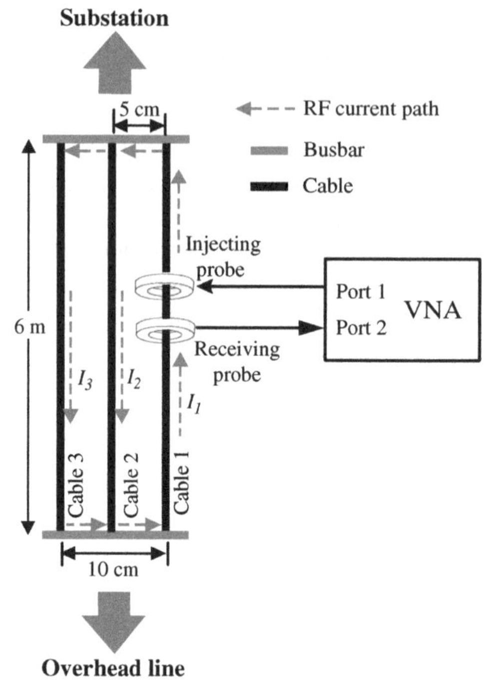

Figure 9.15 Configuration of the emulated DC feeder and the online impedance monitoring setup based on the FD-TPS.

respectively. After traversing Cables 2 and 3, the RF currents return to Cable 1 and are subsequently detected by the receiving probe. In this configuration, the three cables and two busbars are electrically interconnected, and the feeder is considered to be operating under a healthy condition.

9.2.2 Defects Detection Principle for DC Feeders

Under defective conditions, the disconnection of one or more parallel cables from a DC feeder is a common consequence of cable degradation, such as insulation aging, corrosion, or accidental mechanical damage. Since the introduced approach is intended to detect cable defects at an early stage, the defective conditions are emulated by electrically insulating one end of one of the three parallel cables. In this manner, the cable with the insulated end is electrically disconnected from the feeder, while the remaining two cables and the two busbars remain electrically interconnected. Insulating either end of a cable produces an equivalent electrical condition for the feeder circuit. Accordingly, three emulated defective conditions are defined and illustrated in Figure 9.16. To extract the online impedance

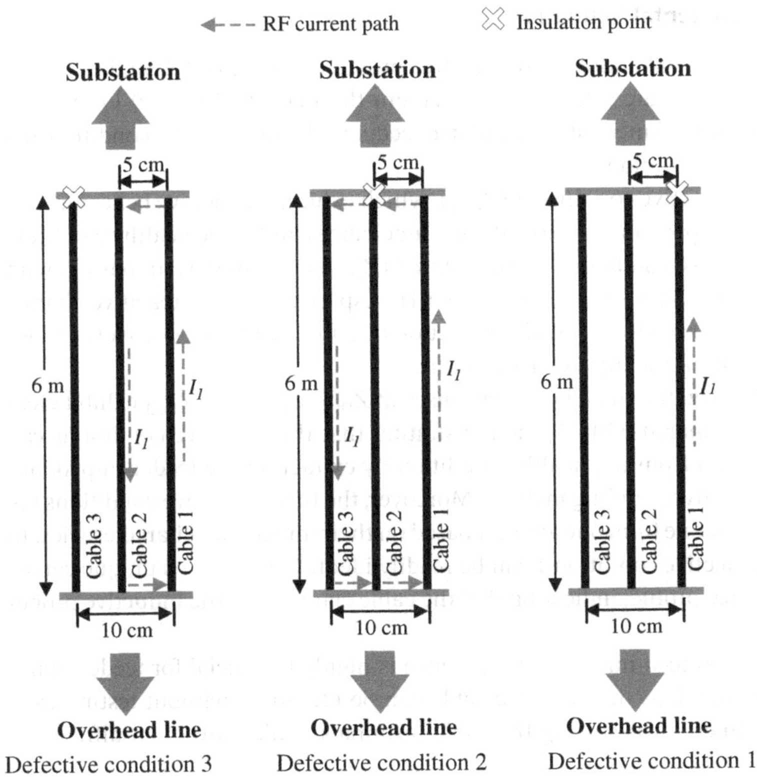

Figure 9.16 Emulated defective conditions of the DC feeders and corresponding RF current paths under single-cable disconnection.

of the feeders under these defective conditions, the injecting and receiving probes remain clamped on Cable 1. The corresponding RF current paths of the injected current I_1 under the three defective conditions are also indicated by dashed arrows in Figure 9.16.

For defective condition 1, Cable 1 is electrically disconnected, and the injected current I_1 flows only along Cable 1 without forming a closed loop through the remaining cables. For defective condition 2, Cable 2 is disconnected, and the injected current I_1 flows through the circuit loop formed by the two busbars together with Cables 1 and 3. For defective condition 3, Cable 3 is disconnected, and the injected current I_1 flows through the circuit loop consisting of the two busbars and Cables 1 and 2. These distinct current paths result in different impedance characteristics, which form the basis for identifying and distinguishing defective feeder conditions.

9.2.3 Experimental Validation

For experimental validation, two inductive probes (Solar 9144-1N) and one VNA (N9917A FieldFox) are selected to implement the FD-TPS. Figure 9.17 presents the measured impedances of the emulated feeder under one healthy condition and three defective conditions.

Except for defective condition 1 (Z_{DC1}), which exhibits capacitive behavior with an equivalent capacitance of 0.08 nF, the impedances under the healthy condition (Z_G) and defective conditions 2 (Z_{DC2}) and 3 (Z_{DC3}) are inductive in nature, with equivalent inductances of 5.1, 7.6, and 6 μH, respectively. The capacitive characteristic of Z_{DC1} arises from the disconnection of Cable 1, which effectively creates an open circuit for the injected RF current.

From 1–7 MHz, the impedance responses of Z_{DC1}, Z_{DC2}, and Z_{DC3} exhibit clear differences compared with Z_G, demonstrating that the defective conditions can be distinguished from the healthy condition by extracting the feeder impedance using the inductive coupling method. Moreover, the three defective conditions are also distinguishable from one another based on their impedance characteristics. In particular, defective condition 1 can be readily identified due to its unique capacitive impedance profile, indicating that the cable clamped by the inductive probes is defective.

Such a pronounced impedance difference is highly beneficial for feeder maintenance personnel, as the defective cable can be identified without testing each cable individually or accessing the cable terminals, which are often difficult to

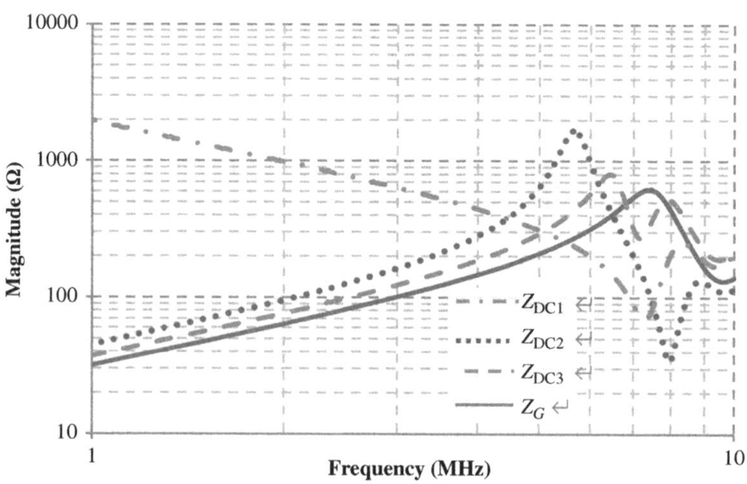

Figure 9.17 Resultant impedance responses of the emulated DC feeders under the healthy condition and three defective conditions.

reach. In practical applications, feeder cables commonly exceed 100 m in length, while only a short section, typically less than 10 m, is accessible and visible within the feeder room. In the frequency range from 5 to 6 MHz, defective conditions 2 and 3 become highly distinguishable due to the significant impedance differences near the respective resonant frequencies of Z_{DC2} and Z_{DC3}. Additionally, Figure 9.17 shows that Z_{DC2} is larger than Z_{DC3}, and Z_{DC3} is larger than Z_G below 5.5 MHz. This relationship among Z_G, Z_{DC2}, and Z_{DC3} is attributed to changes in inductance under different operating conditions, which can be further explained through detailed analysis of the feeder's equivalent circuit under each condition.

This section has demonstrated the feasibility and effectiveness of the proposed inductive-coupling-based online impedance measurement method for defect detection in high-power DC feeders composed of multiple parallel conductors. By extracting the online impedance of the feeder without electrical contact or service interruption, the method enables clear discrimination between healthy and defective operating conditions, as well as between different defect scenarios. Experimental results confirm that cable disconnection-related faults introduce distinct impedance signatures over a wide frequency range, allowing early-stage defect identification without requiring access to cable terminations. These findings highlight the potential of the proposed approach for nonintrusive condition monitoring of high-current DC distribution infrastructures in power and energy applications.

9.3 EMI Estimation of PV Systems

PV systems have been widely deployed as an important renewable energy source in modern power and energy infrastructures. In a typical PV system, electrical energy generated by the photovoltaic effect is processed by high-frequency switching power converters to deliver regulated power to the grid or local loads. While these converters enable efficient energy conversion, their fast-switching operation inevitably generates EMI, particularly on the DC side of the system. Conducted noise can propagate along DC cables and radiate through the PV panel and cable structures, which often act as unintentional antennas, leading to potential interference with nearby electronic equipment and communication systems [16–18].

Existing studies on EMI from PV systems have mainly relied on direct measurement or numerical modeling. Measurement-based approaches provide useful insight into emission characteristics under specific installation and operating conditions [17, 18], whereas simulation-based methods aim to estimate radiated emissions by modeling the PV panel and cable structures [19]. However, these approaches typically require detailed knowledge of internal circuit parameters, geometrical layouts, or installation-specific information, which is often unavailable in practice. Moreover, the impedance characteristics of PV systems vary with operating conditions and converter switching behavior, which are not adequately captured in many conventional EMI estimation methods [20].

These limitations underscore the need for effective EMI estimation that can accurately reflect actual operating conditions while avoiding reliance on detailed internal circuit or geometric information. Accordingly, this section introduces an EMI estimation method for PV systems based on noncontact online impedance measurement. It demonstrates how contactless online impedance extraction can be employed to characterize equivalent noise sources and to estimate both conducted and radiated emissions under realistic operating conditions.

9.3.1 AC Small-signal Model of PV Panel

The electrical characteristics of a PV panel depend strongly on its operating conditions. This dependency is also important from an EMC perspective, as it directly influences noise generation and propagation within the system. Since a PV cell essentially behaves as a p–n junction, it can be modeled using diode-based equivalent circuits with additional parasitic resistance, capacitance, and inductance, as shown in Figure 9.18 [20]. The dynamic PV model consists of a photocurrent source I_{ph}, a forward-biased conduction diode D_f, a variable parallel capacitance C_P, a reverse-bias conducting diode D_r with a breakdown voltage offset V_{bd} (negative value), a shunt resistance R_{sh}, a series resistance R_s, and a series inductance L_s. The diode D_f dominates when the PV cell operates under forward bias, whereas D_r dominates when the junction is reverse biased beyond the breakdown voltage V_{bd}.

In this model, the DC-related parameters include I_{ph}, V_{bd}, D_f, and D_r. For EMI analysis in the frequency range of approximately 150 kHz–30 MHz, these DC parameters can be neglected. Consequently, these elements, together with the shunt resistance R_{sh}, can be replaced by an equivalent parallel resistance R_P. Based on this simplification, the PV panel can be represented by the small-signal AC

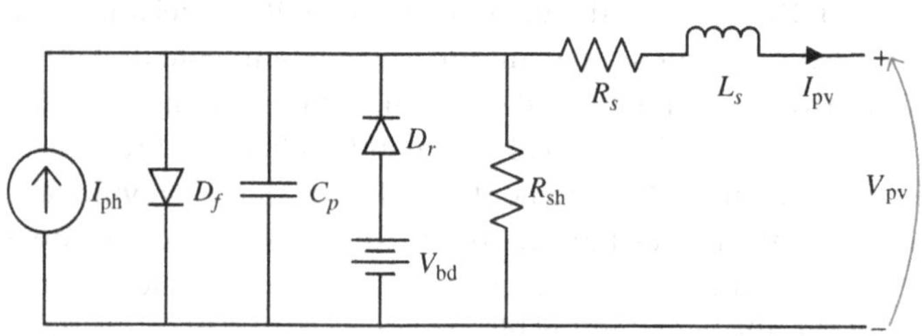

Figure 9.18 Equivalent model of the PV panel.

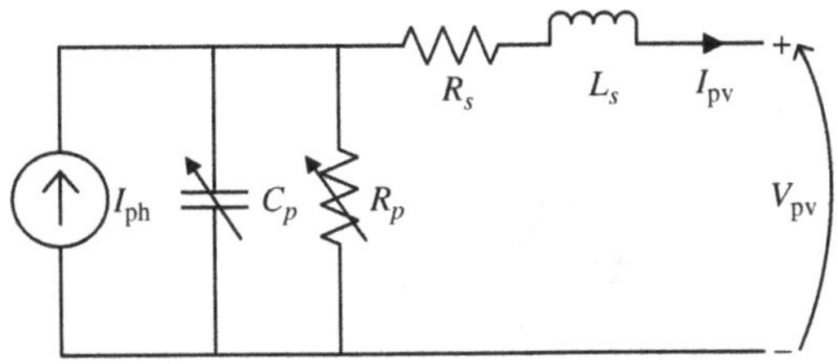

Figure 9.19 AC small-signal model of the PV panel.

model shown in Figure 9.19 [21]. As illustrated in the figure, the remaining parameters in the simplified model include the equivalent parallel capacitance C_P, the series resistanceR_S, and the series inductance L_S.

For the small-signal model, the impedance of the PV panel can be expressed as

$$Z = \left[R_S + \frac{R_P}{\alpha + 1}\right] + j\left[\omega L_S - \frac{\omega R_P^2 C_P}{\alpha + 1}\right] \tag{9.9}$$

where $\omega = 2\pi f$ and $\alpha = \omega^2 R_P^2 C_P^2$.

At low frequencies, the impedance of the PV panel is dominated by R_P and R_S. However, at frequencies of concern for EMI analysis, R_P may no longer be dominant because it is effectively shunted by the parallel capacitance C_P. The series resistance R_S mainly affects the overall impedance around the resonant frequency. Overall, C_P, R_S, and L_S play significant roles in determining the impedance characteristics of the PV panel within the EMI-relevant frequency range. While R_S and L_S are relatively constant, C_P varies with operating conditions. During PV panel operation, C_P is primarily influenced by two components: the junction (or transition) capacitance C_j and the diffusion capacitance C_d [20]. C_j is determined by the charge stored in the depletion region and mainly affects small positive and negative voltages when the junction does not conduct significant current. In contrast, the diffusion capacitance C_d arises from charge stored in the neutral region outside the depletion layer and becomes dominant when the junction conducts a substantial current. As a result of these effects, the impedance of a PV panel is strongly dependent on its operating condition. It is therefore essential to determine the equivalent impedance of the panel at the specific operating point of interest for accurate EMI estimation.

In this section, the equivalent small-signal model parameters of a PV panel are extracted using the online impedance measurement method. The objective of the experiment is to demonstrate both the dependence of the model parameters on operating conditions and the applicability of the impedance-based method for parameter extraction. The experimental setup used to extract the equivalent impedance of the PV panel is illustrated in Figure 9.20.

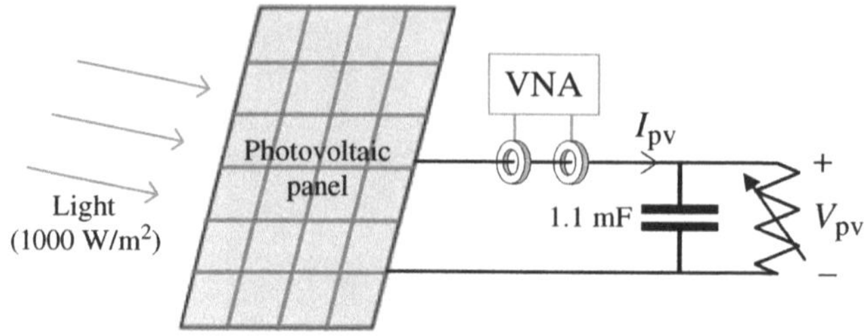

(a)

(b)

Figure 9.20 Extraction of the equivalent impedance of the PV panel. (a) Schematic diagram. (b) Actual experimental setup.

Figure 9.20a shows the schematic diagram of the measurement configuration, in which the PV panel is illuminated by a controlled light source with an irradiance of 1000 W/m^2. An FD-TPS consisting of a VNA (Bode 100) together with two inductive probes (Solar 9144-1N) is used to inject and measure the small-signal excitation. A 1.1 mF capacitor is connected in parallel with a variable resistive load to provide a low-impedance path for the RF signal while allowing the operating point of the PV panel to be adjusted. Figure 9.20b shows the actual experimental setup used in the measurements.

9.3.1.1 Impedance Under Dark and Illuminated Conditions

The magnitude and phase of the extracted PV panel impedance under dark and illuminated conditions are shown in Figure 9.21. To ensure measurement accuracy, the impedance of the short connection wires used to connect the PV panel to the variable resistor and capacitor (1.1 mF) is first measured and subsequently de-embedded from the measured results. Figure 9.21a compares the magnitude

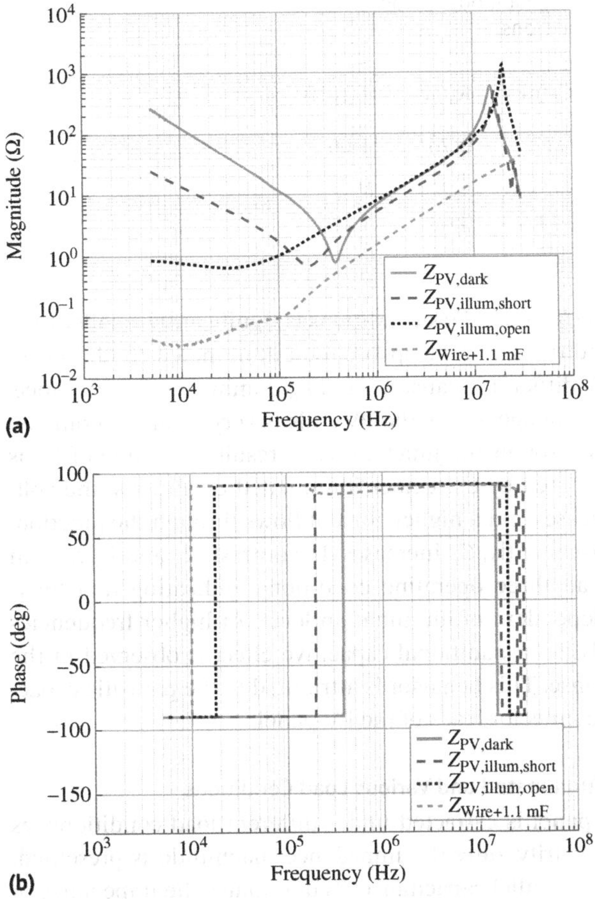

Figure 9.21 Comparison of the PV impedance under dark and illuminated conditions. (a) Magnitude. (b) Phase.

of the PV panel impedance under dark conditions and under two illuminated conditions, corresponding to short-circuit and open-circuit operating points. Figure 9.21b shows the corresponding phase responses. Under illuminated conditions, clear differences in both magnitude and phase are observed, indicating that the impedance characteristics of the PV panel are strongly dependent on its operating state.

The extracted small-signal model parameters under dark and illuminated conditions are summarized in Table 9.2.

Table 9.2 Extracted PV small-signal model parameters under dark and illuminated conditions.

Condition	C_P(μF)	R_S(Ω)	L_S(μH)
Dark	0.12	0.76	1.09
Illuminated (short)	1.01	0.68	0.79
Illuminated (open)	~11.9	0.63	1.13

It shows that the value of C_P under dark conditions is significantly smaller than that under illuminated conditions. For the open-circuit illuminated condition, an accurate estimation of C_P is difficult because R_P and R_S dominate the impedance. Under dark conditions, the voltage across the PV cells is very small and only an insignificant current flows through the junction. As a result, the value of C_P is low and is primarily determined by C_j. Under illuminated conditions, the voltage across the PV cells increases and a higher current flows through the junction, leading to a larger C_d. Consequently, C_P increases. In contrast, R_S and L_S remain relatively constant across all three operating conditions, indicating that these parameters are largely independent of illumination level. At higher frequencies above approximately 10 MHz, an additional capacitive effect is observed in the measured impedance response. This behavior is attributed to the parasitic capacitance associated with the connector leads of the PV panel.

9.3.1.2 Impedance Under Illuminated and Various Load Conditions

The impedance of the PV panel is extracted under different load conditions, as shown in Figure 9.22. For clarity, only the impedance magnitude is presented. The results indicate that the parallel capacitance C_P dominates the impedance of the PV panel at frequencies below approximately 100 kHz. Above 100 kHz, the impedance becomes inductive as the effect of the series inductance L_S becomes dominant. Near the resonance around 100 kHz, the impedance is primarily determined by the series resistance R_S. At higher frequencies above approximately 10 MHz, connectors of the PV panel introduce another capacitive effect due to the parasitic capacitance.

The estimated values of the small-signal model parameters (C_P, R_S, and L_S) obtained from the measurements are summarized in Table 9.3. As shown, C_P increases as the panel voltage V_{PV} increases. This behavior can be explained by the transition from C_j, which varies only slightly with voltage, to C_d, which becomes dominant as the operating voltage increases. As V_{PV} increases, the voltage bias across the junction also increases, leading to higher carrier injection and, consequently, an increase in C_d. The impedance of the PV panel under various

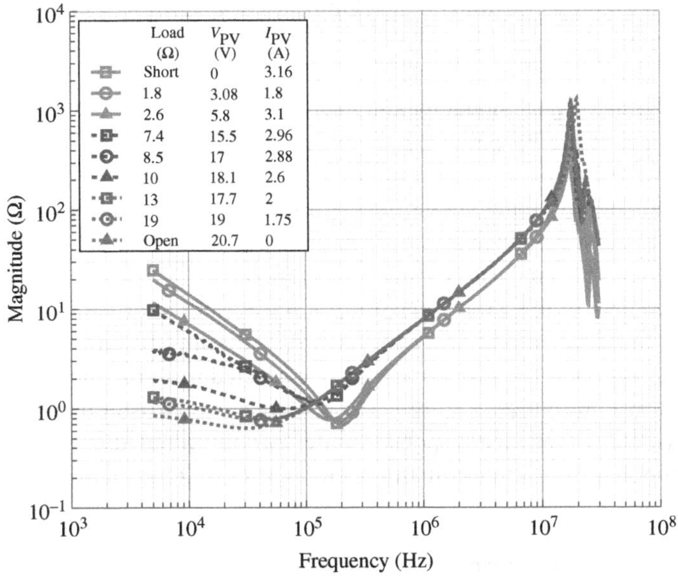

Figure 9.22 Comparison of PV impedance under illuminated and various load conditions.

Table 9.3 Extracted PV small-signal model parameters under illuminated and different load conditions.

Operation Condition		**Model Parameters**			
Load	V_{PV} (V)	I_{PV} (A)	C_P(μF)	R_S(Ω)	L_S(μH)
Short	0	3.16	1.01	0.68	0.79
1.8	3.08	1.8	1.19	0.72	0.80
2.6	5.8	3.1	1.91	0.76	0.79
7.4	15.5	2.96	2.29	1.10	1.13
8.5	17	2.88	~2.82	1.19	1.15
10	18.1	2.6	~5.71	0.99	1.15
13	17.7	2	~8.30	0.79	1.15
19	19	1.75	~9-10	0.76	1.16
Open	20.7	0	~11.90	0.63	1.13

operating conditions is therefore effectively extracted using the online impedance measurement method. If the resistive load is replaced by a power converter, the same approach can be applied to extract the PV panel impedance under actual operating conditions, which is essential for developing accurate EMI noise models in practical PV systems.

9.3.1.3 Modified PV Small-signal Model for EMI Analysis

The small-signal model of the PV panel can be constructed for a specific operating condition based on the measurement results discussed earlier. However, modifications to the small-signal model shown in Figure 9.19 are required in order to accurately estimate the impedance at higher frequencies. The modified model used for high-frequency EMI analysis is shown in Figure 9.23.

To account for the effect of the PV panel connection leads at high frequencies, an external parallel capacitance C_{ext} and resistance R_{ext} are introduced into the modified model. The values of these parameters are extracted from the measurement results shown in Figure 9.22. Figure 9.24 presents a comparison between simulated and measured impedance responses under two illuminated operating conditions, corresponding to short-circuit and open-circuit load terminations.

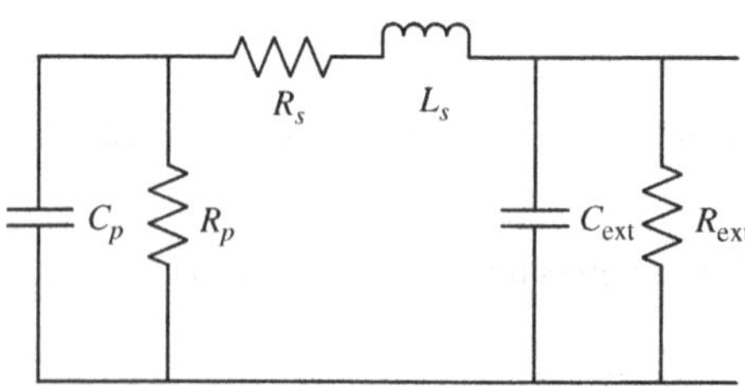

Figure 9.23 Modified PV small-signal model for EMI analysis.

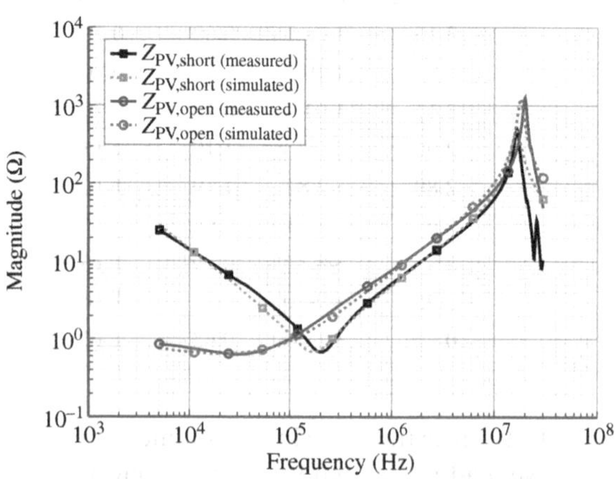

Figure 9.24 Comparison of simulated and measured impedance of PV under illumination.

For the short-circuit condition, the model parameters C_P, L_S, and R_S are obtained directly, as shown in Table 9.3. R_P is not included in this case, as it does not significantly influence the impedance within the frequency range of interest. $C_{ext} = 0.12$ nF and $R_{ext} = 498.9\ \Omega$ are extracted from the measured results shown in Figure 9.22. For the open-circuit condition, L_S and R_S are again obtained as shown in Table 9.3. R_P is determined from DC measurements, while C_P is estimated analytically. The parameters used for the open-circuit case are $C_P =$ 1.3 μF, $R_p = 0.3\ \Omega$, $C_{ext} = 70$ pF, and $R_{ext} = 1.3$ kΩ in this case study. As shown in Figure 9.24, good agreement is achieved between the simulated and measured impedance results.

This section employs an online impedance measurement method to extract the equivalent circuit model of the PV panel under its actual operating conditions. Accurate extraction of the panel impedance is essential, as it directly determines the DM impedance of the PV system. In contrast, the CM impedance is strongly influenced by factors such as grounding configuration and panel placement. It is therefore necessary to extract both DM- and CM-related impedance characteristics under realistic operating conditions. The extracted impedance over the frequency range of interest can subsequently be used to evaluate the EMI impact of a PV system when it is connected to a high-frequency switching power converter.

9.3.2 Conducted Emissions Estimation

The experimental setup of the PV system with an actual PV panel, operating under variable conditions, is shown in Figure 9.25. To characterize the influence of the PV panel and the DC cables at the input side, the AC grid connection at the output side is intentionally omitted. For simplicity and relevance, the system is configured using a DC–DC converter, which constitutes the input stage of most PV power conversion systems.

In the setup, a DC–DC buck converter is employed, with its output connected to a resistive load, while its input is supplied by a 32-cell 60 W PV panel. The PV panel and the buck converter are placed 5 cm above a copper ground plate with dimensions of 3 m × 1 m. This metal plate serves as the system ground and provides a well-defined return path for CM noise currents. The actual experimental setup is shown in Figure 9.25b, and the device specifications and operating parameters are summarized in Table 9.4. The entire setup is placed inside a shielded room to facilitate accurate radiated emission measurements.

For system operation, the PV panel is illuminated using halogen lamps. The panel is operated at a light intensity of 1000 W/m^2, with the ambient temperature maintained consistent. DC power is supplied by the PV panel to the buck converter, which operates at a switching frequency of 50 kHz. The duty cycle of the switch is manually set and fixed at 0.55 to maintain the system at a specific operating point. Under this operating condition, the output voltage and current of the PV panel remain stable during operation. To model the EMI noise in the system, the noise source impedance and termination impedance are extracted under operating conditions using the TD-TPS. The noise current in the system is then measured using current probes to estimate the noise source current. In addition, an accurate model of the DC cable is required to represent the noise propagation path. These aspects are discussed in detail in the following sections.

The measurement setups used to extract the CM and DM impedances of the noise sources with an actual PV panel are shown in Figure 9.26a,b, respectively. For the FD-TP, one Bode 100 VNA and two Solar 9144-1N inductive probes are used. Capacitors with values of 1 and 22 μF are connected to provide low-impedance paths for RF signals.

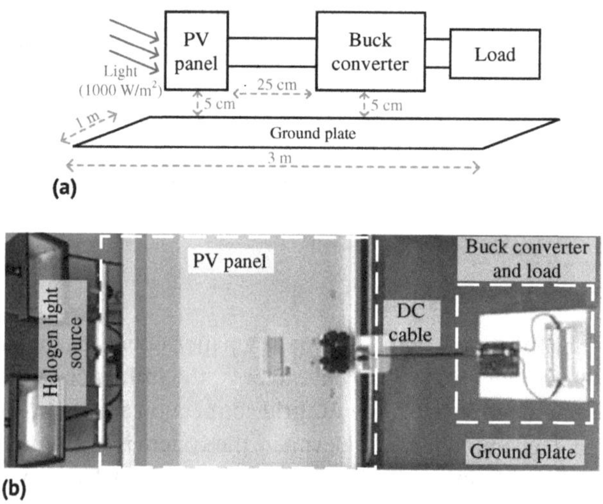

Figure 9.25 Measurement setup of the PV system. (a) Block diagram. (b) Actual setup.

Table 9.4 Equipment specifications and settings.

Equipment	Specifications
PV panel	Nominal peak power (W_p): 60 W Nominal voltage (V_{mp}): 18.3 V Nominal current (I_{mp}): 3.27 A Open-circuit voltage (V_{oc}): 22 V Short-circuit current (I_{sc}): 3.67 A Dimensions: 685 × 670 × 35 mm^3
Buck converter	Input: 10–40 V DC, 5 A Switching frequency: 50 kHz
Load	2.2 Ω

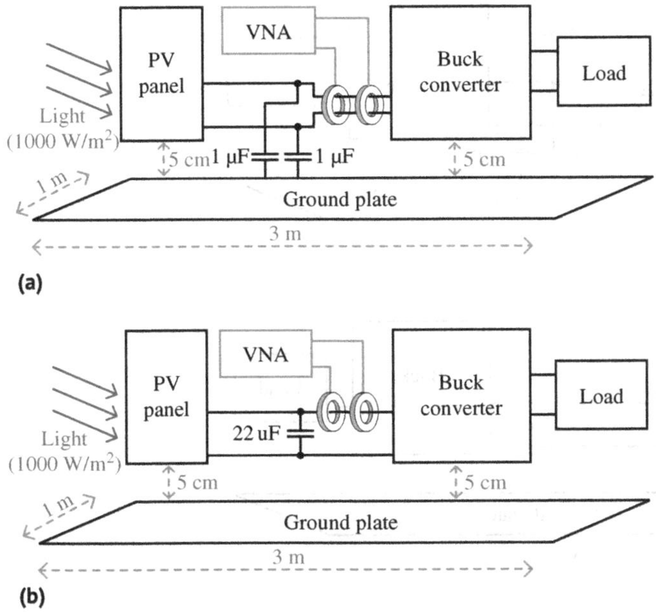

Figure 9.26 Measurement setup for extracting the noise source online impedance. (a) CM. (b) DM.

The total impedances measured for the two configurations are given as

$$Z_{\text{total,CM}} = Z_{\text{noise,CM}} + Z_{\text{wire+2}\mu\text{F}} \tag{9.10}$$

$$Z_{\text{total,DM}} = Z_{\text{noise,DM}} + Z_{\text{wire+22}\mu\text{F}} \tag{9.11}$$

where $Z_{\text{noise,CM}}$ and $Z_{\text{noise,DM}}$ denote the actual CM and DM noise source impedances, respectively, and $Z_{\text{wire+2}\mu\text{F}}$ and $Z_{\text{wire+22}\mu\text{F}}$ represent the combined impedances of the connection wires and bypass capacitors for the two cases.

After extracting the noise source impedance, the online impedance of the PV panel is measured by clamping the inductive probes on the PV panel side of the system, as shown in Figure 9.27. Since the electrical characteristics of the PV panel depend on its operating point, as demonstrated in Section 9.3.1, its impedance must be extracted under the actual operating condition. Although the DM impedance of the PV panel is more strongly influenced by operating conditions, both CM and DM impedances are extracted at the operating point to ensure accurate EMI modeling. The total impedances extracted for the PV panel are given by

$$Z_{\text{total,CM}} = Z_{\text{PV,CM}} + Z_{\text{wire+2}\mu\text{F}} \tag{9.12}$$

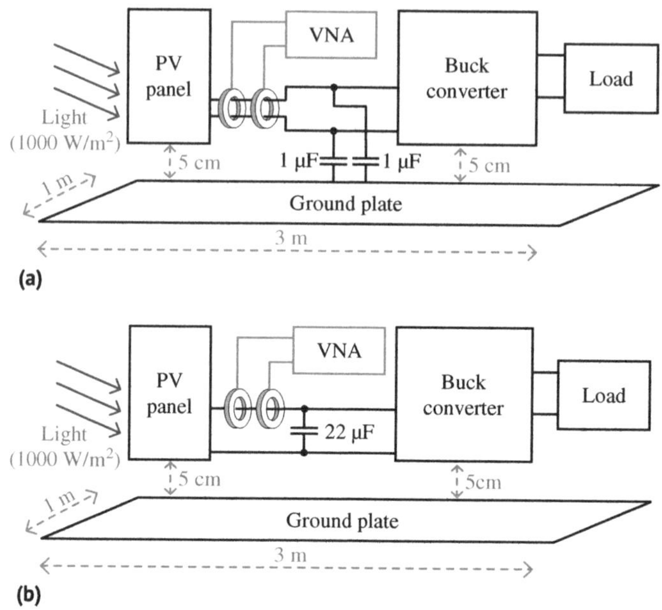

Figure 9.27 Measurement setup for extracting the PV panel online impedances. (a) CM. (b) DM.

$$Z_{\text{total,DM}} = Z_{\text{PV,DM}} + Z_{\text{wire+22}\mu\text{F}} \tag{9.13}$$

where $Z_{\text{PV,CM}}$ and $Z_{\text{PV,DM}}$ denote the actual CM and DM impedances of the PV panel, respectively.

The extracted magnitude and phase responses of the CM impedances are shown in Figure 9.28a,b, respectively. These results indicate that the bypass capacitors operate effectively over the frequency range of interest. Consequently, the total impedances measured in Equations (9.10) and (9.12) are dominated by $Z_{\text{noise,CM}}$ and $Z_{\text{PV,CM}}$, respectively. The results further show that $Z_{\text{noise,CM}}$ can be well approximated by a 0.02 nF capacitor, which is expected since the CM impedance of the noise source is dominated by the parasitic capacitance between the converter and ground. Similarly, $Z_{\text{PV,CM}}$ can be approximated by a 0.07 nF capacitor. The larger capacitance value is attributed to the parasitic capacitance between the PV panel structure and ground, which is physically larger in extent.

Figure 9.29a,b show the extracted magnitude and phase responses of the DM impedances, respectively. The results confirm that the bypass capacitor also operates effectively for the DM case. Under these conditions, the total impedance in Equation (9.13) is dominated by $Z_{\text{PV,DM}}$, and the contribution of $Z_{\text{wire+22}\mu\text{F}}$ can be neglected. In contrast, for the total impedance in Equation (9.11), the contribution

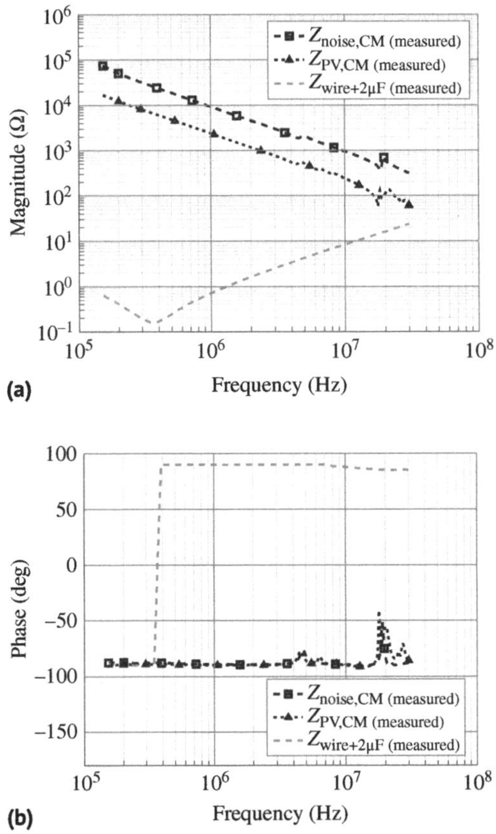

Figure 9.28 Extracted CM impedances. (a) Magnitude. (b) Phase.

of $Z_{wire+22\mu F}$ cannot be ignored, as it is comparable to $Z_{noise,DM}$ and must therefore be de-embedded. After de-embedding, it is observed that $Z_{noise,DM}$ can be effectively represented by an inductor with an inductance of 0.055 μH.

In the case of $Z_{PV,DM}$, the extracted results exhibit behavior similar to that observed in Figure 9.22. The equivalent PV model illustrated in Figure 9.23 can therefore be constructed based on the obtained results. At the operating point of $V_{PV} = 18.1$ V and $I_{PV} = 2.6$ A, the parallel capacitance C_P, parallel resistance R_P, and series resistance R_S are insignificant within the frequency range of interest and can be neglected. Under these conditions, the equivalent PV model can be represented by a series inductance $L_S = 1.5$ μH, an external parallel capacitance $C_{ext} =$ 50 pF, and an external resistance $R_{ext} = 840$ Ω. The CM and DM impedances of both the noise source and the PV panel are thus extracted under operating conditions using the FD-TPS. These impedances enable estimation of the noise sources

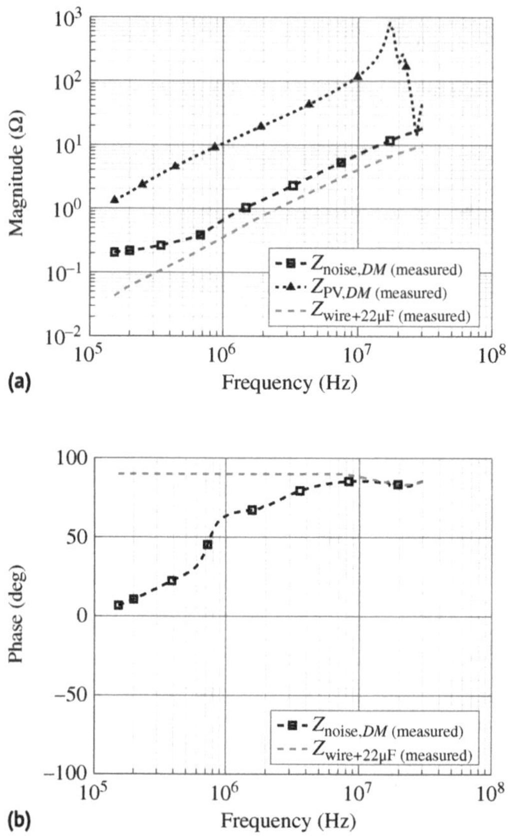

Figure 9.29 Extracted DM impedances. (a) Magnitude. (b) Phase.

in the PV system, which is essential for developing a complete noise model. The estimation of the noise source is discussed in the following.

9.3.2.1 Noise Source Estimation

Since the PV system does not employ a LISN, the noise current in the system is measured using a current probe in conjunction with a spectrum analyzer, as shown in Figure 9.30.

The CM noise current $I_{PV,CM}$ is measured by clamping the current probe around both the positive and negative DC conductors, as illustrated in Figure 9.30a. With this configuration, the DM noise currents flowing in the two conductors generate opposing magnetic fluxes in the probe core, which effectively cancel each other. As a result, the measured signal at the spectrum analyzer is attributed primarily

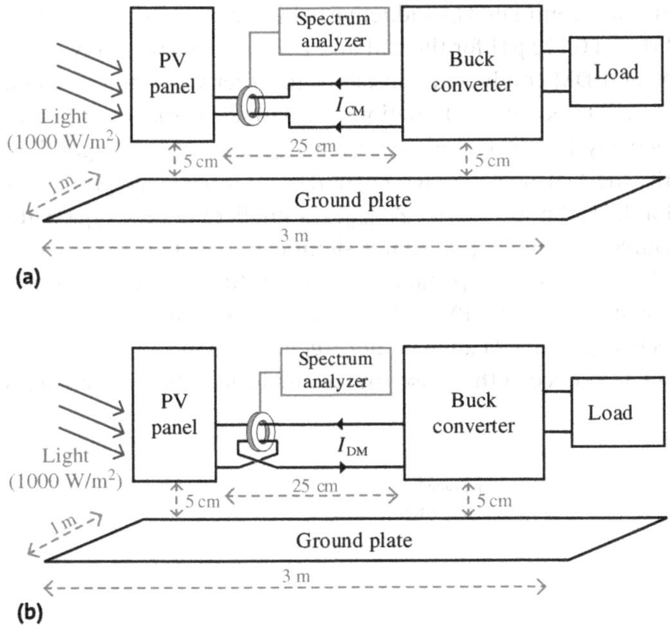

Figure 9.30 Measurement of conducted noise current. (a) CM. (b) DM.

to the CM current. Similarly, the DM noise current $I_{\mathrm{PV,DM}}$ is measured by clamping the probe around both conductors with one conductor wound in the opposite direction, as shown in Figure 9.30b. In this configuration, the magnetic fluxes produced by CM currents cancel each other, allowing the DM current component to be measured by the spectrum analyzer.

The measured current magnitude is obtained as

$$I[\mathrm{dB\mu A}] = V_{\mathrm{receiver}} + K \tag{9.14}$$

where V_{receiver} is the voltage reading from the spectrum analyzer in dBμV, and K is the transfer admittance of the current probe expressed in dBS.

The equivalent noise source currents $I_{\mathrm{noise,CM}}$ and $I_{\mathrm{noise,DM}}$ are derived from the measured currents as

$$I_{\mathrm{noise,CM}} = I_{\mathrm{PV,CM}} \times \frac{Z_{\mathrm{noise,CM}} + Z_{\mathrm{cable,CM}} + Z_{\mathrm{PV,CM}}}{Z_{\mathrm{noise,CM}}} \tag{9.15}$$

$$I_{\mathrm{noise,DM}} = I_{\mathrm{PV,DM}} \times \frac{Z_{\mathrm{noise,DM}} + Z_{\mathrm{cable,DM}} + Z_{\mathrm{PV,DM}}}{Z_{\mathrm{noise,DM}}} \tag{9.16}$$

The short cable between the PV panel and the buck converter is accounted for by measuring the CM cable impedance $Z_{\mathrm{cable,CM}}$ and the DM cable impedance

$Z_{cable,DM}$, both of which are dominated by inductive behavior. The extracted inductance values are 0.351 and 0.172 μH for the CM and DM cases, respectively.

The measured CM and DM conducted emissions, together with the estimated noise source emissions obtained using Equations (9.15) and (9.16), are shown in Figure 9.31a,b, respectively. For the CM noise source, $Z_{noise,CM}$ is much larger than $Z_{PV,CM}$; consequently, the CM noise source current $I_{noise,CM}$ closely approaches $I_{PV,CM}$. In contrast, for the DM noise source, $Z_{noise,DM}$ is smaller than $Z_{PV,DM}$, resulting in a DM noise source current $I_{noise,DM}$ that is significantly higher than $I_{PV,DM}$.

With the extracted CM and DM impedances and the estimated noise sources, the CM and DM noise models of the PV system can be constructed, as shown in Figure 9.32a,b, respectively. Both CM and DM noise generation in a PV system can be significant, and the influence of the noise source and termination impedances

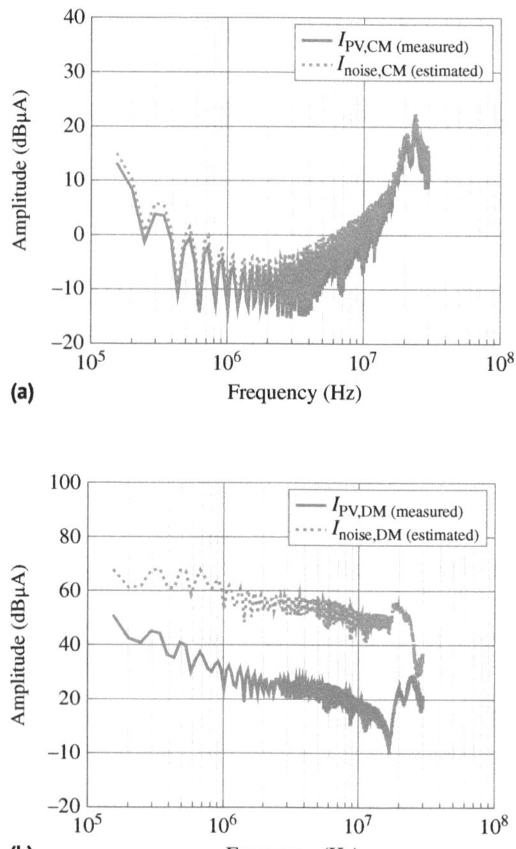

Figure 9.31 Conducted emissions at the PV panel and at the noise source. (a) CM. (b) DM.

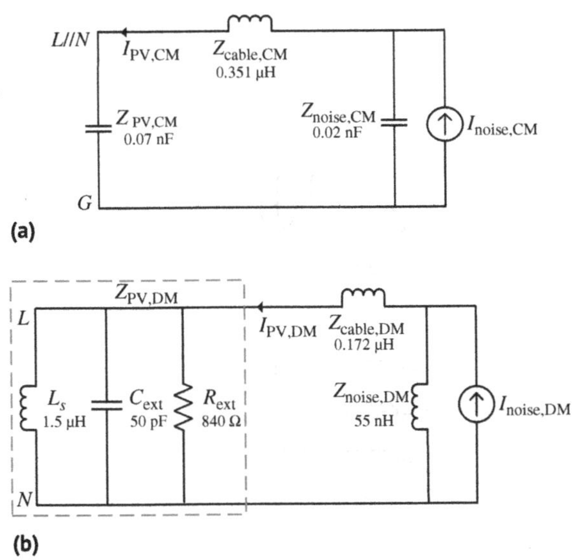

Figure 9.32 Constructed noise models of the PV system. (a) CM. (b) DM.

on noise propagation can be clearly visualized using these models. In the absence of filtering, noise propagates toward the PV panel and can give rise to radiated emissions, which are discussed in Section 9.3.3. In the following analysis, the effect of long DC cables is incorporated into the noise models.

9.3.2.2 Influence of Long Cables

In practical PV installations, long DC cable lengths must be considered, which requires modeling the cables as transmission lines. In the PV system configuration shown in Figure 9.25, a long DC cable is inserted between the PV panel and the buck converter. Figure 9.33a illustrates the cross-section of the DC cable used in this experiment. The conductor radius r is 1.2 mm, and the center-to-center spacing between the conductors d is 2.6 mm. Two cable lengths, 1 and 3 m, are investigated. In both cases, the cable is maintained at a height h of 5 cm above the system ground plane. The equivalent CM and DM noise models incorporating the long DC cable are shown in Figure 9.33b,c, respectively. The remaining elements in the model, including the noise source and termination impedances, are obtained from the previously extracted results.

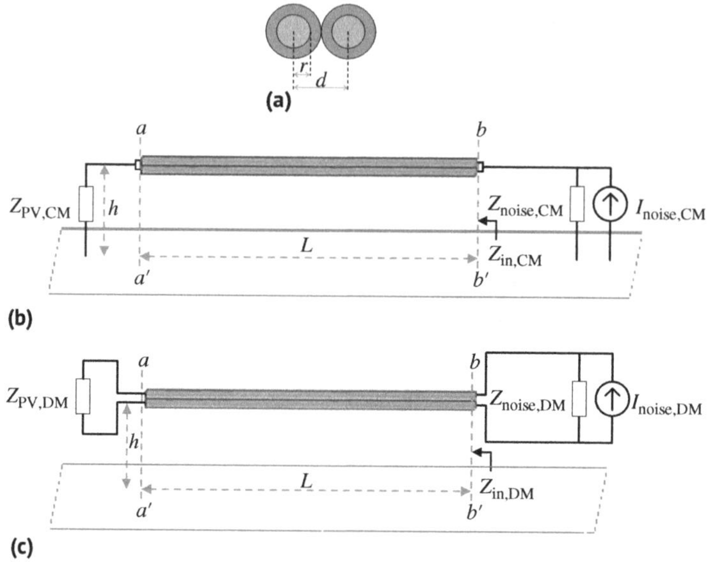

Figure 9.33 PV system model with long DC cable. (a) Cross-section of cable. (b) CM model. (c) DM model.

For a two-conductor cable above a ground plane, the CM characteristic impedance $Z_{C,CM}$ of the transmission line is given by [22]

$$Z_{C,CM} = 30 \ln\left[\frac{2h}{r}\sqrt{1+\left(\frac{2h}{d}\right)^2}\right] \tag{9.17}$$

Similarly, the DM characteristic impedance $Z_{c,DM}$ is given by

$$Z_{C,DM} = 120 \ln\left[\frac{d}{r}\sqrt{\frac{1}{1+\left(\frac{d}{2h}\right)^2}}\right] \tag{9.18}$$

Using Equations (9.17) and (9.18), the calculated values of $Z_{C,CM}$ and $Z_{C,DM}$ for the experimental setup are 218.98 and 92.78 Ω, respectively. Once the characteristic impedances are determined, the noise current $I(L)$ at the end of the transmission line, corresponding to the input of the PV panel, can be calculated as [23]

$$I(L) = \frac{V_0}{Z_C}\left[\frac{1-\Gamma_L}{1+\Gamma_L e^{-j2\beta L}}\right] \tag{9.19}$$

where the phase constant $\beta = 2\pi/\lambda$, and λ is the wavelength. The input voltage of the transmission line is V_0, and the characteristic impedance Z_C is replaced by

$Z_{C,\mathrm{CM}}$ or $Z_{C,\mathrm{DM}}$ for the CM and DM cases, respectively. The load reflection coefficient is defined as $\Gamma_L = (Z_L - Z_C)/(Z_L + Z_C)$, where Z_L is the load impedance at the end of the transmission line.

The estimated and measured noise emissions with a 1 m DC cable are shown in Figure 9.34. The results demonstrate close agreement between the estimated and measured CM and DM noise emissions. Compared with the case of a 25 cm DC cable shown in Figure 9.31, both CM and DM noise emissions increase at higher frequencies when a longer cable is used. The increased conducted noise emission can subsequently lead to higher radiated emissions, particularly in the CM case.

Figure 9.35 shows close agreement between the estimated and measured noise emissions for the system with a 3 m DC cable. By comparing these results with those presented in Figures 9.31 and 9.34, it is evident that the use of a longer DC

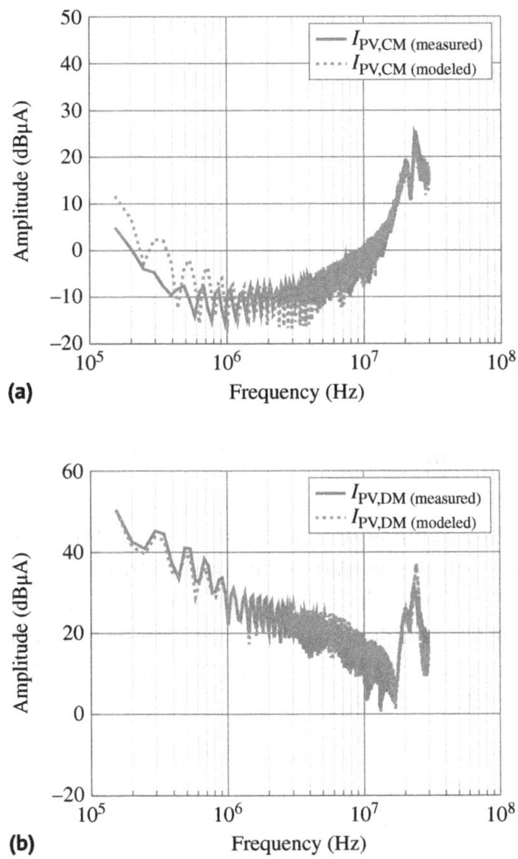

Figure 9.34 Comparison of noise emission with a 1 m DC cable. (a) CM. (b) DM.

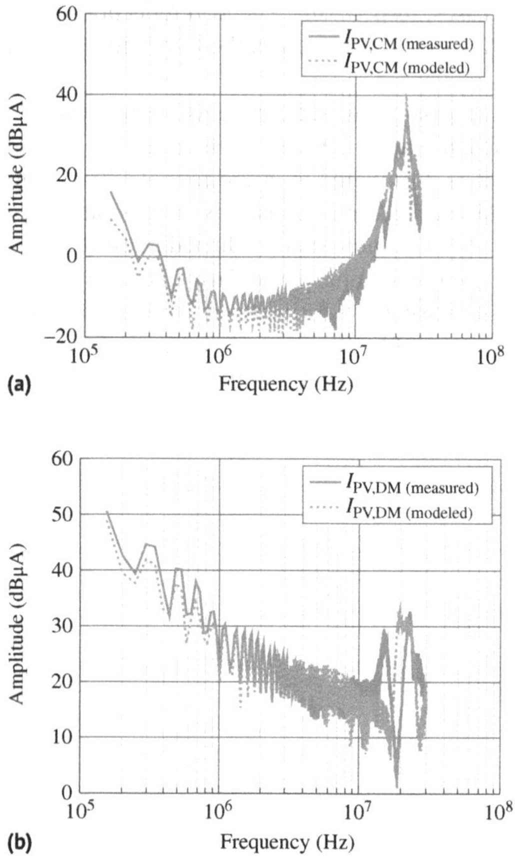

Figure 9.35 Comparison of noise emission with a 3 m DC cable. (a) CM. (b) DM.

cable leads to increased conducted noise emissions in both the CM and DM cases. This increase is primarily caused by multiple resonances introduced by the longer cable length in the system. Figure 9.35a further indicates that the increase in noise emission is more pronounced for the CM component. This behavior is particularly critical from an EMC perspective, as CM noise is more likely to couple to the surrounding environment and result in elevated radiated emissions.

9.3.3 Radiated Emissions Estimation

To measure the radiated emissions from the PV system over the frequency range from 150 kHz to 30 MHz, a loop antenna is employed. The measuring antenna is positioned at a distance of 1 m and a height of 1 m in front of the PV panel, as illustrated in Figure 9.36. As mentioned previously, the entire experimental setup is

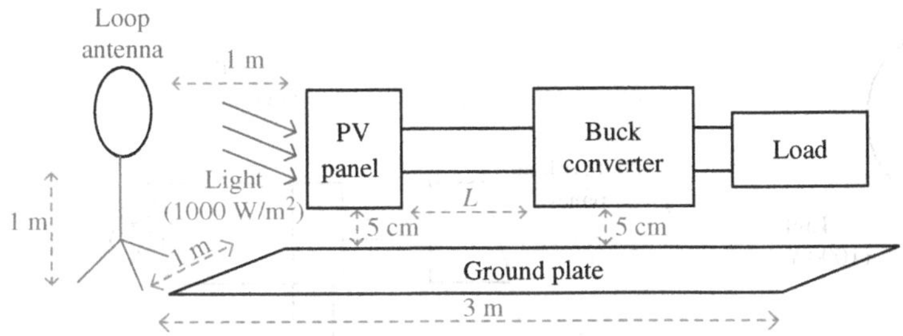

Figure 9.36 Radiated emission measurement setup.

placed inside a shielded room to prevent external electromagnetic noise from contaminating the radiation measurements obtained by the loop antenna. When the PV system is operating, the signal received by the loop antenna is first amplified and then measured using a spectrum analyzer. Both the amplifier and the spectrum analyzer are located outside the shielded room to avoid interference with the measurement environment.

The electric field strength of the radiated emission, E_{noise}, is defined as

$$E_{\text{noise}}[\text{dB}\mu\text{V/m}] = V_{\text{receiver}} - G + \text{AF} \tag{9.20}$$

where V_{receiver} is the voltage measured by the spectrum analyzer in dBμV, G is the amplifier gain in dB, and AF is the antenna factor of the loop antenna in dB/m.

Before measuring the radiated emissions from the complete PV system, the individual contributions of CM and DM conducted noise to radiated emissions are first investigated. This analysis allows the dominant radiation mechanism to be identified and enables the determination of separate AFs for the CM and DM radiation models.

9.3.3.1 Dominant Mode and AF

To characterize the effects of CM and DM conducted noise on radiated emissions, the AF of the PV panel is determined for each case. The AF is defined as the ratio of the radiated electric field intensity to the voltage across the antenna terminals. In decibel form, it is expressed as

$$\text{AF}[\text{dB/m}] = E[\text{dB}\mu\text{V/m}] - V[\text{dB}\mu\text{V}] \tag{9.21}$$

The measurement setups used to determine the AF for CM and DM radiated emissions from the PV panel are shown in Figure 9.37a,b, respectively. Since the electrical characteristics of the PV panel vary with operating conditions, it is necessary to determine the AF at the specific operating point of interest. For both CM and DM measurements, a variable resistor is connected to the PV panel, and the

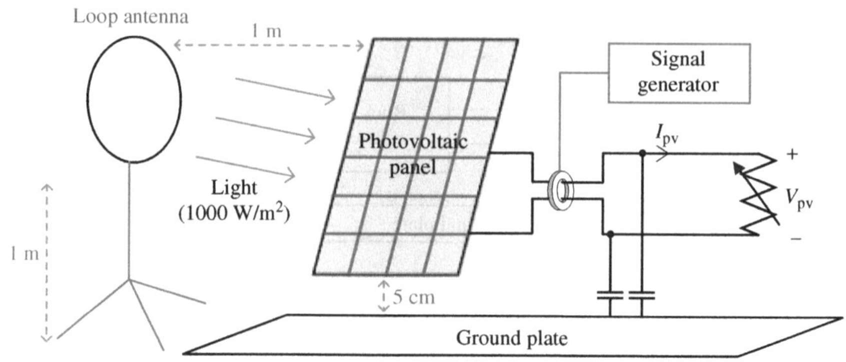

(a)

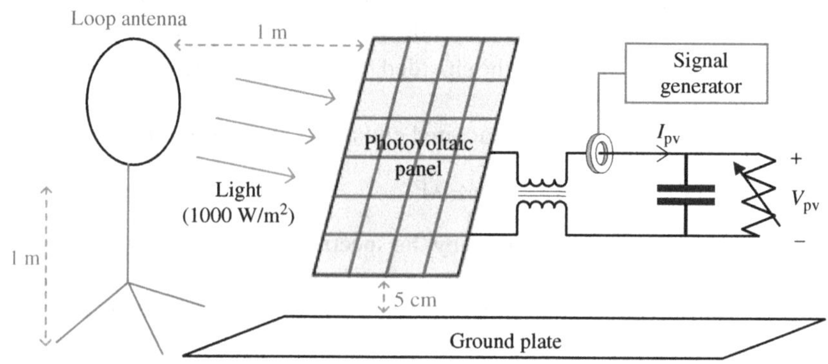

(b)

Figure 9.37 Radiation measurement with: (a) CM source excitation and (b) DM source excitation.

panel is operated at $V_{PV} = 18.1$ V and $I_{PV} = 2.6$ A to match the operating condition used in the previous experimental setups. For the determination of the AF for CM radiated emissions $(\mathrm{AF}_{PV,CM})$, a current probe is clamped around both the line and neutral conductors of the DC cable. Two bypass capacitors are connected from each conductor to the system ground plane. A signal generator is then connected to the probe to inject a CM current into the system. Similarly, for the determination of the AF for DM radiated emissions $(\mathrm{AF}_{PV,DM})$, the injecting probe is clamped onto only one conductor of the DC cable. A bypass capacitor is connected across the line and neutral conductors, and a 30 mH CM choke is inserted between the probe and the PV panel to suppress CM currents. Under this configuration, DM current is effectively injected into the system by the signal generator.

In both cases, the actual amplitudes of the injected noise currents, $I_{PV,CM}$ and $I_{PV,DM}$, are measured using an additional current probe in a configuration similar to that shown in Figure 9.36. Since $Z_{PV,CM}$ and $Z_{PV,DM}$ are known from Section 9.3.2, the voltage across the PV panel can be determined for each case. The electric field strength of the radiated emissions is measured using the loop antenna and calculated using Equation (9.20). Finally, the antenna factors $AF_{PV,CM}$ and $AF_{PV,DM}$ are obtained using Equation (9.21).

The extracted AFs of the PV panel for CM and DM radiated emissions are shown in Figure 9.38. The results clearly indicate that CM noise is the dominant mode responsible for radiated emissions. This implies that, for comparable levels of CM and DM conducted noise, radiated emissions due to CM currents will dominate the overall noise spectrum. Radiated emissions caused by DM conducted noise can become dominant only when the DM conducted noise level is significantly higher than that of CM noise. For low-power PV systems, this situation is typically not encountered, as demonstrated by the preceding results.

Since $AF_{PV,CM} \gg AF_{PV,DM}$, the radiated emission from the PV panel can be reasonably estimated based solely on the CM noise emission, provided that the DM noise level is not substantially higher than the CM noise. Accordingly, the radiated emission from the PV panel (E_{PV}) can be estimated as

$$E_{PV}[\mathrm{dB\mu V/m}] = AF_{PV,CM}[\mathrm{dB/m}] + V_{PV,CM}[\mathrm{dB\mu V}] \tag{9.22}$$

9.3.3.2 Radiated Emission with Different DC Cable Lengths

In this section, the measured radiated emissions from the PV system with different DC cable lengths are presented. The measurement setup follows that shown in Figure 9.36, with DC cables of different lengths inserted between the PV panel and

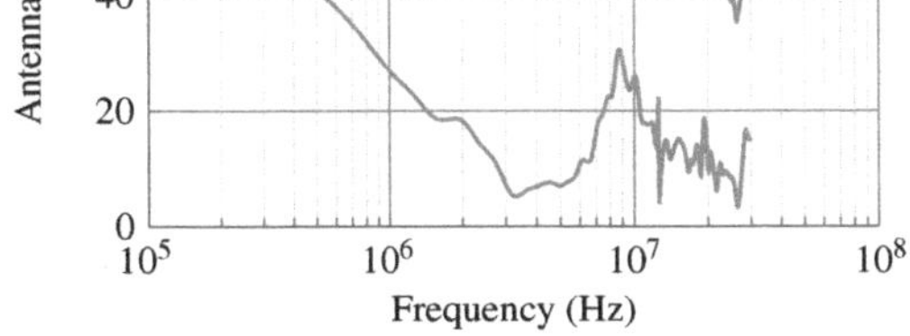

Figure 9.38 CM and DM AFs of the PV system.

the buck converter. In addition to direct measurements, the radiated emissions are also estimated using Equation (9.22).

Firstly, the radiated emission from the PV system with a short 25 cm DC cable is shown in Figure 9.39. The figure also includes the background noise level measured when the halogen lamps are switched on but the PV system is not operating. During the measurements, care is taken to ensure that the background noise remains below the radiated emission levels of interest. The measured radiated emissions are compared with the estimated results derived from the conducted emission data shown in Figure 9.31 using Equation (9.22). The close agreement between the measured and estimated results indicates that the radiated emissions are predominantly governed by CM conducted noise. The results further show that appreciable radiated emissions can occur from the PV system at both low and high frequencies.

Subsequently, Figure 9.40a,b present the radiated emission results for the PV system with DC cable lengths of 1 and 3 m, respectively. In both cases, good agreement is observed between the measured and estimated radiated emissions. The estimated results are obtained using the CM conducted emission data shown in Figures 9.34 and 9.35.

For the case with a 1 m DC cable, the radiated emission increases at higher frequencies, particularly around 25 MHz. This behavior is consistent with the corresponding increase in CM conducted emissions observed for the same cable length. Similarly, for the system with a 3 m DC cable, the radiated emissions worsen at high frequencies, as the longer cable introduces additional resonances and increases CM conducted noise, as shown in Figure 9.35.

This section has presented radiated emission results for PV systems with different DC cable lengths. The results demonstrate that radiated emissions increase

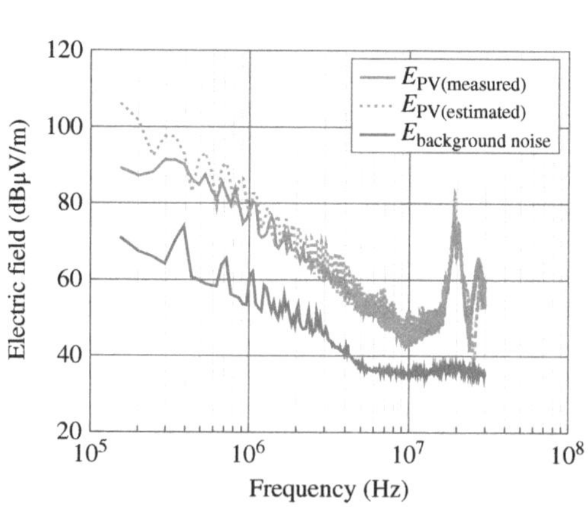

Figure 9.39 Radiated emission from the PV system with a 25 cm DC cable.

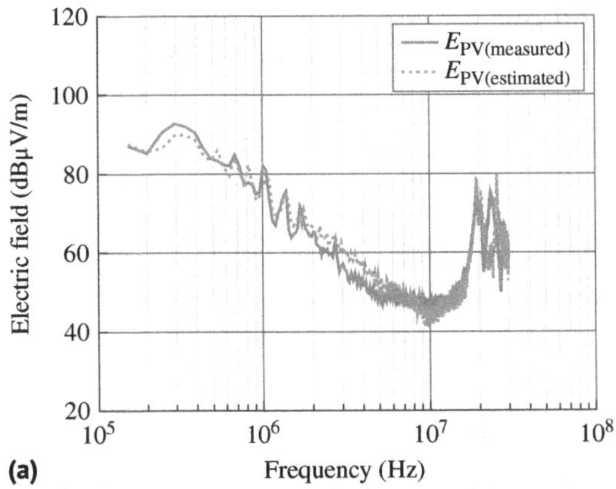

(a)

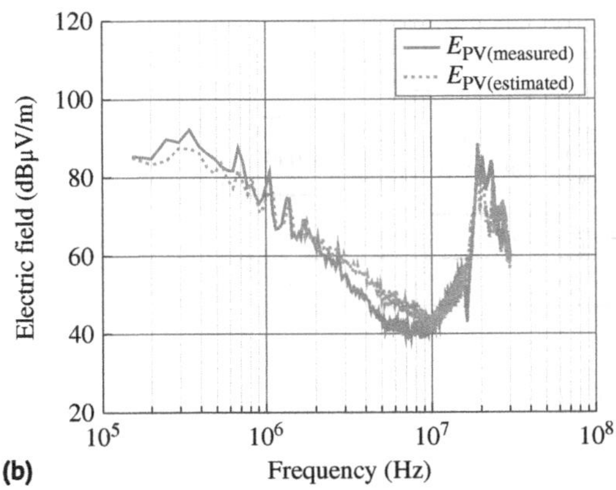

(b)

Figure 9.40 Radiated emission from the PV system with DC cables of different lengths: (a) 1 m and (b) 3 m.

with increasing cable length. It is also shown that radiated emissions can be effectively estimated from conducted emissions using Equation (9.22). The close agreement between measured and estimated results confirms the validity of the proposed approach for analyzing radiated emissions under different system configurations, such as longer DC cables or the inclusion of EMI filters. The use of EMI filters and their impact on noise emissions are discussed in the subsequent section.

9.3.4 EMI Filter Effect on Noise Emissions

The previous sections have shown that both conducted and radiated emissions generated by the PV system can be significant. To reduce these noise emissions,

EMI filters can serve as effective mitigation tools. In this section, the effects of using CM and DM filters on noise emissions from PV systems are investigated.

9.3.4.1 Impact of CM Filter

A CM LC filter consisting of one 10 mH CM choke (L_C) and two 1 nF capacitors (C_Y) is connected to the PV system, as shown in Figure 9.41. The equivalent noise model of the PV system with the CM filter is shown in Figure 9.42.

Using the measured impedances and the extracted noise source model, the conducted emission of the system with the CM filter can be calculated as

$$I_{PV} = I_{noise} \times \frac{Z_{noise} Z_{CY}}{Z_t Z_{CY} + Z_t Z_{noise} + Z_{CY} Z_{noise}} \tag{9.23}$$

where $Z_t = Z_{PV} + Z_{cable} + Z_{LC}$, and all terms are replaced by their respective CM or DM values depending on the mode under consideration.

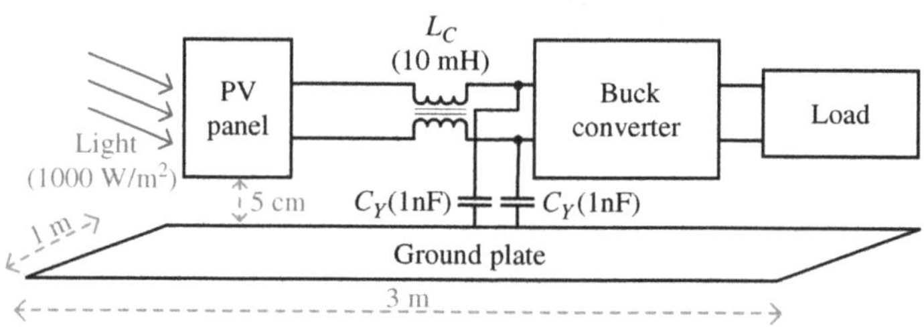

Figure 9.41 PV system with the CM filter.

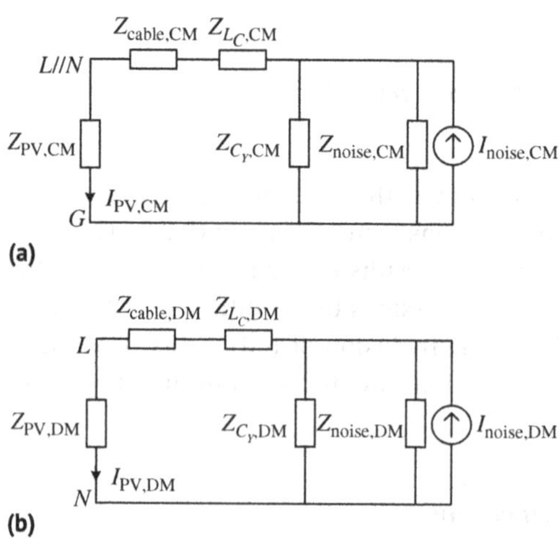

Figure 9.42 Equivalent noise models of Figure 9.41. (a) CM. (b) DM.

The calculated and measured CM and DM conducted emissions after inserting the CM filter are plotted in Figure 9.43a,b, respectively. Comparing these results with the emissions obtained without filters in Figure 9.31 shows that the CM emission is significantly reduced. The DM emission is also reduced in this case due to the effective DM impedance provided by the CM filter.

Radiated emissions are also measured for the PV system with the CM filter. The measured and estimated radiated emissions, shown in Figure 9.44, exhibit good agreement. The estimated radiated emission is obtained from the estimated CM conducted emission shown in Figure 9.43a using Equation (9.22). Since the CM

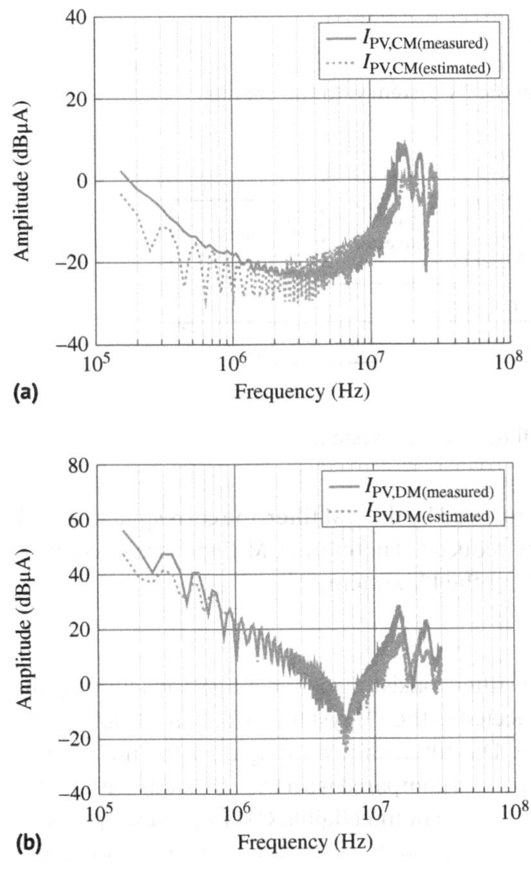

Figure 9.43 Comparison of conducted noise emission with the CM filter. (a) CM emission. (b) DM emission.

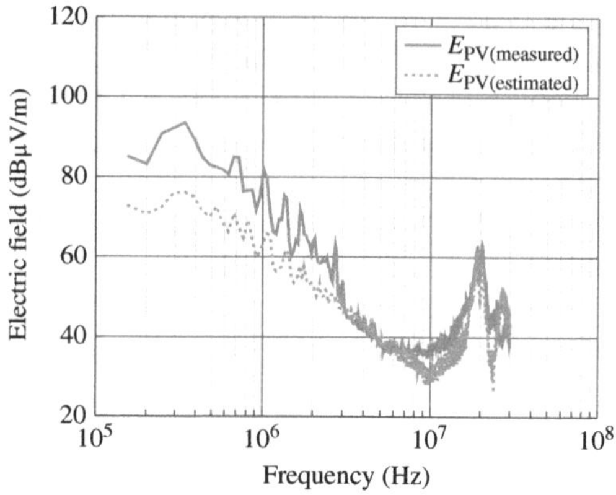

Figure 9.44 Radiated emission from the PV system with the CM filter.

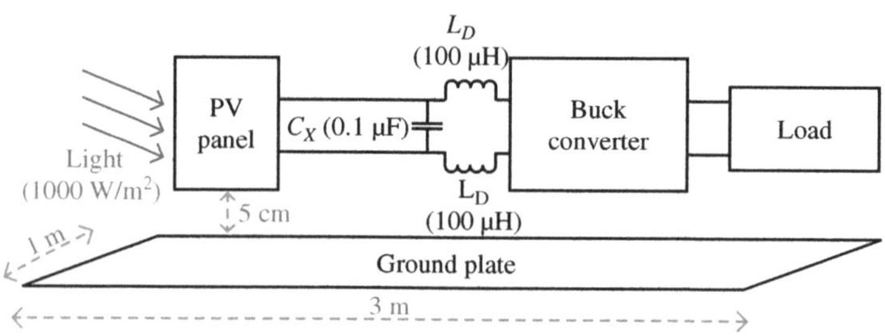

Figure 9.45 Connection of the DM filter in the PV system.

conducted emission is effectively reduced by the CM filter, the corresponding radiated emission is also significantly reduced. Therefore, CM filters are effective in suppressing radiated emissions from the PV system.

9.3.4.2 Impact of DM Filter

For the case of the DM filter, an LC filter consisting of one 0.1 μF capacitor C_X and two 100 μH DM chokes L_D is connected to the PV system, as shown in Figure 9.45. The conducted emissions with the DM filter, shown in Figure 9.46, indicate that the DM emission is significantly reduced compared with the emissions without the filter presented in Figure 9.31. Due to the nonnegligible CM impedance provided by the DM chokes, the CM emission is also reduced to some extent, although this reduction is less pronounced.

The measured impedance magnitude of the DM filter components used in the setup is shown in Figure 9.47. According to the measurement results, the

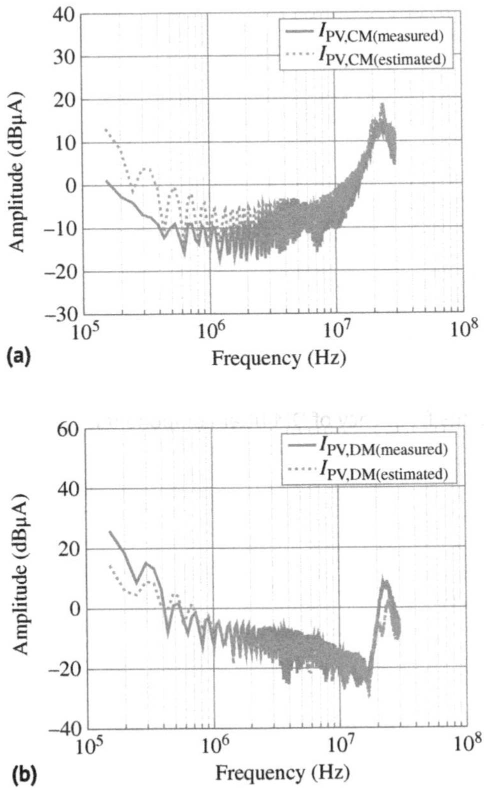

Figure 9.46 Comparison of conducted emission with the DM filter. (a) CM emission. (b) DM emission.

impedance of the DM choke can be modeled by a series *RLC* circuit with parameters $R = 1.08\ \Omega$, $L = 0.09\ \mu\text{H}$, and $C = 0.1\ \mu\text{F}$ while it can be modeled by a parallel *RLC* circuit with parameters $R = 2.82\ \text{k}\Omega$, $L = 100.7\ \mu\text{H}$, and $C = 0.02\ \mu\text{F}$. Considering that the parallel branch has a negligible impact on CM emission, the effective CM and DM impedances of the DM filter correspond to two L_D chokes connected in parallel and two L_D chokes connected in series, respectively. The equivalent noise models of the PV system with the DM filter are shown in Figure 9.48.

The DM emission at the PV system can be computed as

$$I_{\text{PV,DM}} = I_{\text{noise,DM}} \times \frac{Z_{\text{noise,DM}} Z_{Cx}}{Z_t Z_{Cx} + Z_t(Z_{\text{noise,DM}} + Z_{L_D,\text{DM}}) + Z_{Cx}(Z_{\text{noise,DM}} + Z_{L_D,\text{DM}})} \tag{9.24}$$

where $Z_t = Z_{\text{PV,DM}} + Z_{\text{cable,DM}}$.

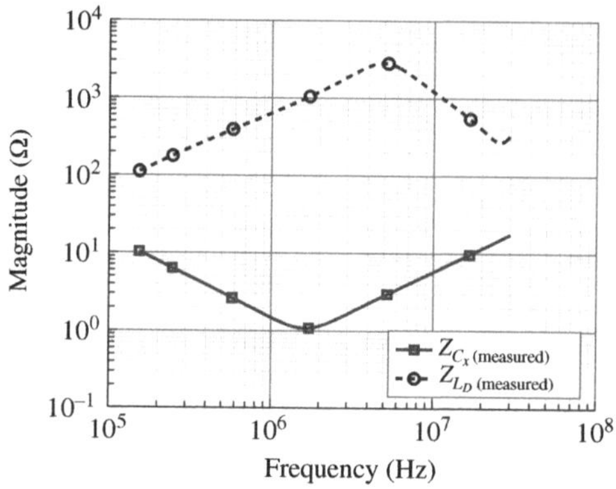

Figure 9.47 Magnitude of impedance versus frequency of DM filter components.

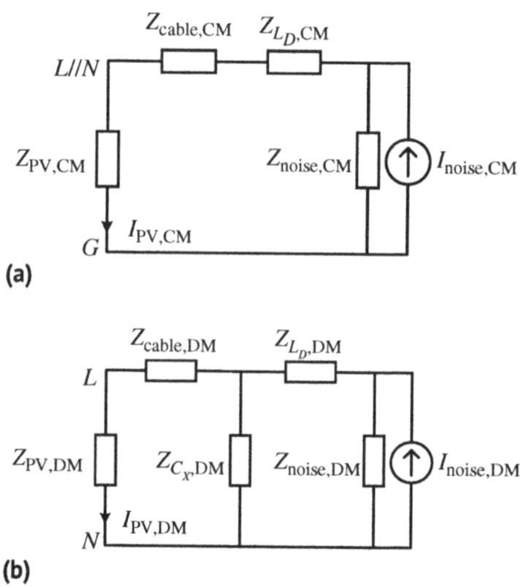

Figure 9.48 Equivalent noise models of the PV system with the DM filter. (a) CM. (b) DM.

In the case of CM noise, the capacitor C_X is ineffective. Therefore, the CM emission is calculated as

$$I_{\text{PV,CM}} = I_{\text{noise,CM}} \times \frac{Z_{\text{noise,CM}}}{Z_{\text{noise,CM}} + Z_{\text{cable,CM}} + Z_{\text{PV,CM}} + Z_{L_D,\text{CM}}} \tag{9.25}$$

The calculated and measured conducted emissions are shown in Figure 9.46. The results demonstrate good agreement between the calculated and measured CM and DM emissions. Accordingly, the measured and estimated radiated

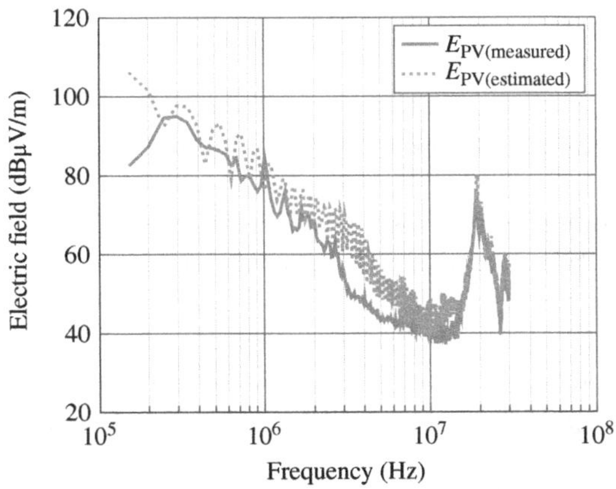

Figure 9.49 Radiated emission from the PV system with the DM filter.

emissions after inserting the DM filter are presented in Figure 9.49. As discussed earlier, the DM filter does not significantly reduce CM conducted emission. Therefore, when compared with the radiated emission shown in Figure 9.39, the radiated emission is not effectively reduced after the insertion of the DM filter.

9.4 Chapter Summary

In summary, this chapter highlights the effectiveness and adaptability of electrically contactless online impedance measurement in addressing key challenges across power and energy systems. By enabling nonintrusive characterization of operating equipment, these methods support accurate modeling, early fault detection, and informed maintenance decision-making in environments where conventional approaches are limited. The presented applications demonstrate that contactless impedance monitoring is not merely a diagnostic tool, but a unifying framework that bridges electromagnetic theory, power electronics, and energy system engineering. These results confirm its strong potential as a scalable sensing and analysis paradigm for enhancing the reliability, efficiency, and safety of next-generation power and energy infrastructures.

References

1 Y. Han and Y. H. Song, "Condition monitoring techniques for electrical equipment-a literature survey," *IEEE Trans. Power Deliv.*, vol. 18, no. 1, pp. 4–13, Jan. 2003.

2 E. P. Dick and C. C. Erven, "Transformer diagnostic testing by frequency response analysis," *IEEE Trans. Power Appar. Syst.*, vol. PAS-97, no. 6, pp. 2144–2153, Nov. 1978.

3 S.B. Rathnayaka, K. Y. See, and K. Li (2018). Inductively coupled on-line impedance measurement for condition monitoring of electrical equipment. IET Science, *Measurement & Technology*, vol. 12, no. 3, pp. 382–387. https://doi.org/10.1049/iet-smt.2017.0083

4 V. Behjat, A. Vahedi, A. Setayeshmehr et al., "Diagnosing shorted turns on the windings of power transformers based upon online FRA using capacitive and inductive couplings," *IEEE Trans. Power Deliv.*, vol. 26, no. 4, pp. 2123–2133, Oct. 2011.

5 M. Bagheri, M. S. Naderi, and T. Blackburn, "Advanced transformer winding deformation diagnosis: moving from off-line to on-line," *IEEE Trans. Dielectr. Electr. Insul.*, vol. 19, no. 6, pp. 1860–1870, Dec. 2012.

6 G. R. Slemon and A. Straughen, *Electric machines*, Addison-Wesley Publishing Company, 1980, pp. 104–142.

7 A. E. Fitzgerald, C. Kingsley, and S. D. Umans, *Electric Machinery*, McGraw-Hill, 1992.

8 P. C. Sen, *Principles of electrical machines & power electronics*, John Wiley & Sons, 1997.

9 J. J. Cathey, *Electric machinery analysis and design applying MATLAB*, McGraw-Hill, 2001.

10 D. Liu and X. Jiang, "High frequency model of common mode inductor for EMI analysis based on measurements," in *Proc. 3rd Int. Symp. Electromagn. Compat.*, China, May 2002.

11 L. Hai Yan, Z. Jian Guo, and S. Y. R. Hui, "Experimental determination of stray capacitances in high frequency transformers," *IEEE Trans. Power Electron.*, vol. 18, no. 5, pp. 1105–1112, Sep. 2003.

12 A. T. Hoang, *Electrical characterization of partial discharge resistant enamel insulation*, Chalmers Tekniska Hogskola (Sweden), 2014.

13 X. J. Shen and X. C. Jiang, "Development of online monitoring system for 1500 V ethylene–propylene–rubber DC feeder cable of Shanghai urban rail transit," *IET Gener. Transm. Distrib.*, vol. 5, no. 7, pp. 720–728, Jul. 2011.

14 K. Gouramanis and C. Demoulias, "Cable overheating in an industrial substation feeder due to untransposed power cables – measurement and simulation," in *Proc. EUROCON 2005 Int. Conf. "Computer as a Tool"*, Nov. 2005.

15 M. Tomita, M. Muralidhar, Y. Fukumoto et al., "Design and development of superconducting DC cable for railway applications," *IEEE Trans. Appl. Supercond.*, vol. 23, no. 3, p. 3601504, 2013.

16 R. Araneo, S. Lammens, M. Grossi et al., "EMC issues in high-power grid-connected photovoltaic plants," *IEEE Trans. Electromagn. Compat.*, vol. 51, no. 3, pp. 639–648, 2009.

17 I. Wu, T. Shinozuka, S. Ishigami et al., "Evaluation of electromagnetic radiation from the DC side of a photovoltaic power generation system," *IEEE Electromagn. Compat. Mag.*, vol. 4, no. 2, pp. 73–81, Aug. 2015.

18 T. Wada, T. Mori, M. Tokuda et al., "Radiated electromagnetic field from a solar cell for CISPR radiated emission measurement method," in *Proc. Int. Symp. Electromagn. Compat.*, Aug. 2005.

19 W. Chen, Y. Duan, L. Guo et al., "Modeling and prediction of radiated emission from solar cell in a photovoltaic generation system," *IEEE J. Photovoltaics*, vol. 6, no. 2, pp. 540–545, Jan. 2016.

20 K. A. Kim, C. Xu, L. Jin et al., "A dynamic photovoltaic model incorporating capacitive and reverse-bias characteristics," *IEEE J. Photovoltaics*, vol. 3, no. 4, pp. 1334–1341, 2013.

21 K. A. Kim, G. S. Seo, B. H. Cho et al., "Photovoltaic hot-spot detection for solar panel substrings using AC parameter characterization," *IEEE Trans. Power Electron.*, vol. 31, no. 2, pp. 1121–1130, Feb. 2016.

22 K. L. Kaiser, *Transmission lines, matching, and crosstalk*, CRC Press, 2005.

23 C. R. Paul, *Electromagnetics for engineers: with applications to digital systems and electromagnetic interference*, Wiley, 2004.

10
Conclusion and Future Direction

This chapter concludes the book by summarizing the key theoretical developments, measurement methodologies, and application insights presented in the preceding chapters, and by outlining important directions for future research and development. Having established a comprehensive framework for electrically contactless online impedance measurement, from fundamental principles and systematic measurement configurations to practical implementations and representative applications, this chapter consolidates the main contributions of the book and places them in a broader engineering context. In addition, it discusses emerging challenges and opportunities related to extending measurement bandwidth, advancing probe design and calibration, developing portable and scalable online impedance analyzers, enhancing signal excitation and processing techniques, integrating artificial intelligence (AI), expanding application domains, and strengthening robustness and standardization for industrial deployment, thereby highlighting how electrically contactless online impedance measurement can continue to evolve and support next-generation electrical and electronic systems.

10.1 Conclusion

This book has presented a comprehensive and systematic treatment of electrically contactless online impedance measurement, with a particular focus on inductive coupling-based techniques. Motivated by the growing demand for safe, nonintrusive, and in situ characterization of energized electrical systems, the book bridges fundamental theory, practical measurement methodologies, and real-world engineering applications.

Starting from the fundamental distinction between offline and online impedance, Chapters 1 and 2 established the motivation and theoretical foundations for contactless measurement approaches. By comparing voltage–current, capacitive coupling, and inductive coupling methods, the inductive coupling approach was identified as a uniquely advantageous solution for applications where direct electrical contact is undesirable or impractical, such as in high-power, high-voltage, and safety-critical environments.

Chapters 3–5 formed the methodological core of the book by systematically developing two-probe, single-probe, and multi-probe inductively coupled measurement setups in both frequency and time domains. These chapters provided detailed analytical formulations, equivalent circuit models, cascaded two-port network representations, calibration strategies, and practical implementation considerations. Special emphasis was placed on addressing key challenges such as probe-to-probe coupling effects, systematic measurement errors, time-variant impedance tracking, and multibranch impedance extraction. Collectively, these developments established a unified and extensible framework for electrically contactless online impedance measurement across a wide range of operating conditions.

Chapter 6 offered a comparative assessment of the different contactless measurement configurations, highlighting their respective advantages, limitations, and suitability for specific application scenarios. Rather than promoting a single "best" solution, this comparison emphasized informed selection based on measurement objectives, system complexity, signal-to-noise constraints, installation conditions, and scalability requirements. This chapter thus served as a practical guide for engineers and researchers seeking to deploy contactless impedance measurement techniques in real systems.

The practical value of the proposed methodologies was demonstrated in Chapters 7–9 through representative applications in power electronics, transport electrification, and power and energy systems. Chapter 7 showed how online impedance information enables a transition from empirical tuning to physics-based power-electronic design and diagnostics, supporting systematic EMI filter design for SMPSs, health diagnosis of power semiconductor devices, and online extraction of voltage-dependent parasitic capacitances, such as those of SiC power MOSFETs, under actual high-voltage biasing conditions without direct electrical contact.

Chapter 8 extended the methodology to safety-critical transportation platforms, demonstrating impedance-informed common-mode EMI filter design for inverter-driven motor systems, motor condition monitoring through online tracking of CM impedance for early detection of stator inter-turn short-circuit faults with robustness to load and speed variations, and scalable rail-transit applications using TBHMS for LRT current collector integrity assessment and MRT railway track condition and defect diagnosis during normal service.

Chapter 9 further demonstrated the effectiveness of the contactless impedance paradigm at infrastructure scale and in high-power and high-voltage environments, enabling online condition monitoring and defect detection of power transformers, nonintrusive fault diagnosis in high-power DC feeder systems with complex parallel-cable infrastructures, and EMI estimation with associated mitigation insights for photovoltaic systems based on online-extracted impedance and noise-source characteristics under realistic operating conditions. Collectively, these studies show that electrically contactless impedance information not only supports real-time diagnostics and monitoring, but also provides a quantitative foundation for model-based design, performance optimization, and proactive risk mitigation in complex and safety-critical electrical systems.

Taken together, the material presented in this book establishes electrically contactless online impedance measurement as a mature, versatile, and enabling technology for a wide spectrum of modern electrical and electronic systems. By eliminating the need for direct electrical contact, these approaches enhance operational safety, simplify deployment in energized and safety-critical environments, and preserve the integrity of the system under test. At the same time, the integration of rigorous theoretical modeling, systematic calibration procedures, and flexible measurement configurations ensures high measurement accuracy, robustness, and practical feasibility across component-, subsystem-, and system-level applications.

As electrical and electronic systems continue to evolve toward higher power density, tighter integration, increased electrification, and greater functional complexity, the demand for nonintrusive, real-time diagnostic, monitoring, and characterization tools will become increasingly critical. The concepts, methods, and application paradigms presented in this book provide a solid foundation for broader adoption in emerging domains such as wide-bandgap power electronics, electrified transportation platforms, renewable energy integration, smart grids, and safety-critical industrial and infrastructure systems. Beyond diagnostics and monitoring, electrically contactless online impedance measurement is expected to play an important role in digital-twin development, predictive maintenance, intelligent control, and reliability-oriented design, thereby supporting the advancement of safer, more reliable, and more intelligent electrical systems in both research and industrial practice.

10.2 Future Direction

Although electrically contactless online impedance measurement has demonstrated strong maturity and broad applicability, several important research and development directions remain open. Addressing these directions will further enhance measurement bandwidth, robustness, portability, scalability, and industrial relevance, thereby extending the impact of the introduced methodologies.

10.2.1 Extension to Wider Measurement Bandwidths

Extending the usable frequency range of contactless online impedance measurement is a fundamental future direction. At present, most implementations operate effectively over a frequency range of approximately 10 kHz to 100 MHz. Extending the upper frequency limit is essential for accurately characterizing the impedance behavior of modern electrical and electronic systems, particularly as switching speeds, integration levels, and parasitic interactions continue to increase, and high-frequency impedance increasingly influences system behavior, performance margins, and reliability. Achieving this goal primarily requires improved inductive probe designs with enhanced high-frequency transfer characteristics, reduced parasitic capacitances, and more accurate high-frequency modeling and calibration.

Equally important is extending the lower frequency limit, which is necessary for applications such as battery impedance estimation and other low-frequency impedance-based evaluation tasks. This direction calls for inductive probes with improved low-frequency sensitivity, optimized magnetic core materials, etc.

10.2.2 Advanced Probe Design Under High-power Current Operation

In some practical applications involving large DC currents or strong low-frequency AC components, inductive probes may be driven into magnetic saturation, which can degrade measurement linearity, sensitivity, and accuracy. Advanced inductive probe designs that are robust against magnetic saturation therefore represent a future research direction. Potential approaches include the use of optimized magnetic core materials with higher saturation flux density, improved magnetic circuit geometries to better distribute magnetic flux, and bias-tolerant probe configurations that mitigate the influence of DC or low-frequency bias currents.

In addition, saturation compensation or online calibration techniques may be explored to extend the usable measurement range without sacrificing accuracy. Such techniques account for nonlinear probe behavior induced by partial saturation and environmental variations, thereby improving measurement effectiveness under such operating conditions.

10.2.3 Advanced Signal Excitation and Signal Processing Techniques

Current electrically contactless online impedance measurement methods primarily employ single-sine, swept-sine, or multi-sine excitation, combined with signal processing techniques such as the fast Fourier transform (FFT), moving-window DFT, or STDFT. These approaches are well established and effective for steady-state impedance extraction and for systems exhibiting temporal variations. However, they inherently involve trade-offs among frequency resolution, temporal resolution, measurement speed, and computational complexity. Future research may therefore focus on the development of more advanced excitation schemes and signal processing algorithms to address these limitations. Promising directions include adaptive excitation strategies that adjust signal content in response to operating conditions, as well as broadband or time–frequency hybrid methods that improve temporal resolution while maintaining sufficient frequency selectivity. Such advancements would enable more accurate tracking of time-varying impedance, improved characterization of transient behaviors, and enhanced sensitivity to small or rapidly evolving impedance changes.

In addition, signal processing techniques with improved robustness to noise are essential for practical deployment. Enhancing noise immunity and stability under harsh electromagnetic environments will further extend the applicability of electrically contactless online impedance measurement to high-power, high-speed, and safety-critical systems.

10.2.4 Development of Portable and Scalable Contactless Online Impedance Analyzers

Wider adoption of electrically contactless online impedance measurement requires moving beyond laboratory-grade instrumentation toward portable and scalable measurement platforms. A key future direction is the development of compact, application-oriented online impedance analyzers that integrate signal generation, data acquisition, calibration, and signal processing into a single platform, while also supporting modular and scalable system architectures.

Such analyzers would enable convenient field deployment for commissioning, maintenance, and embedded monitoring, as well as scalable integration into multi-node or multichannel measurement systems for complex electrical infrastructures. By supporting both standalone operation and coordinated multipoint deployment, portable and scalable impedance analyzers can significantly lower the barrier to practical adoption across a wide range of applications, from component-level diagnostics to system-level monitoring.

10.2.5 Integration with AI

The integration of electrically contactless impedance measurement with AI offers substantial potential. By combining impedance data with machine learning, physics-informed models, and statistical inference, future systems can evolve from direct measurement toward automated fault diagnosis, anomaly detection, predictive maintenance, and remaining useful life estimation. This integration is particularly valuable for large-scale and long-term monitoring scenarios, where manual analysis becomes impractical.

10.2.6 Exploration of Emerging and Expanded Application Domains

Beyond the applications presented in this book, electrically contactless online impedance measurement can be extended to a wide range of emerging domains. Promising areas include electrified aviation and maritime systems, wireless power transfer infrastructures, advanced energy storage systems, safety-critical industrial equipment, etc. In these contexts, the nonintrusive nature of contactless impedance measurement provides unique advantages that are difficult to achieve using conventional techniques.

10.2.7 Robustness Enhancement and Standardization for Industrial Deployment

For widespread industrial adoption, robustness and standardization are essential. Future work should address long-term stability, repeatability across installations, immunity to harsh electromagnetic environments, and reduced calibration effort. The development of standardized measurement procedures, reference setups, and performance benchmarks will facilitate consistent evaluation and accelerate the transition from research prototypes to industrial-grade solutions.

In summary, continued advances along these directions will further establish electrically contactless online impedance measurement as a foundational sensing and characterization technology. By extending bandwidth, improving probe robustness, enhancing signal processing capability, enabling portability and scalability, and integrating intelligence, the methodology is well positioned to support the next generation of high-performance, reliable, and intelligent electrical and electronic systems.

Acronyms

3D	Three-dimensional
AC	Alternating current
ADC	Analog-to-digital converter
AF	Antenna factor
AFB	Abnormal friction bearing
AI	Artificial intelligence
AT	Attenuator
CCA	Current collector assembly
CCS	Current collector shoe
CH1	Channel 1
CH2	Channel 2
CISPR	Comité International Spécial des Perturbations Radioélectriques
CLI	Contact loss index
CM	Common-mode
CQI	Contact quality index
CWT	Continuous wavelet transform
DAQ	Data acquisition
DC	Direct current
DFT	Discrete Fourier transform
DM	Differential-mode
DR	Dummy rail
DSP	Digital signal processing
DUT	Device under test
DWPT	Discrete wavelet packet transform
ECQ	Electrical contact quality
EMC	Electromagnetic compatibility
EMI	Electromagnetic interference
ESR	Equivalent series resistance
EUT	Equipment under test

FCC	Federal Communications Commission
FD-SPS	Frequency-domain single-probe setup
FD-THPS	Frequency-domain three-probe setup
FD-TPS	Frequency-domain two-probe setup
FEM	Finite element method
FFT	Fast Fourier transform
FOI	Frequency of interest
HSP	High-speed ramps
IA	Impedance analyzer
IIP	Injecting inductive probe
IM	Induction motor
ITSC	Inter-turn short-circuit
KCL	Kirchhoff's current law
LISN	Line impedance stabilization network
LRT	Light rail transit
MDS	Motor drive system
MOSFET	Metal–oxide–semiconductor field-effect transistor
MRT	Mass rapid transit
NI	National Instruments
PA	Power amplifier
PCI	Peripheral component interconnect
PV	Photovoltaic
PWM	Pulse-width modulation
RCF	Rotor cage fault
RF	Radio frequency
RIP	Receiving inductive probe
SA	Spectrum analyzer
SAC	Signal acquisition card
SG	Signal generator
SGAS	Signal generation and acquisition system
SGC	Signal generation card
SiC	Silicon carbide
SID	Signal injection device
SMD	Surface-mounted device
SMPS	Switched-mode power supply
SNR	Signal-to-noise ratio
SOC	State of charge
SOH	State of health
SP	Surge protector
SPS	Single-probe setup
ST	Speed transient

STDFT	Short-time discrete Fourier transform
SUT	System under test
TBHMS	Train-borne health monitoring system
TD-MPS	Time-domain multi-probe setup
TD-SPS	Time-domain single-probe setup
TD-TPS	Time-domain two-probe setup
UPS	Uninterruptible power supply
VFD	Variable-frequency drive
VHR	Voltage health ratio
V–I	Voltage–current
VNA	Vector network analyzer
VSWR	Voltage standing wave ratio
WBG	Wide-bandgap

Index

f

g

h

i

Printed in the USA/Agawam, MA
August 28, 2026

909771.003